MATHEMATICAL MODELING IN CONTINUUM MECHANICS
SECOND EDITION

Temam and Miranville present core topics within the general themes of fluid and solid mechanics. The brisk style allows the text to cover a wide range of topics including viscous flow, magnetohydrodynamics, atmospheric flows, shock equations, turbulence, nonlinear solid mechanics, solitons, and the nonlinear Schrödinger equation.

This second edition will be a unique resource for those studying continuum mechanics at the advanced undergraduate and beginning graduate level whether in engineering, mathematics, physics, or the applied sciences. Exercises and hints for solutions have been added to the majority of chapters, and the final part on solid mechanics has been substantially expanded. These additions have now made it appropriate for use as a textbook, but it also remains an ideal reference book for students and anyone interested in continuum mechanics.

MATHEMATICAL MODELING IN CONTINUUM MECHANICS

SECOND EDITION

ROGER TEMAM

Université Paris-Sud, Orsay and Indiana University

ALAIN MIRANVILLE

Université de Poitiers

CAMBRIDGE
UNIVERSITY PRESS

CAMBRIDGE UNIVERSITY PRESS
Cambridge, New York, Melbourne, Madrid, Cape Town, Singapore,
São Paulo, Delhi, Dubai, Tokyo, Mexico City

Cambridge University Press
The Edinburgh Building, Cambridge CB2 8RU, UK

Published in the United States of America by Cambridge University Press, New York

www.cambridge.org
Information on this title: www.cambridge.org/9780521617239

First published 2000
Second edition published 2005

A catalogue record for this publication is available from the British Library

ISBN 978-0-521-61723-9 Paperback

Contents

Preface

This book is an extended version of a course on continuum mechanics taught by the authors to junior graduate students in mathematics. Besides a thorough description of the fundamental parts of continuum mechanics, it contains ramifications in a number of adjacent subjects such as magnetohydrodynamics, combustion, geophysical fluid dynamics, and linear and nonlinear waves. As is, the book should appeal to a broad audience: mathematicians (students and researchers) interested in an introduction to these subjects, engineers, and scientists.

This book can be described as an "interfacial" book: interfaces between mathematics and a number of important areas of sciences. It can also be described by what it is not: it is not a book of mathematics: the mathematical language is simple, only the basic tools of calculus and linear algebra are needed. This book is not a treatise of continuum mechanics: although it contains a thorough but concise description of many subjects, it leaves aside many developments which are fundamental but not needed in practical applications and utilizations of mechanics, e.g., the intrinsic – frame invariance – character of certain quantities or the coherence of certain definitions. The reader interested by these issues is referred to the many excellent mechanics books which are available, such as those quoted in the list of references to Part I. Finally, by its size limitations, this book cannot be encyclopedic, and many choices have been made for the content; a number of subjects introduced in this book can be developed themselves into a full book. All in all, we believe that this book, benefiting from prolonged efforts and teaching experience of the authors, can be very useful to scientists who want to reduce the gap between mathematics and sciences, a gap usually due to the language barrier and the differences in thinking and reasoning.

The core of the book contains the fundamental parts of continuum mechanics: description of the motion of a continuous body, the fundamental law of dynamics, the Cauchy and the Piola-Kirchhoff stress tensors, the constitutive laws, internal energy and the first principle of thermodynamics, shocks and the Rankine–Hugoniot relations, an introduction to fluid mechanics for inviscid and viscous Newtonian fluids, an introduction to linear elasticity and the variational principles in linear elasticity, and an introduction to nonlinear elasticity.

Besides the core of continuum mechanics, this book also contains more or less detailed introductions to several important related fields that could be themselves the subjects of separate books: magnetohydrodynamics, combustion, geophysical fluid dynamics, vibrations, linear acoustics, and nonlinear waves and solitons in the context of the Korteweg–de Vries and the nonlinear Schrödinger equations. The whole book is suitable for a one-year course at the advanced undergraduate or beginning graduate level. Parts of it are suitable for a one-semester course either on the fundamentals of continuum mechanics or on a combination of selected topics.

This second edition of the book has been augmented by the introduction of exercises and hints at solutions making it more suitable for class utilization, by a new chapter on nonlinear elasticity, and by several additions and corrections suggested by the readers of the first edition. In particular it has benefited from the comments of the anonymous and non-anonymous reviewers of the first edition, especially J. Dunwoody and J.J. Telega. The authors want also to thank P. G. Ciarlet for his comments; the new chapter on nonlinear elasticity borrowed very much from his classical book on the subject. Finally they gratefully acknowledge essential help in the production of the volume from Teresa Bunge, Jacques Laminie, Eric Simonnet, and Djoko Wirosoetisno.

Roger Temam
Alain Miranville
June 2004

A few words about notations

The notations in this book are not uniform; this is partly done on purpose and partly because we had no choice. Indeed modelers usually have to comply or at least adapt to the notations common in a given field, and thus they must be trained to some flexibility. Another reason for having non-uniform notations is that different fields are present in this volume, and it was not possible to find notations fitting "all the standards."

Another objective while deciding the notations was to choose notations that can be easily reproduced by handwriting, thus avoiding as much as possible arrows, boldfaced type, and simple and double underlining with bars or tildes; in general, in a given chapter of this book, in a given context, it is clear what a given symbol represents.

Although the notations are not rigid, there are still some repeated patterns in the notations, and we indicate hereafter notations used in several chapters:

Ω or $\mathcal{O}$, possibly with indices: domain in $\mathbb{R}^2$ or $\mathbb{R}^3$

$x = (x_1, x_2)$ or (x_1, x_2, x_3): generic point in $\mathbb{R}^2$ or $\mathbb{R}^3$. Also denoted (x, y) or (x, y, z)

$a = (a_1, a_2)$ or (a_1, a_2, a_3): initial position in Lagrangian variables

t: time

$u = (u_1, u_2)$ or (u_1, u_2, u_3), or v or w: vectors in $\mathbb{R}^2$ or $\mathbb{R}^3$. Also denoted (u, v) or (u, v, w)

AB (or $\overrightarrow{AB}$ to emphasize): vector from A to B

u or U: velocity

u: displacement vector

γ: acceleration

m: mass

f, F: forces; usually f for volume forces and F for surface forces
ρ: density
g: gravity constant. Also used for equation of state for fluids
T or θ: temperature
σ: Cauchy stress tensor (in general)
n: unit outward normal on the boundary of an open set Ω or $\mathcal{O}$, $n = (n_1, n_2)$
  or $n = (n_1, n_2, n_3)$

We will use also the following classical symbols and notations:

δ_{ij}: the Kronecker symbol equal to 1 if $i = j$ and to 0 if $i \neq j$
$\varphi_{,i}$ will denote the partial derivative $\partial\varphi/\partial x_i$.

The Einstein summation convention will be used: when an index (say j) is repeated in a mathematical symbol or within a product of such symbols, we add these expressions for $j = 1, 2, 3$. Hence

$$\sigma_{ij,j} = \sum_{j=1}^{3} \frac{\partial\sigma_{ij}}{\partial x_j}, \qquad \sigma_{ij} \cdot n_j = \sum_{j=1}^{3} \sigma_{ij} n_j.$$

PART I

FUNDAMENTAL CONCEPTS
IN CONTINUUM MECHANICS

CHAPTER ONE

Describing the motion of a system: geometry and kinematics

1.1. Deformations

The purpose of mechanics is to study and describe the motion of material systems. The language of mechanics is very similar to that of set theory in mathematics: we are interested in material bodies or systems, which are made of material points or matter particles. A material system fills some part (a subset) of the ambient space ($\mathbb{R}^3$), and the position of a material point is given by a point in $\mathbb{R}^3$; a part of a material system is called a subsystem.

We will almost exclusively consider material bodies that fill a domain (i.e., a connected open set) of the space. We will not study the mechanically important cases of thin bodies that can be modeled as a surface (plates, shells) or as a line (beams, cables). The modeling of the motion of such systems necessitates hypotheses that are very similar to the ones we will present in this book, but we will not consider these cases here.

A material system fills a domain Ω_0 in $\mathbb{R}^3$ at a given time t_0. After deformation (think, for example, of a fluid or a tennis ball), the system fills a domain Ω in $\mathbb{R}^3$. A material point, whose initial position is given by the point $a \in \Omega_0$, will be, after transformation, at the point $x \in \Omega$.

The deformation can thus be characterized by a mapping as follows (see Figure 1.1):

$$\Phi: a \in \Omega_0 \mapsto x \in \Omega.$$

Assuming that matter is conserved during the deformation, we are led to make the following natural hypothesis:

The function Φ is one – to – one from Ω_0 onto Ω.

We will further assume that the deformation Φ is a smooth application of class $\mathcal{C}^1$ at least, from Ω_0 into Ω, as well as its inverse ($\mathcal{C}^1$ from Ω onto Ω_0). In fact we assume that Φ is as smooth as needed.

3

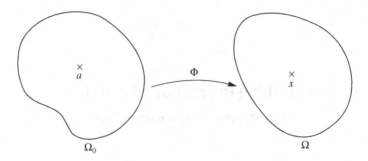

Figure 1.1 The mapping Φ.

Regularity assumption

The regularity assumption made on Φ will actually be general; we will assume that all the functions we introduce are as regular as needed for all the mathematical operations performed to be justified (e.g., integration by parts, differentiation of an integral depending on a parameter, etc.). This hypothesis, which will be constantly assumed in the following, will only be weakened in Chapter 6 for the study of shock waves, which correspond to the appearance of discontinuity surfaces. In that case, we will assume that the map Φ is piecewise C^1. This assumption must be weakened also for the study of other phenomena which will not be considered here, such as singular vortices for fluids, dislocations for solids, or collisions of rigid bodies.

Let grad $\Phi(a) = \nabla\Phi(a)$ be the matrix whose entries are the quantities $(\partial\Phi_i/\partial a_j)(a)$; this is the Jacobian matrix of the mapping $a \mapsto x$ also denoted sometimes Dx/Da. Because Φ^{-1} is differentiable, the Jacobian $\det(\nabla\Phi)$ of the transformation $a \mapsto x$ is necessarily different from zero. We will assume in the following that it is strictly positive; the negative sign corresponds to the nonphysical case of a change of orientation (a left glove becoming a right glove). We will later study the role played by the linear tangent map at point a in relation to the Taylor formula

$$\Phi(a) = \Phi(a_0) + \nabla\Phi(a_0) \cdot (a - a_0) + o(|a - a_0|).$$

We will also introduce the dilation tensor to study the deformation of a "small" tetrahedron.

Displacement

Definition 1.1. *The map* $u : a \mapsto x - a = \Phi(a) - a$ *is called the displacement; $u(a)$ is the displacement of the particle a.*

Elementary deformations

Our aim here is to describe some typical elementary deformations.

a) Rigid deformations

The displacement is called rigid (in this case, we should no longer talk about deformations) when the distance between any pair of points is conserved as follows:

$$d(a, a') = d(x, x'), \qquad \forall\, a, a' \in \Omega_0,$$

where $x = \Phi(a)$, $x' = \Phi(a')$. This is equivalent to assuming that

$$\Phi \text{ is an isometry from } \Omega_0 \text{ onto } \Omega,$$

or, when Ω_0 is not included in an affine subspace of dimension less than or equal to 2,

$$\Phi \text{ is an affine transformation}$$
$$(translation + rotation).$$

In this case

$$x = L \cdot a + c, \quad c \in \mathbb{R}^3, \quad L \in \mathcal{L}_0(\mathbb{R}^3), \quad L^{-1} = L^T,$$

and

$$u(a) = (L - I)a + c,$$

where $\mathcal{L}_0(\mathbb{R}^3)$ is the space of orthogonal matrices on $\mathbb{R}^3$.

b) Linear compression or elongation

A typical example of elongation is given by the linear stretching of an elastic rod or of a linear spring.

Let (e_1, e_2, e_3) be the canonical basis of $\mathbb{R}^3$. The uniform elongation in the direction $e = e_1$ reads

$$x_1 = \lambda a_1, \quad x_2 = a_2, \quad x_3 = a_3,$$

with $\lambda > 1$; $0 < \lambda < 1$ would correspond to the uniform compression of a linear spring or an elastic rod. The displacement is then given by $u(a) = [(\lambda - 1)a_1, 0, 0]$ and

$$\nabla\Phi = \begin{pmatrix} \lambda - 1 & 0 & 0 \\ 0 & 0 & 0 \\ 0 & 0 & 0 \end{pmatrix} + I.$$

c) *Shear*

We consider here simple shear in two orthogonal directions. Such a deformation occurs, for instance, when one tears a sheet of paper.

The shear in the direction e_1 parallel to the direction e_2 reads

$$\begin{cases} x_1 = a_1 + \rho a_2, \\ x_2 = a_2 + \rho a_1, \\ x_3 = a_3, \end{cases}$$

where $\rho > 0$; hence, the displacement is given by

$$u(a) = \begin{pmatrix} \rho a_2 \\ \rho a_1 \\ 0 \end{pmatrix},$$

and

$$\nabla \Phi = \begin{pmatrix} 0 & \rho & 0 \\ \rho & 0 & 0 \\ 0 & 0 & 0 \end{pmatrix} + I.$$

Remark 1.1: We will see in what follows that, in some sense, a general deformation can be decomposed into proper elementary deformations of the types above.

1.2. Motion and its observation (kinematics)

Kinematics is the study of the motion of a system related to an observer, which is called the reference system.

With kinematics, we need to introduce two new elements:

- A privileged continuous parameter t corresponding to time. This implies the choice of a chronology; that is, a way to measure time.[1]
- A given system of linear coordinates, or frame of reference, that is "fixed" with respect to the observer. It is defined, in the affine space, by its origin O and three orthonormal basis vectors e_1, e_2, e_3.

---

[1] From a strictly mathematical point of view, it would not be absurd, for example, to replace t by t^3, but this would change the notion of time interval, and the time $t = 0$ would play a particular role, which it does not.

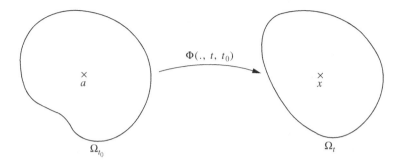

Figure 1.2 The motion of a system.

Definition 1.2. *A reference system is defined by the choice of a chronology and a frame of reference.*[2]

The chronology is fixed once and for all, but, hereafter, we will consider several frames of reference, depending on our objectives.

The motion of the system under consideration is observed during a time interval $I \subset \mathbb{R}$. At each instant $t \in I$, the system fills a domain $\Omega_t \subset \mathbb{R}^3$. The motion is then geometrically defined by a family of deformation mappings, depending on the time $t \in I$ (see Figure 1.2). We denote by $\Phi(t, t_0)$ the diffeomorphism

$$a \in \Omega_{t_0} \mapsto x = \Phi(a, t, t_0) \in \Omega_t,$$

which maps the position a at time t_0 to the position x at time t, and we make the following natural hypotheses:

- $\Phi(t_0, t_0) = I$,
- $\Phi(t', t) \circ \Phi(t, t_0) = \Phi(t', t_0)$, and
- the maps $(t, a) \mapsto \Phi(a, t, t_0)$ are at least of class $\mathcal{C}^1$ (except in the case of shock waves).

Definition 1.3. *A material system is a rigid body if and only if the maps $\Phi(t, t')$ are isometries for every t and t'.*

Explicit representation of the motion

We are given a reference time t_0, and we choose for point O the origin of the frame of reference in the affine space. The position of the material point M

---

[2] We will not emphasize any more these very profound considerations, which can lead, depending on the point of view that is pursued, to nonclassic mechanics (e.g., relativity or quantum mechanics).

is defined at each time t by the vector $\overrightarrow{OM}$; we thus write

$$\overrightarrow{OM} = x = \Phi(a, t, t_0),$$

or more simply, omitting t_0,

$$\overrightarrow{OM} = x = \Phi(a, t).$$

Trajectory of a particle

We consider a particle defined by its position a at time t_0. The trajectory of this particle is the curve $\{\Phi(a, t, t_0)\}_{t \in I}$; I is the interval of time during which the motion is observed.

Velocity of a particle

The velocity of the material point M occupying the position x at time t is the vector

$$U = U(x, t) = \frac{\partial \Phi}{\partial t}(a, t, t_0).$$

Remark 1.2: Of course, we consider the derivatives with respect to the given frame of reference, and the velocity is thus considered with respect to the same system.

Property 1.1. *The vector U is independent of the reference time t_0.*

Proof: Let M be a particle occupying the positions x at time t, a at t_0 and a' at t_0' ($t > t_0 > t_0'$). Thus,

$$\overrightarrow{OM} = x = \Phi(a, t, t_0) = \Phi(a', t, t_0'),$$

and

$$a = \Phi(a', t_0, t_0').$$

Thus,

$$x = \Phi[\Phi(a', t_0, t_0'), t, t_0] = \Phi(a', t, t_0'),$$

where a', t_0, and t_0' are fixed. We set

$$h(t) = \Phi[\Phi(a', t_0, t_0'), t, t_0],$$
$$\ell(t) = \Phi(a', t, t_0').$$

We easily verify that $(dh/dt) = (d\ell/dt)$, which accounts for the result.

Acceleration of a particle

The acceleration of the material particle M occupying the position x at time t is the vector

$$\gamma = \gamma(x, t) = \frac{\partial^2 \Phi}{\partial t^2}(a, t, t_0).$$

Property 1.2. *The vector γ is independent of the reference time t_0 (same proof as for the velocity).*

Remark 1.3: Of course, the vectors U and γ depend on the frame of reference and on the chronology we have chosen.

We will see in Section 1.3 why we prefer to write $U(x, t)$ and $\gamma(x, t)$ instead of $U(a, t)$ and $\gamma(a, t)$.

Stream lines

The stream lines are defined at a time t; they are the lines whose tangent at each point is parallel to the velocity vector at this point.

If $U(x, t)$ is the velocity vector at time t and at $x \in \Omega_t$, computing the stream lines is equivalent to solving the differential system

$$\frac{dx_1}{U_1(x_1, x_2, x_3, t)} = \frac{dx_2}{U_2(x_1, x_2, x_3, t)} = \frac{dx_3}{U_3(x_1, x_2, x_3, t)}.$$

In practice, this system can be solved explicitly by analytic methods only rarely. However, its numerical solution on a computer is easy. For computer simulations of flows, these equations are numerically integrated to obtain the stream lines repeatedly in order to visualize the flow (e.g., movie-type animation).

Remark 1.4: The stream lines are different from the trajectories. However, in the case of a stationary motion (defined below) the stream lines and the trajectories coincide.

1.3. Description of the motion of a system: Eulerian and Lagrangian derivatives

Lagrangian description of the motion of a system (description by the trajectories)

The Lagrangian description of the motion is the one we have considered until now. It consists of giving the trajectory of each particle starting from the initial position, for instance from time $t = 0$:

$$x = \Phi(a, t) = \Phi(a, t, 0), \quad a \in \Omega_0.$$

The velocity and acceleration fields are then, respectively, the vector fields $(\partial\Phi/\partial t)(a, t)$ and $(\partial^2\Phi/\partial t^2)(a, t)$.

This description is too rich for most practical purposes, and it is not used in general. It is, however, very useful for mathematical analysis and for some very specific situations. In general, we prefer to use the Eulerian description of the motion.

Eulerian description of the motion of a system (description by the velocity field)

In this description, we are given at each time t the velocity field $U = U(x, t)$, where $U(x, t)$ is the velocity of the material particle occupying the position x at time t.

Theoretically, we can recover from this velocity field the trajectories and the Lagrangian description of the motion by solving the following differential equations:

$$\begin{cases} \dfrac{dx}{dt} = U(x, t), \\ x(0) = a, \end{cases}$$

the solution of which is $x = x_a(t) = \Phi(a, t)$. Similarly, we can compute the stream lines by solving for x (t being a fixed parameter) the differential system

$$\frac{dx_1}{U_1} = \frac{dx_2}{U_2} = \frac{dx_3}{U_3}.$$

Definition 1.4. *A motion is steady or stationary if and only if $\Omega_t = \Omega_0$, $\forall t$, and the Eulerian velocity field is independent of t, that is, $U(x, t) \equiv U(x)$, $\forall t$.*

We emphasize here the fact that a body undergoing a stationary motion *is not at rest.*

As we said before, the trajectories and streamlines of a stationary flow are the same.

Eulerian and Lagrangian derivatives

Consider a particle M occupying the position a at time $t = 0$ and the position x at time t. A function $f = f(M, t)$ associated with a particle M (e.g., its velocity, acceleration, or other physical quantities to be defined subsequently) can be represented in two different ways during the motion:

$$f(M, t) = g(a, t) \quad \text{or} \quad f(M, t) = h(x, t),$$

where $x = \Phi(a, t)$; that is to say

$$h(\Phi(a, t), t) = g(a, t).$$

Definition 1.5.

1. *The Eulerian derivatives of f are the quantities*

$$\frac{\partial h}{\partial x_i}(x, t), \qquad \frac{\partial h}{\partial t}(x, t).$$

2. *The Lagrangian derivatives of f are the quantities*

$$\frac{\partial g}{\partial a_i}(a, t), \qquad \frac{\partial g}{\partial t}(a, t).$$

In mechanical engineering the derivative $(\partial g / \partial t)(a, t)$ is denoted dh/dt or Dh/Dt (or df/dt or Df/Dt) and is sometimes called the total derivative of f.

We deduce from the relation

$$h(\Phi(a, t), t) = g(a, t),$$

that

$$\frac{\partial g}{\partial a_j}(a, t) = \frac{\partial h}{\partial x_k} \frac{\partial x_k}{\partial a_j} = \frac{\partial h}{\partial x_k}(x, t) \frac{\partial \Phi_k}{\partial a_j}(a, t),$$

where we have used the Einstein summation convention on the indices.

Property 1.3. *We have the following relation:*

$$\boxed{\frac{\partial g}{\partial t} = \frac{Dh}{Dt} = \frac{\partial h}{\partial t} + (U \cdot \nabla) h.}$$

Proof: We see easily that

$$\frac{\partial g}{\partial t}(a,t) = \frac{\partial h}{\partial x_k}(x,t)\frac{\partial \Phi_k}{\partial t}(a,t) + \frac{\partial h}{\partial t}(x,t)$$

$$= \left(U_k \cdot \frac{\partial h}{\partial x_k} + \frac{\partial h}{\partial t}\right)(x,t).$$

Application to the computation of acceleration in the Eulerian representation

Here, we consider the case in which the function f above is the velocity. Thus,

$$h(x,t) = U(x,t) = U[\Phi(a,t),t] = g(a,t) = \frac{\partial \Phi}{\partial t}(a,t).$$

Consequently,

$$\gamma(x,t) = \frac{\partial^2 \Phi}{\partial t^2}(a,t) = \frac{\partial g}{\partial t}(a,t) = \frac{Dh}{Dt}(a,t)$$

$$= \frac{\partial}{\partial t}[U(\Phi(a,t),t)],$$

which yields

$$\gamma(x,t) = \frac{\partial U}{\partial t}(x,t) + \sum_{i=1}^{3} U_i(x,t)\frac{\partial U}{\partial x_i}(x,t),$$

or

$$\gamma = \frac{\partial U}{\partial t} + (U \cdot \nabla)U.$$

This formula is a fundamental one for fluids: the Eulerian expression of the acceleration is constantly used in fluid mechanics.

Remark 1.5: Even though it is far from being an unbreakable rule, we generally (or more often) use the Eulerian representation for fluids and the Lagrangian representation for solids.

1.4. Velocity field of a rigid body: helicoidal vector fields

Proposition 1.1. *A necessary and sufficient condition for $U(x, t)$, $x \in \Omega_t$, to be the velocity field at time t of a rigid body is that*

$$(x - x') \cdot [U(x, t) - U(x', t)] = 0, \tag{1.1}$$

for every x and $x' \in \Omega_t$, and for the whole interval of time I under consideration.

Proof: Let us consider two material points of a system occupying the positions a and a' at time t_0, and the positions x and x' at time t, $t \in I$.

We have

$$\|x - x'\|^2 = \|\Phi(a, t) - \Phi(a', t)\|^2,$$

and this quantity remains constant (equal to $\|a - a'\|^2$) for $t \in I$, if and only if

$$\frac{d}{dt}\|x - x'\|^2 = 0;$$

that is to say, because $x = \Phi(a, t)$ and $x' = \Phi(a', t)$:

$$[\Phi(a, t) - \Phi(a', t)] \cdot \left[\frac{\partial \Phi}{\partial t}(a, t) - \frac{\partial \Phi}{\partial t}(a', t)\right] = 0,$$

$$(x - x') \cdot [U(x, t) - U(x', t)] = 0$$

and hence Eq. (1.1).

Proposition 1.2 hereafter gives a useful characterization of the vector fields $U(x) = U(x, t)$, verifying Eq. (1.1) at a given time.

Proposition 1.2. *Equation (1.1) is satisfied for every x and $x' \in \Omega$ if and only if there exist a vector $b \in \mathbb{R}^3$ and an antisymmetric matrix B such that*

$$U(x) = Bx + b, \quad \forall x \in \Omega, \tag{1.2}$$

where

$$B = \begin{pmatrix} 0 & -\omega_3 & \omega_2 \\ \omega_3 & 0 & -\omega_1 \\ -\omega_2 & \omega_1 & 0 \end{pmatrix},$$

which is equivalent to

$$U(x) = \omega \wedge x + b, \quad \forall x \in \Omega, \tag{1.3}$$

where ω is the vector of components $(\omega_1, \omega_2, \omega_3)$.

Proof: Let us show that Eq. (1.2) implies Eq. (1.1). We first notice that

$$U(x') = U(x) + B(x' - x) = U(x) + \omega \wedge (x' - x).$$

Consequently,

$$[U(x) - U(x')] \cdot (x - x') = [B(x - x')] \cdot (x - x')$$
$$= 0$$

because $By \cdot y = 0$, $\forall\, y \in \mathbb{R}^3$.

Conversely, let us prove that Eq. (1.1) implies Eq. (1.2). Let us assume for simplicity that $0 \in \Omega$, and let $b = U(0)$. We set

$$V(x) = U(x) - U(0) = U(x) - b.$$

Then,

$$V(x) \cdot x = 0, \quad \forall\, x \in \Omega. \tag{1.4}$$

Because

$$[U(x) - U(e_i)] \cdot (x - e_i) = 0, \quad \forall\, x \in \Omega,$$

we find

$$[V(x) - V(e_i)] \cdot (x - e_i) = 0, \quad \forall\, x \in \Omega,$$

which yields, thanks to Eq. (1.4),

$$V(x) \cdot e_i = -V(e_i) \cdot x, \quad \forall\, x \in \Omega, \quad i = 1, 2, 3. \tag{1.5}$$

Therefore,

$$V(x) = \sum_{i=1}^{3}[V(x) \cdot e_i]\, e_i = -\sum_{i=1}^{3}[V(e_i) \cdot x]\, e_i;$$

hence,

$$V_i(x) = \sum_{j=1}^{3} b_{i,j} x_j,$$

where

$$b_{i,j} = -V(e_i) \cdot e_j = -V_j(e_i).$$

We then infer that

$$b_{i,i} = -V(e_i) \cdot e_i = 0,$$
$$b_{i,j} = -V(e_i) \cdot e_j = V(e_j) \cdot e_i = -b_{j,i},$$

and the result follows.

Remark 1.6: For a given vector field U, ω, and b are unique. Indeed

$$\omega \wedge x + b = \omega' \wedge x + b', \quad \forall x \in \Omega,$$

yields, for $x = 0$, $b = b'$. Similarly, $(\omega - \omega') \wedge x = 0, \forall x \in \Omega$, implies $\omega - \omega' = 0$.

We now introduce a special class of vector fields on $\mathbb{R}^3$ that play an important role in this part of mechanics. The terminology hereafter is not commonly used by engineers, but it is a convenient one, as we will see later on: such vector fields appear in relation to velocities of rigid bodies, forces, momentum, and quantities of acceleration (concepts subsequently defined).

Definition 1.6. *A helicoidal vector field (HVF) is a vector field on $\mathbb{R}^3$ satisfying the conditions of Eq. (1.1), or equivalently Eq. (1.2). The corresponding vector ω is called the resultant (or linear resultant) of the helicoidal vector field, and $U(x_0)$ is its resulting (or angular) momentum at x_0; $\{U(x_0), \omega\}$ are the reduction elements at the point x_0.*

A helicoidal vector field is characterized by its reduction elements at any given point. Indeed, for every x and x'

$$U(x') = U(x) + \omega \wedge (x' - x). \tag{1.6}$$

Note that, by Eq. (1.6), $U(x)$ is constant along a line parallel to ω.

A *uniform HVF* is an HVF for which $\omega = 0$, and then, by Eq. (1.6), $U(x') = U(x), \forall x', x$. Another particular HVF is the *torque*: this is a HVF for which the scalar product $U(x) \cdot \omega$ (which is independent of x, see the next subsection) vanishes. One can show (this is left as an exercise) that, for any HVF that is not uniform, there is a unique line Δ (called the *axis of the HVF*), along which $U(x)$ is parallel to ω. A torque is a HVF for which $U(x) = 0$, $\forall x \in \Delta$.

Operations on helicoidal vector fields

We are given two helicoidal vector fields, namely $[\mathcal{T}] = \{U(x)\}_{x\in\Omega}$, and $[\tilde{\mathcal{T}}] = \{\tilde{U}(x)\}_{x\in\Omega}$:

$$U(x') = U(x) + \omega \wedge (x' - x),$$
$$\tilde{U}(x') = \tilde{U}(x) + \tilde{\omega} \wedge (x' - x).$$

a) Sum and product by a scalar

We define respectively the sum $[T] + [T']$ and the product $\lambda[T]$ by their reduction elements at any point x_0:

$$[T] + [\tilde{T}]: \{U(x_0) + \tilde{U}(x_0),\, \omega + \tilde{\omega}\},$$
$$\lambda[T]: \{\lambda U(x_0),\, \lambda\omega\}.$$

It is easily seen that this definition is independent of x_0. Also, using the linear mapping

$$[T] \xrightarrow{\varphi_x} \text{(reduction elements at a point } x),$$

we see that the space of HVFs is isomorphic to $\mathbb{R}^6$. One can show that any HVF is the sum of a uniform HVF and a torque (hint: use the axis of the HVF).

b) Topology

We define the convergence of a sequence of helicoidal vector fields by the convergence of their reduction elements at one point. We notice that the convergence of the reduction elements at one point yields the convergence of the reduction elements at every point; this topology is essentially that of $\mathbb{R}^6$.

c) Differentiation

If $[T(t)]$, $t \in I$, is a family of HVFs depending on a parameter $t \in I$ and defined by their reduction elements at a point x_0: $\{U(x_0, t), \omega(t)\}$, we define

$$\frac{d[T]}{dt}(t) = \lim_{\Delta t \to 0} \frac{[T(t + \Delta t)] - [T(t)]}{\Delta t};$$

this is the helicoidal vector field whose reduction elements at x_0 are

$$\left\{ \frac{\partial U}{\partial t}(x_0, t),\, \frac{d\omega}{dt}(t) \right\}.$$

d) Scalar product of two helicoidal vector fields

Let $[T]$ and $[\tilde{T}]$ be two helicoidal vector fields whose reduction elements at x are respectively $[U(x), \omega]$ and $[\tilde{U}(x), \tilde{\omega}]$. We set

$$[T] \cdot [\tilde{T}] = U(x) \cdot \tilde{\omega} + \tilde{U}(x) \cdot \omega.$$

This quantity is independent of the choice of the point x. Indeed, if x' is another point,

$$[U(x') - U(x)] \cdot \tilde{\omega} + [\tilde{U}(x') - \tilde{U}(x)] \cdot \omega$$
$$= [\omega \wedge (x' - x)] \cdot \tilde{\omega} + [\tilde{\omega} \wedge (x' - x)] \cdot \omega$$
$$= [x' - x, \tilde{\omega}, \omega] + [x' - x, \omega, \tilde{\omega}]$$
$$= 0.$$

Remark 1.7: The scalar product above of two helicoidal vector fields is a bilinear product but not a positive definite form on the space of *HVFs* $\sim \mathbb{R}^6$. Indeed,

$$[T] \cdot [T] = 2U(0) \cdot \omega,$$

which can be positive, negative, or even equal to 0 for a suitable $[T]$.

Remark 1.8: We have defined the helicoidal vector field associated with the velocity field of a rigid body. In the sequel we will introduce other HVFs corresponding to different physical quantities such as forces, momentum and quantities of acceleration.

Structure of a helicoidal vector field

When ω vanishes, the vector field $U = U(x)$ is constant.

It is interesting to describe the structure of the vector field $U(x)$ when the resultant ω of the HVF does not vanish. To do so, we look for the axis Δ, called axis of the HVF, along which $U(x)$ is parallel to ω. We start by searching for such a point x_0 in the plane containing 0 and orthogonal to ω. It follows that

$$0 = U(x_0) \wedge \omega = U(0) \wedge \omega + (\omega \wedge x_0) \wedge \omega$$
$$0 = U(0) \wedge \omega + |\omega|^2 x_0 - (\omega \cdot x_0)\omega;$$

since $\omega \cdot x_0 = 0$, there remains

$$x_0 = |\omega|^{-2}[\omega \wedge U(0)].$$

For any other point $x \in \Delta$, we then have

$$0 = U(x) \wedge \omega = U(x_0) \wedge \omega + [\omega \wedge (x - x_0)] \wedge \omega$$
$$0 = U(x_0) \wedge \omega + |\omega|^2(x - x_0) - [\omega \cdot (x - x_0)]\omega;$$

hence, since $U(x_0) \wedge \omega = 0$

$$x - x_0 = |\omega|^{-2} \{[\omega \cdot (x - x_0)]\omega\},$$

and the points x belong to an axis Δ parallel to ω, the axis Δ of the HVF.

For any point x in space, we then compare $U(x)$ to $U(x')$, where x' is the orthogonal projection of x on Δ; we deduce that the vector field $U(x)$ possesses a helicoidal symmetry, that is to say it is invariant by rotation around Δ and by translation parallel to Δ, that is the application (rotation, translation), mapping x onto x', also maps $U(x)$ onto $U(x')$.

1.5. Differentiation of a volume integral depending on a parameter

When considering a material system whose motion is as described earlier in this chapter, we will often need to compute the time derivative of integrals of the form

$$K(t) = \int_{\Omega_t} C(x, t) \, dx,$$

where $C = C(x, t)$ is a given scalar function.

Here and henceforth, the expression *a set that moves with the flow* denotes a family of sets Ω_t, where t belongs to a time interval $I \subset \mathbb{R}$, and $\Omega_t = \Phi(\Omega_0, t)$, and Φ, Ω_0, and Ω_t satisfy the assumptions introduced earlier.

To compute the derivative of $K(t)$, we first recall the following result:

Lemma 1.1 (Differentiation of a Determinant). *We are given a family of linear operators $S = S(t) \in \mathcal{L}(\mathbb{R}^n)$, $t \in I$, such that $\det S(t) \neq 0$, for every $t \in I$. Then,*

$$\frac{d}{dt}[\det(S(t))] = \det[S(t)] \cdot \operatorname{tr}\left[\frac{dS}{dt}(t)S^{-1}(t)\right].$$

Proof:

1. We first recall the differentiation of the determinant function.

   We set $f(S) = \det S$. The function f is a polynomial function of (the coefficients of) S; therefore, it is differentiable and, by definition, its differential $f'(S)$ at S satisfies

   $$f'(S) \cdot T = \lim_{\lambda \to 0}\left\{\frac{f(S + \lambda T) - f(S)}{\lambda}\right\}$$

   $$= \lim_{\lambda \to 0}\left\{\frac{\det(S + \lambda T) - \det(S)}{\lambda}\right\},$$

   for all test matrices T. To compute this limit, write

   $$\det(S + \lambda T) = \det[S_1 + \lambda T_1, \ldots, S_n + \lambda T_n],$$

   where S_i, T_i is the ith column vector of S, T. Then, expand this determinant in the form

   $$\det(S_1, \ldots, S_n) + \lambda k + O(\lambda^2) \, (as \, \lambda \to O).$$

   It is left as an exercise for the reader to check that

   $$k = f'(S) \cdot T = \sum_{i=1}^{n}\sum_{j=1}^{n} T_{ij} A_{ij} = \operatorname{tr}(T \cdot (\operatorname{cof} S)^T),$$

where $A = (\text{cof } S)$ is the matrix, the entries of which are the cofactors[3] of S, and $(\text{cof } S)^T$ is the transposed matrix. Recalling that

$$(\text{cof } S)^T = (\det S) S^{-1},$$

we see that, at a point S where $\det S \neq 0$, the differential of the determinant can be written in the form

$$f'(S) \cdot T = \text{tr}(TS^{-1}) \det S.$$

2. Application
   We have, by the chain differentiation rule:

$$\frac{d}{dt} \det[S(t)] = f'[S(t)] \cdot \frac{dS}{dt}(t)$$

$$= \det[S(t)] \cdot \text{tr}\left[\frac{dS}{dt}(t) S^{-1}(t)\right].$$

We are now in position to prove the following result.

Proposition 1.3. *We assume that $C = C(x, t)$ is a function of class $\mathcal{C}^1$, for $x \in \Omega_t$ and $t \in I$, and that (this is a standing hypothesis) U (resp. Φ) is $\mathcal{C}^1$ with respect to x (resp. a) and t. Then,*

$$\frac{dK}{dt}(t) = \int_{\Omega_t} \frac{\partial C}{\partial t}(x, t)\, dx + \int_{\Omega_t} \text{div}(CU)(x, t)\, dx, \qquad (1.7)$$

$$\frac{dK}{dt}(t) = \int_{\Omega_t} \frac{\partial C}{\partial t}\, dx + \int_{\Gamma_t} CU \cdot n\, d\Gamma, \qquad (1.8)$$

where Γ_t is the boundary of Ω_t and n the unit outward normal vector on Γ_t.

Proof:
1. We first consider the case in which $C \equiv 1$ and $K(t) = \text{vol } \Omega_t$. We perform the change of variables $x = \Phi(a, t)$ in the integral; because we assume that $\det \nabla_a \Phi > 0$, we then have

$$\int_{\Omega_t} dx = \int_{\Omega_0} \frac{D\Phi}{Da}\, da$$

$$= \int_{\Omega_0} \det(\nabla_a \Phi)\, da.$$

---

[3] We recall that the cofactor S_{ij} of S is the determinant, multiplied by $(-1)^{i+j}$, of the matrix obtained by removing the ith line and the jth column of S.

Hence,

$$\frac{d}{dt}\int_{\Omega_t} dx = \int_{\Omega_0} \frac{\partial}{\partial t}\det(\nabla_a\Phi)\,da$$

$$= \int_{\Omega_0} \mathrm{tr}\left[\nabla_a\frac{\partial\Phi}{\partial t}\cdot(\nabla_a\Phi)^{-1}\right]\det(\nabla_a\Phi)\,da$$

$$= \int_{\Omega_t} \mathrm{tr}\left[\nabla_a\frac{\partial\Phi}{\partial t}\cdot(\nabla_a\Phi)^{-1}\right]dx.$$

On the other hand,

$$U_i(x,t) = \frac{\partial\Phi_i}{\partial t}(a,t),$$

$$\frac{\partial U_i}{\partial x_j}(x,t) = \frac{\partial^2\Phi_i}{\partial t\partial a_k}\cdot\frac{\partial a_k}{\partial x_j},$$

so that

$$(\nabla_x U)_{i,j} = \frac{\partial U_i}{\partial x_j} = \left(\nabla_a\frac{\partial\Phi}{\partial t}\right)_{i,k}\cdot(\nabla_a\Phi)^{-1}_{k,j},$$

and thus

$$\mathrm{div}\,U = \mathrm{tr}(\nabla_x U) = \mathrm{tr}\left[\left(\nabla_a\frac{\partial\Phi}{\partial t}\right)\cdot(\nabla_a\Phi)^{-1}\right],$$

which yields

$$\frac{d}{dt}\int_{\Omega_t} dx = \int_{\Omega_t} \mathrm{div}\,U\,dx.$$

2. The general case
   A computation similar to that performed above yields

$$\frac{d}{dt}\int_{\Omega_t} C(x,t)\,dx = \int_{\Omega_0} \frac{\partial}{\partial t}(C(\Phi(a,t),t)\det(\nabla_a\Phi))\,da$$

$$= \int_{\Omega_0} \frac{\partial C}{\partial t}\cdot\det(\nabla_a\Phi)\,da$$

$$+ \int_{\Omega_0} \frac{\partial C}{\partial x_j}\frac{\partial\Phi_j}{\partial t}\cdot\det(\nabla_a\Phi)\,da$$

$$+ \int_{\Omega_0} C\frac{\partial}{\partial t}\det(\nabla_a\Phi)\,da$$

$$= \int_{\Omega_t} \left(\frac{\partial C}{\partial t} + \frac{\partial C}{\partial x_j} U_j + C \operatorname{div} U \right) dx$$

$$= \int_{\Omega_t} \left(\frac{\partial C}{\partial t} + \operatorname{div} (CU) \right) dx.$$

This completes the proof of the proposition.

Remark 1.9: Proposition 1.3 is relative to a domain Ω_t limited by a closed surface Γ_t and that moves with the flow; $U = U(x, t)$ is then the velocity of the boundary Γ_t. Equation (1.8) can be extended to a time-dependent domain Ω_t, which does not necessarily move with the flow (i.e., Ω_t is not $\Phi(\cdot, t)\Omega_0$, at each time $t > 0$). In that case, Eq. (1.8) applies, $U = U(x, t)$ being then replaced by the (given) velocity of the boundary Γ_t of Ω_t. The proof is very similar; this generalization of Eq. (1.8) will be used in Chapter 6.

Remark 1.10: The expression *set that moves with the flow* introduced at the beginning of Section 1.5 applies also to surfaces, curves, points, and all sorts of sets. Sometimes the expression *comoving* is used in this context.

Exercises

1. We consider the velocity field of a continuum given in Eulerian description

$$U = (U_1, U_2, U_3) = (\sin t, \cos t, \alpha), \alpha \geq 0.$$

Compute the streamlines and the trajectories.
2. Show that the acceleration field γ in Eulerian representation can be written as

$$\gamma = \frac{\partial U}{\partial t} + \frac{1}{2} \nabla |U|^2 + (\operatorname{curl} U) \wedge U.$$

3. We consider the velocity field of a material given in Eulerian representation

$$U = (x_2 x_3, -x_1 x_3, kt + k_0), \quad k, k_0 \in \mathbb{R}.$$

   a) Compute the trajectories and specify their shape when $k = 0$ and $k_0 \neq 0$.
   b) Give the Lagrangian representation of the motion.
   c) For $k = 0$, compute the acceleration in Lagrangian and Eulerian representations.

4. We consider the motion of a continuum medium given in Lagrangian representation between times 0 and t by:

$$x_1 = a_1 \cos \omega t - a_2 \sin \omega t$$
$$x_2 = a_1 \sin \omega t + a_2 \cos \omega t$$
$$x_3 = a_3,$$

$\omega \in \mathbb{R}$.

a) Check that the Jacobian of the mapping $a \mapsto x = \Phi(a, t)$ does not vanish.

b) Give the Eulerian representation of the motion.

c) Compute the trajectories and the streamlines.

5. (*Complements of Vector Analysis.*) Show that

(i) $\det[\vec{A}, \vec{B}, \vec{C}] = \vec{A} \cdot (\vec{B} \wedge \vec{C}) = (\vec{A} \wedge \vec{B}) \cdot \vec{C}$
$= (\vec{C} \wedge \vec{A}) \cdot \vec{B} = \vec{C} \cdot (\vec{A} \wedge \vec{B}) = \vec{B} \cdot (\vec{C} \wedge \vec{A}) = (\vec{B} \wedge \vec{C}) \cdot \vec{A}.$

(ii) $\vec{A} \wedge (\vec{B} \wedge \vec{C}) = (\vec{A} \cdot \vec{C})\vec{B} - (\vec{A} \cdot \vec{B})\vec{C}.$

Here, $\vec{A}, \vec{B}$, and $\vec{C}$ denote three vectors in $\mathbb{R}^3$.

6. Show that

$$\text{div} (\Psi u) = \nabla \Psi \cdot u + \Psi \text{ div } u,$$
$$\text{curl} (\Psi u) = \nabla \Psi \wedge u + \Psi \text{ curl } u,$$
$$\text{curl curl } u = -\Delta u + \nabla \text{ div } u$$
$$\text{div} (u \wedge v) = v \cdot \text{curl } u - u \cdot \text{curl } v,$$
$$\text{curl} (u \wedge v) = u \text{ div } v - v \text{ div } u + (v \cdot \nabla)u - (u \cdot \nabla)v,$$

where Ψ is a scalar function (on $\mathbb{R}^3$) and u and v are vector functions.

7. We recall Green's formula

$$\int_\Omega \text{div } u \, dx = \int_{\partial\Omega} u \cdot n \, d\Gamma,$$

where Ω is an open set of $\mathbb{R}^d$ with a regular boundary $\partial\Omega, dx = dx_1 \dots dx_d$, and $d\Gamma$ is the surface measure on $\partial\Omega$.

Deduce from Green's formula that

(i) $\int_\Omega \frac{\partial u}{\partial x_i} dx = \int_{\partial\Omega} u \cdot n_i \, d\Gamma,$

(ii) $\int_\Omega \Delta u \cdot v \, dx = \int_{\partial\Omega} \frac{\partial u}{\partial n} v \, d\Gamma - \int_\Omega \nabla u \cdot \nabla v \, dx,$

where $\dfrac{\partial u}{\partial n} = \displaystyle\sum_{i=1}^{d} \dfrac{\partial u}{\partial x_i} n_i,$

(iii) $\displaystyle\int_{\Omega} (\Delta u \cdot v - u \cdot \Delta v) \, dx = \int_{\partial\Omega} (\dfrac{\partial u}{\partial n} v - \dfrac{\partial v}{\partial n} u) \, d\Gamma,$

(iv) $\displaystyle\int_{\Omega} u \cdot \text{curl } v \, dx = \int_{\Omega} v \cdot \text{curl } u \, dx - \int_{\partial\Omega} (u \wedge v) \cdot n \, d\Gamma,$

where u and v are scalar functions on $\mathbb{R}^d$ in (i), (ii), and (iii), and vector functions in (iv).

CHAPTER TWO

The fundamental law of dynamics

2.1. The concept of mass

The aim of this chapter is to allow to express and to generalize to continuum media the well-known law in mechanics (for one material point) $f = m\gamma$. Having introduced the acceleration, we now need to introduce the concepts of mass and force.

In this section, we consider a material system S in motion. This system fills the domain Ω_{t_0} at time t_0 and the domain Ω_t at time t. We define the concept of mass through the following hypothesis:

> *For every material system S and at each time t, there exists a positive measure μ_t carried by Ω_t and called the mass distribution.*

From here on, we will, most of the time, assume that μ_t is regular (with respect to the Lebesgue measure dx), that is to say, there exists a function $\rho = \rho(x, t)$ such that

$$d\mu_t(x) = \rho(x, t) \, dx;$$

ρ is called the volumic mass (or density) of the system at point x and at time t (we use here the Eulerian description of the motion). As usual, we assume that the function ρ is as smooth as necessary – at least of class $\mathcal{C}^1$ in x and t.

Other cases, which we will not consider here, are important in mechanics; for plates and shells the measure μ_t is carried by a surface Σ_t; for beams and strings the measure μ_t is carried by a curve Γ_t and is regular with respect to the arc-length measure, and finally for point masses, the measure μ_t is a linear combination of Dirac masses. All theses cases can be studied by methods that are similar to those we will present below.

24

The mathematician reader is in general familiar with measure's theory; the non-mathematician reader may simply consider $d\mu_t(x)$ as a convenient notation for one of the above-mentioned measures $\rho(x, t)dx$, $\rho(x, t)d\Sigma_t(x)$, $\rho(x, t)dl_t(x)$, or Dirac masses. Only elementary results of measure's theory will be used hereafter.

Conservation of mass

Here, we make the following assumption (namely, the conservation of mass hypothesis):

For every t and t', we have

$$\mu_{t'} = \Phi(\cdot, t', t)\mu_t,$$

which means that the mass distribution at time t' is the image of the mass distribution at time t by the mapping $\Phi(t', t) = \Phi(\cdot, t', t)$ (in the sense of the image of a measure by a mapping). In terms of volumic densities, this reads:

$$\rho'(x', t') = \rho(\Phi(x, t', t), t) \cdot \frac{Dx'}{Dx}. \tag{2.1}$$

Definition 2.1. *The (total) mass of the system at time t is the integral $m = \int_{\Omega_t} d\mu_t(x)$. If $d\mu_t(x) = \rho(x, t)\, dx$, then*

$$m = \int_{\Omega_t} d\mu_t(x) = \int_{\Omega_t} \rho(x, t)\, dx.$$

Similarly, if a material subsystem S' fills the domain $\Omega'_t \subset \Omega_t$ at time t, then the mass of S' is the integral

$$\int_{\Omega'_t} d\mu_t(x).$$

Remark 2.1: Of course, according to the mass conservation assumption, the mass of a system is constant:

$$\int_{\Omega_t} d\mu_t(x) = \int_{\Omega_{t'}} d\mu_{t'}(x')$$

for every t and t'. More generally, let us recall that if $\varphi = \varphi(x')$ is a sufficiently regular function defined on $\Omega_{t'}$, then, by definition of the image of a measure

$$\int_{\Omega_{t'}} \varphi(x')\, d\mu_{t'}(x') = \int_{\Omega_t} \varphi[\Phi(t', t)(x)]\, d\mu_t(x).$$

Definition 2.2. *The center of mass or center of inertia of the material system S at time t is the point G_t defined by*

$$OG_t = \frac{\int_{\Omega_t} x \, d\mu_t}{\int_{\Omega_t} d\mu_t}.$$

Remark 2.2: One can easily check that the definition of G_t is independent of the choice of the point O.

The point $G = G_t$ is not necessarily a material point that moves with the flow. Nevertheless, if $x_G(t) = OG_t$, we can define, as usual, the velocity and acceleration of the center of mass as

$$v(G) = \frac{d}{dt} x_G(t), \tag{2.2}$$

$$\gamma(G) = \frac{d^2}{dt^2} x_G(t) = \frac{dv(G)}{dt}. \tag{2.3}$$

It is clear that $x_G(t)$ is the center of mass (or barycenter) of the vectors x for the measure $d\mu_t(x)$. We will see hereafter that, similarly, $v(G_t)$ and $\gamma(G_t)$ are the centers of mass (or barycenters) of the velocities and accelerations of the points of S (of Ω_t) for the same measure.

Consequence of mass conservation: the continuity equation in Eulerian variables

Proposition 2.1. *We consider a material system S that fills the domain Ω_t at time $t \in I$. Then, for every $x \in \Omega_t$ and for every $t \in I$,*

$$\boxed{\frac{\partial \rho}{\partial t} + \text{div}(\rho U) = 0,}$$

where $\rho(x, t)$ is the volumic mass density and $U(x, t)$ the velocity of the particle occupying the position x at time t.

Proof: If $\Omega_t' \subset \Omega_t$, then, by the conservation of mass hypothesis,

$$0 = \frac{d}{dt} \int_{\Omega_t'} d\mu_t(x)$$

$$= \frac{d}{dt} \int_{\Omega_t'} \rho(x, t) \, dx$$

(thanks to Proposition 1.3)

$$= \int_{\Omega_t'} \left[\frac{\partial \rho}{\partial t} + \text{div}(\rho U) \right] dx.$$

Thus, for every $\Omega'_t \subset \Omega_t$,

$$\int_{\Omega'_t} \left[\frac{\partial \rho}{\partial t} + \text{div}(\rho U) \right] dx = 0,$$

which yields necessarily[1]

$$\frac{\partial \rho}{\partial t} + \text{div}(\rho U) = 0,$$

for every $(x, t) \in \Omega_t \times I$.

Remark 2.3: For certain materials (e.g., incompressible homogeneous fluids, see Part 2), ρ is constant in space and time, and the continuity equation then reads

$$\boxed{\text{div } U = 0.}$$

Proposition 2.2. *Let S be a three-dimensional material system that fills the domain Ω_t at time t, and let C be a function of class C^1 in x and t. Then,*

$$\frac{d}{dt} \int_{\Omega_t} C(x, t)\rho(x, t)\, dx = \int_{\Omega_t} \frac{DC}{Dt}(x, t)\rho(x, t)\, dx,$$

where $(D/Dt) = (\partial/\partial t) + U \cdot \nabla$ is the convective (or material, or total) derivative.

Proof: Thanks to Proposition 1.3, we see that

$$\frac{d}{dt} \int_{\Omega_t} C(x, t)\rho(x, t)\, dx$$

$$= \int_{\Omega_t} \left(\frac{\partial C}{\partial t}\rho + C\frac{\partial \rho}{\partial t} + \nabla C \cdot \rho U + C(\nabla \rho) \cdot U + C\rho \text{ div } U \right)(x, t)\, dx.$$

Because (see Proposition 2.1)

$$\frac{\partial \rho}{\partial t} + (\nabla \rho) \cdot U + \rho \text{ div } U = 0,$$

there remains

$$\frac{d}{dt} \int_{\Omega_t} C(x, t)\rho(x, t)\, dx = \int_{\Omega_t} \left(\frac{\partial C}{\partial t} + U \cdot \nabla C \right) \rho\, dx$$

$$= \int_{\Omega_t} \frac{DC}{Dt}\rho\, dx.$$

---

[1] If g is a continuous function on $\bar{\Omega}$ such that $\int_{\Omega'} g(x)\, dx = 0$, $\forall \Omega' \subset \Omega$, then g vanishes. This property, which is easy to verify, will be used many times.

Proposition 2.2 can be rewritten also as follows:

Corollary 2.1. *We have the following relation:*

$$\frac{d}{dt}\int_{\Omega_t} C(x,t)\,d\mu_t(x) = \int_{\Omega_t}\frac{DC}{Dt}d\mu_t(x).$$

Remark 2.4: A consequence of Corollary 2.1 is that the velocity of the center of mass of a material system S is given by the relation

$$v(G_t) = \frac{1}{\int_{\Omega_t} d\mu_t(x)}\int_{\Omega_t}\left[\frac{\partial x}{\partial t} + (U\cdot\nabla)x\right]d\mu_t(x);$$

that is to say, because $\partial x/\partial t = 0$ and $(U\cdot\nabla)x = U$,

$$v(G_t) = \frac{\int_{\Omega_t} U\,d\mu_t(x)}{\int_{\Omega_t} d\mu_t(x)}. \tag{2.3}$$

Similarly, differentiating a second time, we find the acceleration of the center of mass

$$\gamma(G_t) = \frac{\int_{\Omega_t}\gamma\,d\mu_t(x)}{\int_{\Omega_t} d\mu_t(x)}. \tag{2.4}$$

Hence, as we said before, although $x_G(t)$ is the center of mass (barycenter) of x for the measure $d\mu_t(x)$, $v(G_t)$ and $\gamma(G_t)$ are, respectively, the centers of mass (barycenters) of the velocity and acceleration functions $U(x,t)$ and $\gamma(x,t)$ for the same measure $d\mu_t(x)$.

Conservation of mass in Lagrangian variables

We now write the equation of conservation of mass for the Lagrangian description of the motion of the system. We have

$$\rho(x,t) = \rho[\Phi(a,t),t] = \sigma(a,t),$$

σ denoting the volumic mass density in Lagrangian variables. Furthermore, if $\Omega'_t \subset \Omega_t$,

$$\int_{\Omega'_t}\rho(x,t)\,dx = \int_{\Omega'_{t_0}}\sigma(a,t)J(a,t)\,da,$$

where $J = \det\nabla_a\Phi$. Because

$$\frac{d}{dt}\int_{\Omega'_{t_0}}\sigma(a,t)J(a,t)\,da = 0, \quad \forall\,\Omega'_{t_0}\subset\Omega_{t_0},$$

it follows that

$$\sigma(a, t)J(a, t) = \text{Const.} = \sigma(a, 0),$$

which expresses the conservation of mass in Lagrangian variables. When the fluid is homogeneous and incompressible (see Remark 2.3), we then find

$$J(a, t) \equiv 1.$$

We observe that this condition is less convenient than the divergence-free condition derived in Eulerian variables and is seldom used; the same remark applies to the conservation of mass equation in the compressible case.

Kinetic energy of a system with respect to a frame of reference

We consider a material system S that fills the domain Ω_t at time t. The kinetic energy of the system at time t is given by[1]

$$E_c = \frac{1}{2} \int_{\Omega_t} |U(x, t)|^2 \, d\mu_t(x),$$

where U denotes the velocity field in the corresponding frame of reference.

Linear and angular momentum and the corresponding HVF (Helicoidal Vector Field)

We now introduce two important HVFs associated with a material system, the *momentum HVF* and the *quantities of acceleration HVF*.

Given a material system S in motion, we define, at each instant of time t, the momentum HVF: its elements of reduction at the origin O of the frame of reference are

- The *linear momentum* of the system: $R = \int_{\Omega_t} U(x, t) \, d\mu_t(x)$,
- The *angular momentum* at 0 of the system: $\sigma(0) = \int_{\Omega_t} x \wedge U(x, t) \, d\mu_t(x)$.

The resulting or angular momentum at a point z is then given by

$$\sigma(z) = \sigma(0) + R \wedge z,$$

where $R = mv(G_t)$ is the linear momentum; $\{\sigma(z)\}_z$ is the momentum HVF of the system and

$$\sigma(z) = \int_{\Omega_t} (x - z) \wedge U(x, t) d\mu_t(x). \tag{2.5}$$

---

[1] Here and in other places, $|U|$ denotes the Euclidian norm of the vector U.

Quantities of acceleration and the corresponding HVF

Similarly, we define the quantities of acceleration HVF of a material system
S, at time t, as the HVF whose elements of reduction at 0 are

- The *dynamical resultant* of the system: $R' = \int_{\Omega_t} \gamma(x,t)\,d\mu_t(x)$,
- The *dynamical momentum* at 0: $\delta(0) = \int_{\Omega_t} x \wedge \gamma(x,t)\,d\mu_t(x)$.

The dynamical momentum at a point z is then given by

$$\delta(z) = \delta(0) + R' \wedge z = \int_{r_t} (x - z) \wedge \gamma(x,t)\,d\mu_t(x),$$

where $R' = m\gamma(G_t)$ is the dynamical resultant; $\{\delta(z)\}_z$ is the quantities of
acceleration HVF of the system at time t.

Remark 2.5: Consider two disjoint systems S and S'; then, the momentum
and quantities of acceleration HVFs for the system $S \cup S'$ are the sum of the
momentum and quantities of acceleration HVFs of S and S', respectively (this
result is an immediate consequence of the additive property of the integrals
with respect to the domain).

Proposition 2.3. *In a given frame of reference, the quantities of acceleration
HVF of a system is equal to the time derivative of its momentum HVF.*

Proof: It follows from Proposition 2.2 that $R' = (dR/dt)$. Hence, it suffices
to check that $\delta(0) = (d\sigma(0)/dt)$. Indeed, using the conservation of mass and
the Lagrangian variables, we have

$$\delta(0) = \int_{\Omega_t} x \wedge \gamma(x,t)\,d\mu_t(x) = \int_{\Omega_0} \Phi(a,t) \wedge \frac{\partial^2}{\partial t^2}\Phi(a,t)\,d\mu_0(a)$$
$$= \frac{d}{dt}\int_{\Omega_0} \Phi(a,t) \wedge \frac{\partial\Phi}{\partial t}(a,t)\,d\mu_0(a) = \frac{d}{dt}\int_{\Omega_t} x \wedge U(x,t)\,d\mu_t(x)$$
$$= \frac{d}{dt}\sigma(0).$$

2.2. Forces

We are given two disjoint material systems S and S' in motion; concerning the
actions (or forces) exerted by S' on S, we make the following assumptions:

*At each time t, the actions or forces exerted by S' on S are represented
by a vector measure $d\varphi_t(x)$ (or, equivalently, by three scalar measures)
carried by Ω_t.*

As in the case of the mass, we consider in practice four types of measures:

- Measures that are regular with respect to dx. In that case,

$$d\varphi_t(x) = f(x, t)\,dx, x \in \Omega_t,$$

  where $f = (f_1, f_2, f_3)$. We then say that f is the volumic density of the forces exerted (by S' on S at time t).
- Measures that are carried by a surface Σ_t and regular with respect to the surface measure $d\Sigma_t$;
- Measures that are carried by a curve Γ_t and regular with respect to the arc-length measure $d\ell_t$;
- Measures concentrated at a point that occur in the case of point forces.

In this study, we will consider all four cases that may occur in tridimensional continuum media.

As for the mass, the reader who is not familiar with measure's theory will be able to consider $d\varphi_t(x)$ as a convenient notation for one of these force fields or combinations of such force fields.

The forces that we encounter in mechanics are of various types: gravity, electromagnetism, contact forces (i.e., forces resulting from the contact of two bodies and concentrated on the surface of contact), and so forth.

The corresponding HVF

There is a natural way to associate an HVF with the forces exerted by S' on S; this is the HVF whose reduction elements at point 0 are $\int_{\Omega_t} d\varphi_t(x)$ and $\int_{\Omega_t} x \wedge d\varphi_t(x)$. The reduction elements at any other point z are then $\int_{\Omega_t} d\varphi_t(x)$ and $\int_{\Omega_t} (x - z) \wedge d\varphi_t(x)$. We say that $\int_{\Omega_t} d\varphi_t(x)$ is the *resultant (or linear resultant) of the forces* exerted by S' on S, and $\int_{\Omega_t} (x - z) \wedge d\varphi_t(x)$ is their *resulting momentum* at z.

Remark 2.6 (Additivity Assumption): Let S_1, S_2, and S be three disjoint material systems in motion. We make the natural assumption that the measure associated with the forces exerted by $S_1 \cup S_2$ on S is the sum of the measures associated with the forces exerted by S_1 and S_2 on S. Similarly, we assume that the measure associated with the forces exerted by S on $S_1 \cup S_2$ is the sum of the measures associated with the forces exerted by S on S_1 and by S on S_2.

A consequence of Remark 2.6 is that the HVF for the forces exerted by $S_1 \cup S_2$ on S is the sum of the HVF for the forces exerted by S_1 on S and of the HVF for the forces exerted by S_2 on S. Similarly, the HVF for the forces exerted by S on $S_1 \cup S_2$ is the sum of the HVF for the forces exerted by S on S_1 and of the HVF for the forces exerted by S on S_2 (this result is a mere consequence of the additivity of the integrals with respect to the domains).

Remark 2.7: In the case of a material point occupying the position M_t at time t, the momentum at M_t of the helicoidal vector field of the external forces applied to this point obviously vanishes. Thus, the corresponding HVF is fully characterized by its general resultant $F = F_t$ and its momentum equal to 0 at M_t.

2.3. The fundamental law of dynamics and its first consequences

We define the external forces of a material system S as the forces exerted by the whole universe (the complement of S) on S.

The fundamental law of dynamics. *There exists at least one frame of reference $\mathcal{R}$, called Galilean, such that at each time t and for every material system S, the two helicoidal vector fields associated with the quantities of acceleration of S, and with the external forces of S, are equal.*

In other words, for any system S, at every time t, and in a Galilean frame of reference, we have, keeping the notation above

$$\int_{\Omega_t} \gamma(x, t)\, d\mu_t(x) = \frac{d}{dt} \int_{\Omega_t} U(x, t)\, d\mu_t(x) = \int_{\Omega_t} d\varphi_t(x), \qquad (2.6)$$

$$\int_{\Omega_t} x \wedge \gamma(x, t)\, d\mu_t(x) = \frac{d}{dt} \int_{\Omega_t} x \wedge U(x, t)\, d\mu_t(x) = \int_{\Omega_t} x \wedge d\varphi_t(x).$$
$$(2.7)$$

For Eqs. (2.6) and (2.7) we have also used the fact that the HVF of quantities of acceleration is the time derivative of the momentum HVF, 0 being fixed in the Galilean frame of reference.

The fundamental equations (2.6) and (2.7) are also called the **Laws** or **Theorems** of **Conservation of Linear Momentum** and of **Conservation of Angular Momentum.** Theorem 2.1 below is a rephrasing of the law of conservation of linear momentum.

Remark 2.8: The fundamental law is no longer true when the frame of reference is not Galilean. Non-Galilean frames of reference are not much used in continuum mechanics but are often considered in celestial mechanics, in meteorology or oceanography, or for rigid body mechanics. In this case, it is necessary to add to the external force HVFs some other suitable HVFs related to the accelerations (see Section 2.5).

Several important results are deduced from the fundamental law of dynamics:

Corollary 2.2 (Action and Reaction Principle). *Let S_1 and S_2 be two disjoint material systems. Then, the HVF associated with the forces exerted by S_1 on S_2 (denoted by $[\mathcal{F}_{12}]$) is opposed to the HVF corresponding to the forces exerted by S_2 on S_1 (denoted by $[\mathcal{F}_{21}]$), that is to say*

$$[\mathcal{F}_{12}] = -[\mathcal{F}_{21}].$$

Proof: We set $S = S_1 \cup S_2$. By Remark 2.6, the external forces exerted on S_1 consist on the one hand of the forces exerted by S_2 on S_1, whose helicoidal vector field is $[\mathcal{F}_{21}]$ and, on the other hand, of the forces exerted by the complement of S on S_1, whose helicoidal vector field is denoted by $[\mathcal{F}_{e1}]$. Similarly, the external forces exerted on S_2 consist of the forces exerted by S_1 on S_2, with helicoidal vector field $[\mathcal{F}_{12}]$, and of the forces exerted by the complementary of S on S_2, whose helicoidal vector field is denoted by $[\mathcal{F}_{e2}]$.

Using the results derived in Section 2.2, we obtain, thanks to the fundamental law

$$[\mathcal{A}_S] = [\mathcal{F}_{e1}] + [\mathcal{F}_{e2}],$$

where $[\mathcal{A}]$ denotes the quantities of acceleration helicoidal vector field. Similarly

$$\left[\mathcal{A}_{S_1}\right] = [\mathcal{F}_{e1}] + [\mathcal{F}_{21}],$$
$$\left[\mathcal{A}_{S_2}\right] = [\mathcal{F}_{e2}] + [\mathcal{F}_{12}].$$

Because $[\mathcal{A}_S] = [\mathcal{A}_{S_1}] + [\mathcal{A}_{S_2}]$ (see Remark 2.5), we obtain, after summing these two equations:

$$[\mathcal{F}_{12}] + [\mathcal{F}_{21}] = 0.$$

Corollary 2.3 (The Fundamental Law of Statics). *If a material system S is at rest (in equilibrium) with respect to a Galilean frame of reference, then the HVF corresponding to the external forces applied to S vanishes.*

Remark 2.9: This last result is still valid when the motion of the system is rectilinear and uniform with respect to a Galilean frame of reference.

Remark 2.10: For a material point M, the fundamental law of dynamics implies

$$m\gamma = F,$$

where γ is the acceleration of M and F the external force exerted on M (see Remark 2.7). Because all momenta vanish at M, the fundamental law of dynamics does not provide any further equation; that is, this equation is equivalent to the fundamental law in the case of a material point.

Theorem 2.1 (Motion of the Center of Mass). *The motion of the center of mass of a material system S is the same as that of a material point whose mass is the total mass of the system and that would be subjected to a force equal to the resultant of the external forces exerted on S.*

Proof: We set $m = \int_{\Omega_t} d\mu_t(x)$. Then, according to Eq. (2.4)

$$m\gamma(G_t) = \int_{\Omega_t} \gamma(x,t)\,d\mu_t(x),$$

$$= \text{The dynamical resultant of } S,$$

$$= \text{The resultant of the external forces on } S,$$

owing to the fundamental law of dynamics.

A consequence of this theorem is the following corollary, which is important for the characterization of Galilean systems:

Corollary 2.4. *In a Galilean system, the motion of the center of mass of an isolated system is rectilinear and uniform.*

This corollary will be used in Section 2.5 for the description of some Galilean systems. An isolated system is a system S on which no external forces are applied.

2.4. Application to systems of material points and to rigid bodies

Systems of material points

We are given a set of material points M_i, $i = 1, \ldots, n$ (such a situation occurs for instance in the study of electromagnetic forces or for the motion of planets). The fundamental law applied to M_i reads

$$m_i\,\gamma_i = F_i + \sum_{k\neq i} F_{ki}, \tag{2.8}$$

where F_i denotes the external force on M_i, and F_{ki} is the force exerted by M_k on M_i. The fundamental law applied to M_j and to $M_i \cup M_j$ gives

$$m_i \gamma_i = F_i + \sum_{k \neq i} F_{ki}, \quad m_j \gamma_j = F_j + \sum_{k \neq j} F_{kj},$$

$$m_i \gamma_i + m_j \gamma_j = F_i + F_j + \sum_{k \neq i, k \neq j} F_{ki} + \sum_{k \neq i, k \neq j} F_{kj}.$$

Comparing these relations, we see that

$$F_{ji} + F_{ij} = 0. \tag{2.9}$$

Similarly, concerning the resulting momentum at a point O, we have

$$OM_i \wedge F_{ji} + OM_j \wedge F_{ij} = 0, \tag{2.10}$$

which yields, thanks to Eq. (2.8), that F_{ij} is necessarily parallel to $M_i M_j$:

$$M_i M_j \wedge F_{ij} = 0. \tag{2.11}$$

Equation (2.9) is proved exactly as is Eq. (2.8) by applying the fundamental law to M_i, M_j and $M_i \cup M_j$. This gives, for the resulting momentum at O,

$$OM_i \wedge m_i \gamma_i = OM_i \wedge F_i + OM_i \wedge \sum_{k \neq i} F_{ki},$$

$$OM_j \wedge m_j \gamma_j = OM_j \wedge F_j + OM_j \wedge \sum_{k \neq j} F_{kj},$$

$$OM_i \wedge m_i \gamma_i + OM_j \wedge m_j \gamma_j = OM_i \wedge F_i + \sum_{k \neq i, k \neq j} OM_i \wedge F_{ki}$$

$$+ OM_j \wedge F_j + \sum_{k \neq i, k \neq j} OM_j \wedge F_{kj};$$

Equation (2.9) follows by comparison of these three relations. Of course Eqs. (2.8) and (2.9) follow as well from the principle of action and reaction.

Analytical expression of the fundamental law

In the case of a single material point M, Eq. (2.7) reads

$$m\gamma = F. \tag{2.12}$$

Assuming, for instance, that F is a function of x, $\dot{x} = dx/dt$, and t, $F = F(x, \dot{x}, t)$, we obtain

$$m \frac{d^2}{dt^2} x(t) = F(x(t), \dot{x}(t), t), \tag{2.13}$$

which is equivalent to a system of three ordinary differential equations for the three components of $x(t) = OM(t)$. A classical problem in mechanics is the following: given the position and velocity of the point M at time 0 (that is to say $x(0)$ and $\dot{x}(0)$), determine the subsequent motion of the point M, that is to say determine $x(t)$ for $t > 0$. When the function F is smooth enough (which we always assume), the existence and uniqueness of solutions for Eq. (2.12) for given $x(0)$ and $\dot{x}(0)$ result from classical theorems on ordinary differential equations (solution of the Cauchy problem).

Similarly, for two material points M_1 and M_2 (and we could actually generalize to several material points), if we assume that the forces F_1, F_2, and F_{12} are, respectively, functions of $x_1, \dot{x}_1$, and t, of $x_2, \dot{x}_2$, and t, of $x_1, x_2, \dot{x}_1$, $\dot{x}_2$, and t, then Eq. (2.7) is equivalent to

$$\begin{cases} m_1\ddot{x}_1 = F_1(x_1, \dot{x}_1, t) - F_{12}(x_1, \dot{x}_1, x_2, \dot{x}_2, t), \\ m_2\ddot{x}_2 = F_2(x_2, \dot{x}_2, t) + F_{12}(x_1, \dot{x}_1, x_2, \dot{x}_2, t). \end{cases} \quad (2.14)$$

Here, $x_i(t) = OM_i(t)$, $i = 1, 2$. Similarly, given $x_1(0), \dot{x}_1(0), x_2(0)$, and $\dot{x}_2(0)$, and if the functions F_1, F_2, and F_{12} are sufficiently smooth, the existence and uniqueness of solutions for Eq. (2.13) result from the classical theorems for the Cauchy problem for differential systems.

In conclusion, for a finite number of material points, and under reasonable assumptions on the forces, *the fundamental law allows us to show that the motion of such systems is entirely determined by the knowledge of the external and internal forces, the initial position of the system, and the initial velocity distribution.* It is the ultimate purpose of the following chapters to prove similar results for continuous systems, but this problem is much more complicated for continuum media and is far beyond the scope of this book. In particular, the *ordinary differential equations* (ODEs) must then be replaced by partial differential equations (PDEs). The PDEs are generally considerably more difficult to study and solve than ODEs, and some of the PDEs that appear subsequently still lead to many open mathematical problems (see some remarks on the PDEs of mechanics in the Appendix to this book).

Rigid bodies

We now move to the case of a rigid body S whose center of mass or center of inertia is denoted by G. A first consequence of the fundamental law is the following theorem.

Theorem 2.2 (Angular Momentum Theorem). *Let Δ be an axis fixed in the Galilean frame of reference $\mathcal{R}$. Then,*

$$\frac{d\sigma_\Delta}{dt} = \delta_\Delta = m_\Delta,$$

where $\sigma_\Delta = \sigma(A) \cdot k$ is the angular momentum with respect to Δ and $m_\Delta = m_A \cdot k$, m_A being the momentum of the external forces at any point A, $A \in \Delta$, and k is the unit vector of Δ; $\sigma(A)$ denotes, as before, the angular momentum at A.

Proof: We saw (Proposition 2.3) that

$$m_O = \frac{d}{dt}\sigma(O) = \frac{d}{dt}\int_{\Omega_t} x \wedge U(x,t)\,d\mu_t.$$

Similarly, at the point A $(OA = x_A)$:

$$m_A = \frac{d}{dt}\int_{\Omega_t}(x - x_A) \wedge U(x,t)\,d\mu_t(x);$$

hence,

$$m_\Delta = m(A) \cdot k = \frac{d}{dt}\int_{\Omega_t}[(x - x_A), U(x,t), k]\,d\mu_t(x),$$

and the result follows.

The motion of a rigid body is described by six parameters, and it is governed by six scalar equations corresponding to the theorem of center of inertia motion and to the angular momentum theorem projected on three noncoplanar axes.

Inertia tensor at a point A

Let S be a rigid body, and let A be a fixed point of S. Then, the angular momentum at A reads

$$\sigma_A = \int_{\Omega_t}(x - x_A) \wedge U(x,t)\,d\mu_t(x),$$

$(OA = x_A)$ and, because S is a rigid body,

$$U(x,t) = U(x_A,t) + \omega \wedge (x - x_A),$$

for some $\omega \in \mathbb{R}^3$. Consequently,

$$\sigma_A = \int_{\Omega_t}(x - x_A) \wedge U(x_A,t)\,d\mu_t(x)$$

$$+ \int_{\Omega_t}(x - x_A) \wedge [\omega \wedge (x - x_A)]\,d\mu_t(x)$$

$$= mAG \wedge U(x_A,t) + \int_{\Omega_t}(x - x_A) \wedge [\omega \wedge (x - x_A)]\,d\mu_t(x).$$

The mapping

$$J_A : \omega \mapsto \int_{\Omega_t} (x - x_A) \wedge [\omega \wedge (x - x_A)]\, d\mu_t(x) \qquad (2.15)$$

is a linear operator called the inertia tensor of S at the point A (this notion is independent of the choice of the frame of reference).

If S is an homogeneous body, that is, if the mass measure $d\mu_t$ is proportional to the volume, $d\mu_t(x) = \rho dx$, where ρ is constant, and if $O \equiv A$ is fixed in $S = \Omega_0$, then

$$J_0(u) \cdot v = \rho \int_{\Omega_0} [x \wedge (u \wedge x)] \cdot v\, dx$$

$$= \rho \int_{\Omega_0} [|x|^2 u - (x \cdot u)x] \cdot v\, dx$$

$$= \rho \int_{\Omega_0} [|x|^2(u \cdot v) - (x \cdot u)(x \cdot v)]\, dx$$

$$= J_0(v) \cdot u. \qquad (2.16)$$

Thus, $(u, v) \mapsto J_0(u) \cdot v$ is a bilinear symmetric form. Furthermore, J_0 is entirely defined by the quantities

$$(J_0)_{ij} = J_0(e_i) \cdot e_j,$$

(e_1, e_2, e_3) being the canonical basis of $\mathbb{R}^3$. If follows immediately that

$$(J_0)_{ii} = \rho \int_{\Omega_0} \left(|x|^2 - x_i^2\right) dx,$$

$$(J_0)_{ij} = -\rho \int_{\Omega_0} x_i x_j dx, \quad i \neq j.$$

In particular, if G is the center of mass of S and Δ is an axis containing G, with unit vector k,

$$\sigma_G = J_G(\omega), \quad \sigma_\Delta = J_G(\omega) \cdot k.$$

2.5. Galilean frames: the fundamental law of dynamics expressed in a non-Galilean frame

We start with the following remark pertaining just to kinematics:

Comparison of velocities and accelerations in two different frames

We consider two different frames of reference $\mathcal{R}_f$ and $\mathcal{R}_m$ with the same chronologies; we want to compare the velocities and accelerations with respect to $\mathcal{R}_f$ and $\mathcal{R}_m$ of a moving point $M = M(t)$.

The two frames of reference play a somehow symmetric role, but for the sake of convenience, we will say that $\mathcal{R}_f$ is the fixed frame and $\mathcal{R}_m$ is the moving one. The velocities and accelerations of M with respect to $\mathcal{R}_f$ and $\mathcal{R}_m$ will be called, respectively, absolute and relative and will be denoted by $U_f, \gamma_f, U_m, \gamma_m$.

We are thus given the frame $\mathcal{R}_f$ with origin O_f and basis e_i, $i = 1, 2, 3$, and the frame $\mathcal{R}_m$ with origin O_m and basis $f_i = f_i(t)$, $i = 1, 2, 3$; these bases are connected by the relations

$$f_i(t) = a_{ij}(t)e_j,$$

where the Einstein convention of summation of repeated indices is understood throughout. The position of $M = M(t)$ is determined in the frame $\mathcal{R}_m$ by the vector

$$O_m M(t) = x_i(t) f_i(t).$$

The velocity of M with respect to $\mathcal{R}_f$ is given by the relation

$$U_f = \frac{dO_f M(t)}{dt} = U_f(O_m) + \dot{x}_i(t) f_i(t) + x_i(t) \dot{f}_i(t),$$

where $U_f(O_m) = (dO_f O_m)/(dt)$ is the absolute velocity of O_m and where $\dot{u}$ denotes the derivative of u with respect to t. We can rewrite this relation in the form

$$U_f = U_m + U_e, \tag{2.17}$$

where $U_m = \dot{x}_i(t) f_i(t)$ is the relative velocity of M in $\mathcal{R}_m$ and $U_e = U_f(O_m) + x_i(t) \dot{f}_i(t)$. It is easy to verify that U_e is the velocity in $\mathcal{R}_f$ of the point, fixed in $\mathcal{R}_m$, which coincides with M at time t; it is called the transport velocity of M (in the motion of $\mathcal{R}_m$ with respect to $\mathcal{R}_f$). Similarly, the acceleration of M in $\mathcal{R}_f$ is given by

$$\gamma_f = \frac{d^2 O_f M}{dt^2} = \gamma_f(O_m) + \ddot{x}_i(t) f_i(t) + 2\dot{x}_i(t) \dot{f}_i(t) + x_i(t) \ddot{f}_i(t),$$

$\gamma_f(O_m) = (d^2 O_f O_m)/(dt^2)$ being the absolute acceleration of O_m; we rewrite this last relation as

$$\gamma_f = \gamma_m + \gamma_c + \gamma_e, \tag{2.18}$$

where $\gamma_m = \ddot{x}_i(t) f_i(t)$ is the relative acceleration of M in $\mathcal{R}_m$, $\gamma_c = 2\dot{x}_i(t)\dot{f}_i(t)$ is called the Coriolis acceleration, and $\gamma_e = \gamma_f(O_m) + x_i(t)\ddot{f}_i(t)$ is called the transport acceleration of M (in the motion of $\mathcal{R}_m$ with respect to $\mathcal{R}_f$). This is the acceleration in $\mathcal{R}_f$ of the point, fixed in $\mathcal{R}_m$, which coincides with M at time t.

If $\{U(x_0), \omega\}$ is the HVF corresponding to the velocity field of $\mathcal{R}_m$ in its motion with respect to $\mathcal{R}_f$, we have

$$\dot{f}_i = \omega \wedge f_i,$$

and

$$\ddot{f}_i = \dot{\omega} \wedge f_i + \omega \wedge (\omega \wedge f_i).$$

Hence,

$$\gamma_c = 2\omega \wedge U_m,$$

$$\gamma_e = \gamma_f(O_m) + \dot{\omega} \wedge O_m M + \omega \wedge (\omega \wedge O_m M). \tag{2.19}$$

Among other cases of interest, let us mention the case in which $\mathcal{R}_m$ is moving in a uniform translation with respect to $\mathcal{R}_f$, in which case $\gamma_f = \gamma_m$.

The fundamental law in a non-Galilean frame

Let us call this frame of reference $\mathcal{R}_m$ and let us assume that $\mathcal{R}_f$ is Galilean. We can write the fundamental law in $\mathcal{R}_m$, the acceleration being the relative accelerations in $\mathcal{R}_m$, provided that we add to the external forces the forces corresponding to the opposite of the accelerations γ_c and γ_e, namely the forces defined by the measures $-\gamma_c d\mu_t$ and $-\gamma_e d\mu_t$. Then:

> *The HVF corresponding to the accelerations of a system in a non-Galilean frame is the sum of the HVFs corresponding to the external forces of the system and to the inertial forces corresponding to the Coriolis acceleration and the frozen point acceleration.*

Galilean frames

We use Corollary 2.4 to determine Galilean frames of reference. Furthermore, we infer from the previous discussion that if a frame of reference is Galilean, then every other frame moving in a rectilinear uniform translation with respect to it is Galilean as well because the accelerations γ_c and γ_e vanish (by Eq. (2.18)).

On the basis of Corollary 2.4, here are examples of frames that can be considered as Galilean, depending on the problems under consideration, up to an acceptable level of approximation.

If we assume that the forces exerted by the stars on the solar system are negligible, then the solar system is isolated, and a frame centered at the center of mass of the solar system, whose axes have fixed directions with respect to the solar system, is Galilean; such Galilean frames are used in celestial mechanics. Similarly, if we assume that the forces exerted by the stars and the planets on the earth are negligible, then a frame whose center is the center of the earth and whose directions are fixed with respect to the stars is Galilean at this level of precision; such Galilean frames are used, for instance, in meteorology and oceanography. Finally, for engineering problems of mechanics on earth (e.g., motion of a vehicle or of a robot), we assume that the forces due to the rotation of the earth (see Chapter 12, Section 12.1) and the forces exerted by the stars and the planets are negligible; therefore, a frame connected to the earth (fixed with respect to the earth) is considered as Galilean for such problems.

Exercises

1. Determine the motion of a material point of mass m under the action of gravitation, knowing its initial position and velocity.
2. Compute the center of mass of a half-disc D of radius R.
3. Study the motion of a material point of mass m which moves, without friction, on a vertical circle of radius R. The position of the material point will be determined by the angle θ made by the corresponding radius with the vertical.
4. We consider a body Ω submitted to forces with surface density $-\rho \vec{n} d\Sigma$, on its boundary $\partial \Omega$, where $\vec{n}$ is the unit outer normal vector and $\rho = \rho(x_3)$ is a linear function of x_3, $x = (x_1, x_2, x_3)$. Compute the HVF corresponding to these forces.

CHAPTER THREE

The Cauchy stress tensor and the
Piola-Kirchhoff tensor. Applications

This chapter is central to continuum mechanics. Our aim is to model and study the cohesion forces (or internal forces) of a system, that is to say the actions exerted by part of a system S on another part of S. Our study leads to the definition of the Cauchy stress tensor and to the equations of statics and dynamics that then follow by application of the fundamental law of dynamics.

The Cauchy stress tensor is expressed in the Eulerian variable; its analogue in the Lagrangian variable is the Piola-Kirchhoff tensor introduced in the last section of this chapter.

3.1. Hypotheses on the cohesion forces

We are given a material system S. Let $S = S_1 \cup S_2$ be a partition of S, Ω_1 and Ω_2 being the domains occupied by S_1 and S_2 at a given time. In Sections 3.1 and 3.2, the time will be fixed and will not appear explicitly. We assume that the common boundary Σ of Ω_1 and Ω_2 (see Figure 3.1) is sufficiently regular.

Concerning the actions exerted by S_2 on S_1, we make the following assumptions introduced by Cauchy:

(H1) *The forces exerted by S_2 on S_1 are contact forces, which means that they can be represented by a vector measure $d\varphi$ concentrated on $\Sigma = \partial\Omega_1 \cap \partial\Omega_2$.*

(H2) *The measure $d\varphi$ is absolutely continuous with respect to the surface measure $d\Sigma$, that is,*

$$d\varphi = T\,d\Sigma,$$

where T is the (vector) surface density of the forces.

42

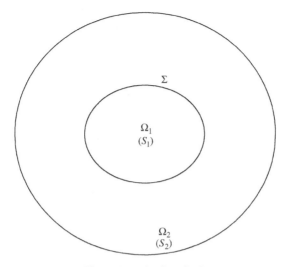

Figure 3.1 The domain Ω.

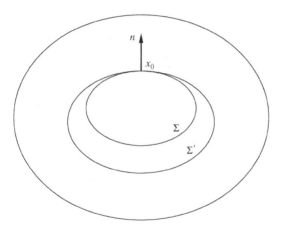

Figure 3.2 T depends only on n.

(H3) *The function T depends only on the point x of Σ and on the unit normal n to Σ at point x:*

$$T = T(x, n).$$

Thus, under the conditions of Figure 3.2 corresponding to two partitions of S, $S_1 \cup S_2$ and $S_1' \cup S_2'$, $T(x_0, n)$ is the same at the point x_0 for both partitions.

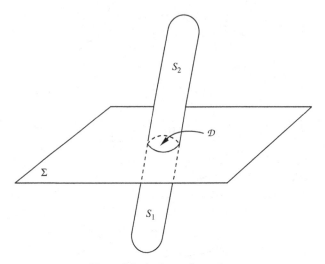

Figure 3.3 Action and reaction.

Considering general partitions, we see that the function T is defined for every $x \in \Omega$ and every $n \in \mathbb{R}^3$ such that $|n| = 1$. We then make the following minimal regularity assumption:

(H4) *For fixed n, the function $x \mapsto T(x, n)$ is continuous.*

A first consequence of the fundamental law (and more precisely of the action and reaction principle) is the following:

Proposition 3.1. *For every $x \in \Omega$ and for every $n \in \mathbb{R}^3$, $|n| = 1$, we have the following:*

$$T(x, n) = -T(x, -n).$$

Proof: Let $x \in \Omega$ and $n \in \mathbb{R}^3$ satisfy $|n| = 1$. We consider two systems S_1 and S_2 contained in S and separated by a plane Σ perpendicular to n (see Figure 3.3).

Then, according to the action–reaction principle, $[\mathcal{F}_{ij}]$ denoting as in Chapter 2 the helicoidal vector field for the actions exerted by $S_i(\Omega_i)$ on $S_j(\Omega_j)$, we have

$$[\mathcal{F}_{12}] + [\mathcal{F}_{21}] = 0.$$

Consequently (hypotheses (H1) and (H2)), if $\mathcal{D} = \partial\Omega_1 \cap \partial\Omega_2$,

$$\int_{\mathcal{D}} T(x, n) \, d\Sigma + \int_{\mathcal{D}} T(x, -n) \, d\Sigma = 0.$$

We thus deduce that

$$\int_{\mathcal{D}} [T(x, n) + T(x, -n)] \, d\Sigma = 0, \quad \forall \, \mathcal{D} \subset \Sigma,$$

which yields

$$T(x, n) + T(x, -n) = 0, \quad \forall \, x \in \Sigma.$$

Definition 3.1. *The vector $T(x, n)$ is called the stress vector at x for the direction n. Furthermore, $T_n(x, n) = T(x, n) \cdot n$ is the normal stress at x for the direction n and $T_t(x, n) = T(x, n) - nT_n(x, n)$ is the tangential stress or shear stress at x for the direction n.*

We say that the material is subjected to a tension when $T_n > 0$ and to a compression when $T_n < 0$.

Remark 3.1: An immediate consequence of Proposition 3.1 is that

$$T(x, -n) \cdot (-n) = T(x, n) \cdot n,$$

$\forall \, x \in \Omega, \, \forall \, n \in \mathbb{R}^3, \, |n| = 1$. Thus, $T_n(x, n)$ only depends on the direction of n.

3.2. The Cauchy stress tensor

To study $T(x, n)$, $x \in \Omega$, $n = (n_1, n_2, n_3)$, $|n| = 1$, we assume for simplicity that $x = 0$ is the origin of the orthonormal system of coordinates. Let e_1, e_2, and e_3 be the unit vectors on $0x_1$, $0x_2$, and $0x_3$, respectively. We set

$$T(0, e_i) = \sigma_{ji} e_j.$$

Thus, at every point x ($x = 0$ here), we define a matrix whose entries are $\sigma_{ij}(x)$.

We then consider the domain S_1 consisting of a tetrahedron with summit 0 and with bases A_1, A_2, A_3 (see Figure 3.4), $A_1 A_2 A_3$ being perpendicular to n. If $h > 0$ is the distance from 0 to the plane $A_1 A_2 A_3$, the equation of the plane reads

$$n_1 x_1 + n_2 x_2 + n_3 x_3 = h.$$

The tetrahedron $A_1 A_2 A_3$ is chosen so that $n_i > 0$, $i = 1, 2, 3$.

Let $d\varphi$ be the measure associated with the external forces on S, and let γ be the acceleration in a Galilean frame and $d\mu$ the mass distribution of S_1.

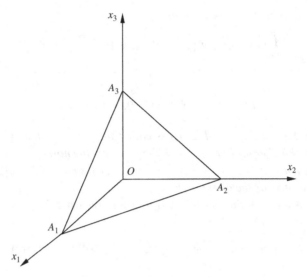

Figure 3.4 The system S_1.

The fundamental law of dynamics applied to the system S_1 reads

$$\int_{\Omega_1} \gamma(x)\,d\mu(x) = \int_{\Omega_1} d\varphi(x) + R,$$

where Ω_1 is the domain filled by S_1, and R is the resultant of the forces exerted by $S_2 = S \backslash S_1$ on S_1. Thus, according to Cauchy's hypotheses:

$$R = \int_{0A_2A_3} T(x, -e_1)\,dx_2\,dx_3 + \int_{0A_1A_2} T(x, -e_3)\,dx_1\,dx_2$$

$$+ \int_{0A_1A_3} T(x, -e_2)\,dx_1\,dx_3 + \int_{A_1A_2A_3} T(x, n)\,d\Sigma.$$

We have

$$\int_{0A_2A_3} T(x, -e_1)\,dx_2\,dx_3 = -\int_{0A_2A_3} T(x, e_1)\,dx_2\,dx_3,$$

and, because the functions $x \mapsto \sigma_{ij}(x)$ are continuous,

$$T(x, e_1) = \sigma_{j1}(x)e_j = [\sigma_{j1}(0) + o(1)]e_j,$$

where $o(1)$ tends to 0 as h tends to 0. Therefore,

$$\int_{0A_2A_3} T(x, e_1)\,dx_2\,dx_3 = \text{area}(0A_2A_3)\,[\sigma_{j1}(0)e_j + o(1)].$$

Because area$(0A_2A_3) = n_1 \cdot$ area$(A_1A_2A_3)$, where n_1 is the cosine of the angle of the normals n and e_1 to the triangles $A_1A_2A_3$ and $0A_2A_3$, it follows that

$$\frac{1}{\text{area}(A_1A_2A_3)} \int_{0A_2A_3} T(x, e_1)\, dx_2\, dx_3 = n_1[\sigma_{j1}(0)e_j + o(1)].$$

We proceed similarly for the faces $0A_1A_3$ and $0A_1A_2$ of the tetrahedron $\Omega_1 = \Omega_1(h)$ and then, for $A_1A_2A_3$,

$$\int_{A_1A_2A_3} T(x, n)\, d\Sigma = \text{area}(A_1A_2A_3)[T(0, n) + o(1)].$$

On the whole,

$$R = R(h) = \text{area}(A_1A_2A_3)[T(0, n) - n_i\sigma_{ji}(0)e_j + o(1)]$$
$$= O(h^2)[T(0, n) - n_i\sigma_{ji}(0)e_j + o(1)].$$

On the other hand, all the functions being regular,

$$\int_{\Omega_1(h)} \gamma(x)\, d\mu(x) - \int_{\Omega_1(h)} d\varphi(x) = \int_{\Omega_1(h)} [\gamma(x)\rho(x) - f(x)]\, dx$$
$$= [\gamma(0)\rho(0) - f(0)]O(h^3),$$

For $h \to 0$, there remains, after dividing by h^2,

$$T(0, n) = \sigma_{ji}(0)n_i e_j. \tag{3.1}$$

This relation, valid for every vector n when $|n| = 1$ and when the n_i are > 0 can be easily generalized to the case where the n_i have arbitrary signs. This formula then suggests to define a linear operator $\sigma (= \sigma(x)$ or $\sigma(x, t)$ to be more precise):

$$\sigma : n = (n_1, n_2, n_3) \mapsto \sigma \cdot n = \sigma_{ji}n_i e_j.$$

The fact that this definition is intrinsic (independent of the given frame e_1, e_2, e_3) is assumed and we will not prove it here (we recall that $\sigma_{ji} = T(0, e_i) \cdot e_j$). We then obtain

$$T(x, n) = \sigma_{ij}(x)n_j e_i, \quad \text{for every } x, \text{ and for every } n. \tag{3.2}$$

A consequence of Eq. (3.2) is the following result.

Theorem 3.1. *The stress vector at x for the direction n is a linear function of the components of n.*

We thus have a linear operator defined even when n is not a unit vector:

$$n \mapsto T(x, n) = \sum_{i,j=1}^{3} \sigma_{ji}(x) n_i e_j$$

$$= \sum_{i,j=1}^{3} [T(x, e_i) \cdot e_j] e_j n_i,$$

where $n = \sum_{i=1}^{3} n_i e_i$. This operator is called the *Cauchy stress tensor* at x of the continuum and is denoted by $\sigma(x)$ or, reintroducing time, by $\sigma(x, t)$.

As usual, we assume that the stress tensor $\sigma = \sigma(x, t)$ is a regular function of class at least $\mathcal{C}^1$ of x and t.

Remark 3.2: Just as we assumed the intrinsic nature of the definition of σ (independence with respect to the orthogonal frame), we will not discuss the invariance of σ under a change of frame in relation with the invariance of forces (change of Galilean frame in particular). This is very important for the next section, but we refer the reader to more specialized books for this point.

3.3. General equations of motion

We reintroduce the time variable $\sigma = \sigma(x, t)$. Our aim in this section and the next is to write the fundamental law of dynamics using the stress tensor. In this section we write the linear momentum equations; in Section 3.4 we will write the angular momentum equations. We have

Theorem 3.2. *Given a body whose mass density is $\rho(x, t)$ and that is subjected to external forces with volumetric density $f(x, t)$, we have*

$$\boxed{\rho\, \gamma_i = f_i + \sigma_{ij,j},} \tag{3.3}$$

where

$$\sigma_{ij,j} = \sigma_{ij,j}(x, t) = \sum_{j=1}^{3} \frac{\partial \sigma_{ij}(x, t)}{\partial x_j}, \quad i = 1, 2, 3, \ x \in \Omega_t, \ t \in I.$$

Proof: Consider $S' \subset S$ filling the volume Ω'_t at time t, $\bar{\Omega}'_t \subset \Omega_t$, and let $\Gamma'_t = \partial \Omega'_t$. The external forces on S' are the volume forces defined by the measure $f\, dx$ and the contact forces exerted by $S \backslash S'$ on S' and defined by the measure $T\,d\Gamma$ concentrated on Γ'_t (according to the hypotheses in Section 3.2). Hence, by the fundamental law, with $T_i = \sigma_{ij} \cdot n_j$,

$$\int_{\Omega'_t} \rho(x,t)\gamma_i(x,t)\,dx = \int_{\partial\Omega'_t} T_i(x,n)\,d\Gamma + \int_{\Omega'_t} f_i(x,t)\,dx$$

$$= \int_{\partial\Omega'_t} \sigma_{ij}(x)\cdot n_j\,d\Gamma + \int_{\Omega'_t} f_i(x,t)\,dx$$

$$= \text{(using the Stokes formula)}$$

$$= \int_{\Omega'_t} \sigma_{ij,j}(x)\,dx + \int_{\Omega'_t} f_i(x,t)\,dx.$$

Finally,

$$\int_{\Omega'_t} (\rho(x,t)\gamma_i(x,t) - \sigma_{ij,j}(x,t) - f_i(x,t))\,dx = 0,$$

$\forall\,\Omega'_t \subset \Omega_t$, and thus the integrand vanishes.

We saw that in Eulerian coordinates

$$\gamma_i = \frac{\partial U_i}{\partial t} + \sum_{j=1}^{3} U_j \frac{\partial U_i}{\partial x_j}.$$

We deduce from the preceding relations the following fundamental equations of continuum mechanics:

$$\rho\left(\frac{\partial U_i}{\partial t} + \sum_{j=1}^{3} U_j \frac{\partial U_i}{\partial x_j}\right) = \sigma_{ij,j} + f_i,$$
$$i = 1, 2, 3.$$

(3.4)

Equilibrium equations

When the system is in equilibrium, $U = 0$ and Eqs. (3.4) reduce to

$$\sigma_{ij,j}(x) + f_i(x) = 0, \quad \text{in } \Omega, \quad i = 1, 2, 3.$$

(3.5)

Remark 3.2 (Continuity of the Stress Vector and Boundary Conditions): As we will see (in Chapter 5), the stress tensor is not necessarily continuous at the boundary between two media. However, the stress vector $T(x,n)$ is continuous provided we assume that the accelerations and forces remain

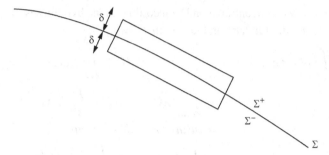

Figure 3.5 The domain Ω_δ.

bounded at the boundary between two continua (which is true except in the case of shock waves). To prove this continuity of $T(x, n)$, we apply the fundamental law of dynamics to the domain Ω_δ (shown in Figure 3.5), and we let δ tend to 0. We omit the details.

In continuum mechanics, the continuity of $T(x, n)$ yields interesting boundary conditions. For instance, if S is subjected to a surface density of forces F on its boundary Γ_t, then

$$F(x) = T(x, n) = \sigma(x) \cdot n(x), \quad \text{on } \Gamma_t = \partial\Omega_t.$$

3.4. Symmetry of the stress tensor

In Section 3.3, we have used the linear momentum equations resulting from the fundamental law of dynamics to establish Theorem 3.2. We now use the angular momentum equations, which lead to the following result.

Theorem 3.3. *The stress tensor σ is symmetric, $\sigma_{ij} = \sigma_{ji}$, $\forall\, i, j$.*

Proof: We consider an arbitrary domain $\Omega' \subset \Omega$ and write the equality of the angular momentum resulting from the fundamental law as follows:

$$\int_{\Omega'} x \wedge (\rho\gamma - f)\,dx = \int_{\partial\Omega'} x \wedge T(x, n)\,d\Gamma$$

$$= \int_{\partial\Omega'} x \wedge [\sigma(x) \cdot n]\,d\Gamma.$$

We assume here that the external forces are volume forces defined by the measure $f\,dx$.

We write the first component of the previous equation as follows:

$$\int_{\Omega'} [x_2(\rho\gamma_3 - f_3) - x_3(\rho\gamma_2 - f_2)]\, dx$$

$$= \int_{\partial\Omega'} (x_2\sigma_{3j}n_j - x_3\sigma_{2j}n_j)\, d\Gamma$$

$$= \int_{\Omega'} \frac{\partial}{\partial x_j}(x_2\sigma_{3j} - x_3\sigma_{2j})\, dx$$

$$= \int_{\Omega'} (\delta_{2j}\sigma_{3j} + x_2\sigma_{3j,j} - \delta_{3j}\sigma_{2j} - x_3\sigma_{2j,j})\, dx,$$

where δ_{ij} denotes the Kroenecker symbol equal to 1 if $i = j$ and to 0 if $i \neq j$. Consequently,

$$\int_{\Omega'} [\sigma_{32} - \sigma_{23} + x_2(\sigma_{3j,j} - \rho\gamma_3 + f_3) - x_3(\sigma_{2j,j} - \rho\gamma_2 + f_2)]\, dx = 0.$$

Thus,

$$\int_{\Omega'} (\sigma_{32} - \sigma_{23})\, dx = 0, \quad \forall\, \Omega' \subset \Omega,$$

and

$$\sigma_{32} = \sigma_{23}.$$

If the last two components are considered similarly, it follows that

$$\sigma_{13} = \sigma_{31}, \qquad \sigma_{12} = \sigma_{21}.$$

Consequences

1. For every n, n', $T(x, n) \cdot n' = T(x, n') \cdot n$.
2. The two quadrics of equation $\sigma_{ij}\xi_i\xi_j = \pm 1$ are called the stress qua-drics. We note that $\sigma_{ij}(x)\xi_i\xi_j$ is not a positive definite quadratic form; for instance, we already observed that $T_n = (\sigma \cdot n) \cdot n$ can be either positive or negative.
3. The eigenvectors of $\sigma(x)$ are called the principal directions of stresses at x. They consist of the vectors n such that $T(x, n)$ is parallel to n,

$$\sigma(x) \cdot n = \lambda n.$$

In other words, $n = n(x)$ is an eigenvector of $\sigma(x)$. For such a direction, the stress is purely normal. Owing to the symmetry of the tensor $\sigma = \sigma(x)$, there exist at every point x at least three principal directions for the stresses that are mutually orthogonal (diagonalization of a real symmetric matrix).

4. The deviator of a tensor σ (in particular the stress tensor) is the tensor $\sigma^D = \sigma - rI$, where r is such that $Tr\,\sigma^D = 0$; hence,

$$\sigma^D = \sigma - \frac{1}{3}(Tr\,\sigma)I,$$

$$\sigma_{ij}^D = \sigma_{ij} - \frac{1}{3}\sigma_{kk}\delta_{ij}.$$

This equation's spherical part σ^S is the difference $\sigma^S = \sigma - \sigma^D = \frac{1}{3}\sigma_{kk}I$.

Examples

We end this section by a few classical elementary examples of stress tensors.

1. A spherical stress tensor is a tensor σ proportional to identity, which we write in the form

$$\sigma = -pI.$$

2. The uniaxial stress tensor in direction e_1 at x is defined by

$$\sigma_{11} \neq 0;$$
$$\sigma_{ij} = 0 \quad \text{otherwise.}$$

3. The shear stress tensor in the two orthogonal directions $0x_1$ and $0x_2$ is defined by

$$\sigma_{12} \neq 0;$$
$$\sigma_{ij} = 0 \quad \text{otherwise.}$$

3.5. The Piola-Kirchhoff tensor

We indicated in Remark 1.5 of Chapter 1 that the Eulerian representation is more often used for fluids while the Lagrangian representation is more often used for solids, although this is not an absolutely unbreakable rule.

Our aim is now to express the stress tensor, and equations (3.3) to (3.5), in the Lagrangian variables. We consider a material system which occupies the domain Ω_0 at time t_0 and the domain Ω_t at time t, and we write, as in Chapter 1,

$$x = \Phi(a, t), \quad a\epsilon\Omega_0, \quad x\epsilon\Omega_t.$$

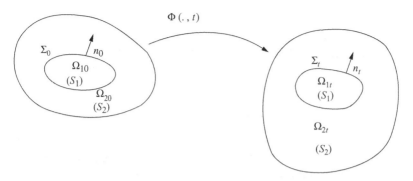

Figure 3.6 Partition $S = S_1 \cup S_2$ of system S at times 0 and t.

Anticipating a notation which will be re-introduced and used systematically in Chapter 5, we call $\mathbf{F}$ the Jacobian matrix $\nabla_a \Phi = Dx/Da$, which already appeared in Chapter 1.

The Cauchy stress tensor $\sigma = \sigma(x, t)$ was defined above in the Eulerian variable x; for the Lagrangian variable a, we can of course introduce the tensor $\bar{\sigma}$,

$$\bar{\sigma}(a, t) = \sigma(\Phi(a, t), t), \tag{3.6}$$

but $\bar{\sigma}$ does not lead to a convenient form of equations (3.3) to (3.5).

The tensor Π that we introduce, called the Piola-Kirchhoff tensor, corresponds to the following concern: we come back to the situation depicted in Figure 3.1, specifying the time t, and we introduce the pre-image by $\Phi(\cdot, t)$, which gives a similar geometry in Ω_0 (see Figure 3.6); n_t is the unit normal vector on Σ_t pointing from Ω_{1t} towards Ω_{2t} and n_0 is the unit normal vector on Σ_0 pointing from Ω_{10} to Ω_{20}. At time t, the resultant of the forces exerted by S_2 on S_1 is:

$$\int_{\Sigma_t} \sigma(x, t) \cdot n_t(x, t) \ d\Gamma_t(x). \tag{3.7}$$

We look for a tensor (if it exists) $\Pi = \Pi(a, t)$, defined at each time t, such that, at time t, the cohesion forces exerted by S_2 on S_1 can be expressed in the form

$$\int_{\Sigma_0} \Pi(a, t) \cdot n_0(a, t) \ d\Gamma_0(a), \tag{3.8}$$

this expression being valid for every partition $S = S_1 \bigcup S_2$ (and thus for every closed regular surface Σ_0 in the interior of Ω_0).

In order to answer this question, we need the following formula of change of variable for integral (3.7), a formula which is assumed here and proved in the appendix of Chapter 5:

$$\int_{\Sigma_t} \sigma(x, t) \cdot n_t(x, t) \, d\Gamma_t(x) = \int_{\Sigma_0} \sigma(\Phi(a, t), t) \cdot (\mathbf{F}^{-1})^T$$
$$\cdot n_0(a, t)[\det \mathbf{F}(a, t)] \, d\Gamma_0(a), \tag{3.9}$$

where the dots denote products of operators (or products of matrices in a given basis). We conclude from this that the following tensor answers the question asked (the variables a and t are omitted and we use notation (3.6)):

$$\Pi = (\det \mathbf{F}) \, \bar{\sigma} \cdot (\mathbf{F}^{-1})^T. \tag{3.10}$$

This is the Piola-Kirchhoff tensor. Since σ and $\bar{\sigma}$ are symmetric tensors, we see that Π is not symmetric in general, but it satisfies

$$\Pi \cdot \mathbf{F}^T = \mathbf{F} \cdot \Pi^T. \tag{3.11}$$

We also introduce sometimes the second Piola-Kirchhoff tensor which is symmetric

$$\mathbf{P} = \mathbf{F}^{-1} \cdot \Pi = (\det \mathbf{F})\mathbf{F}^{-1} \cdot \sigma \cdot (\mathbf{F}^{-1})^T;$$

Π is then called the first Piola-Kirchhoff tensor.

Instead of expressing equations (3.3) to (3.5) in the variables a, t, via the change of variable $x = \Phi(a, t)$ (and by using the tensor Π), we write directly the fundamental law for $S_1, \forall S_1 \subset S$, assuming, as in Section 3, the existence of forces with volumic density $f = f(x, t)$ on S. Then, for the resultants, we have

$$\int_{\Omega_{1t}} \rho(x, t)\gamma(x, t) \, dx = \int_{\Omega_{1t}} f(x, t) \, dx + \int_{\Sigma_t} \sigma(x, t) \cdot n_t(x, t) \, d\Gamma_t(x).$$

We have $\gamma = \partial^2 \Phi / \partial t^2$ and set

$$\rho_a(a, t) = \rho(\Phi(a, t), t) \det \mathbf{F}(a, t),$$
$$f_a(a, t) = f(\Phi(a, t), t) \det \mathbf{F}(a, t).$$

The change of variable is elementary for the volume integrals and, for the surface integral on Σ_t, we use (3.9). It follows that

$$\int_{\Omega_{10}} \rho_a \frac{\partial^2 \Phi}{\partial t^2} da = \int_{\Omega_{10}} f_a da + \int_{\Sigma_0} \Pi \cdot n_0 \, d\Gamma_0,$$

$$\int_{\Omega_{10}} \rho_a \frac{\partial^2 \Phi}{\partial t^2} da = \int_{\Omega_{10}} f_a da + \int_{\Omega_0} \mathrm{Div}_a \Pi \, da,$$

where $\text{Div}_a \Pi$ is the vector whose i^{th} component is $\sum_{j=1}^{3} \partial \Pi_{ij}/\partial a_j$. Hence the analogue of (3.3) in Lagrangian variable ($\Omega_{10} \subset \Omega_0$ being arbitrary):

$$\rho_a \frac{\partial^2 \Phi}{\partial t^2} = f_a + \text{Div}_a \Pi, \quad \text{in } \Omega_0. \tag{3.12}$$

In statics, the analogue of (3.5) reads

$$f_a + \text{Div}_a \Pi = 0, \quad \text{in } \Omega_0. \tag{3.13}$$

There is no need to write the analogue of (3.4), the acceleration being very simple to express in the Lagrangian variables.

We could also write the fundamental law for S_1 regarding the equality of momentum, and we would find (3.11) again (an equation for which we used the symmetry of σ, obtained itself in Section 4 thanks to the equality of angular momentums); this is left as an exercise for the reader.

Equations (3.12) and (3.13) express the fundamental law in its more general form for a continuum medium in the Lagrangian variables.

Exercises

1. We assume that the stress tensor of a continuum is given by

$$\sigma_{ij} = \lambda + \mu \delta_{ij}, \quad \lambda \neq 0, \mu \neq 0.$$

   Compute the normal stresses and the principal directions.

2. Same question as above for

$$\sigma = \begin{pmatrix} 15 & 0 & 5 \\ 0 & 15 & 5 \\ 5 & 5 & 15 \end{pmatrix}$$

3. Show that the fundamental law of equilibrium for a continuum with stress tensor σ can be written as

$$\int_A f_{ext} \cdot \theta \, dx + \int_{\partial A} (\sigma n) \cdot \theta \, d\Sigma = 0,$$

   for every subdomain $A \subset \Omega$ and every vector field $\theta(M) = u + v \wedge \overrightarrow{OM}$, $(u, v) \in (\mathbb{R}^3)^2$, f_{ext} being the external forces.

4. What relations must the components σ_{ij} of the tensor σ given in an arbitrary basis satisfy for σ to be an uniaxial stress tensor?

5. We consider a mechanical body having the shape of a cylinder of axis $(0; \vec{e}_3)$ and radius a, limited by the planes $x_3 = 0$ and $x_3 = h$. This body is at equilibrium under the action of forces exerted on the bases $S_0(x_3 = 0)$ and $S_1(x_3 = h)$ only.

a) We assume that, at each point of the cylinder, the stress tensor is of the form

$$\sigma_{11}^{(1)} = \sigma_{12}^{(1)} = \sigma_{22}^{(1)} = 0, \quad \sigma_{13}^{(1)} = k(7x_1^2 + x_2^2 - c_1 a^2),$$
$$\sigma_{23}^{(1)} = 6kx_1x_2, \quad \sigma_{33}^{(1)} = c_2 k(h - x_3)x_1,$$

where k, c_1 and c_2 are constants.
- (i) Determine the dimension of k. Compute the constants c_1 and c_2.
- (ii) Describe the volume forces exerted on the cylinder.

b) We assume that, at each point of the cylinder, the stress tensor is of the form

$$\sigma_{11}^{(2)} = \sigma_{12}^{(2)} = \sigma_{22}^{(2)} = 0, \, \sigma_{13}^{(2)} = \sigma_{13}^{(1)} + c_3 x_2,$$
$$\sigma_{23}^{(2)} = \sigma_{23}^{(1)} + c_4 x_1, \, \sigma_{33}^{(2)} = \sigma_{33}^{(1)},$$

where c_3 and c_4 are constants. Describe the volume forces exerted on the cylinder.

CHAPTER FOUR

Real and virtual powers

Our aim is now to introduce the concepts of real power and virtual power produced by forces and to present some applications. We first consider the very simple cases of a material point and of a system of material points (Section 4.1). We then study more complex situations (Section 4.2) before finally defining and studying the power of internal forces for a continuum medium in Section 4.3. This eventually leads to the virtual power theorem and to the kinetic energy theorem.

From the standpoint of mechanics, this chapter does not present much new material, but it gives very useful and different perspectives on the concepts and notions already introduced.

4.1. Study of a system of material points

Before considering the case of a system of material points, we start by considering that of a single material point. All that we say for a point or even for a system of points is simple and sometimes naive; it is, however, instructive.

The case of a material point

Definition 4.1. *For a force F applied to a material point M, the (real) power produced by F at time t and for the given frame of reference is the scalar product F · U (U being the velocity of M at time t in the considered frame of reference).*

If the frame of reference is Galilean, and if F denotes the total force applied to M, and m is the mass of M, then, thanks to the fundamental law, we have

$$m\gamma \cdot U = F \cdot U;$$

that is to say,

$$\frac{d}{dt}\frac{m|U|^2}{2} = F \cdot U. \tag{4.1}$$

We thus deduce the following result, which is easy in this case:

Theorem 4.1 (The Kinetic Energy Theorem). *In a Galilean frame of reference, the derivative with respect to time of the kinetic energy of a material point M is, at each time, equal to the power developed by the resultant of the forces applied to M.*

Definition 4.2. *Let V be a vector of* $\mathbb{R}^3$ *called the virtual velocity of M. The virtual power produced by the force F at time t and for this velocity V, is the scalar product* $F \cdot V$.

We easily see that the relation $m\gamma = F$ holds if and only if $m\gamma \cdot V = F \cdot V, \forall V \in \mathbb{R}^3$. Thus, the fundamental law of dynamics is equivalent to the following principle:

In a Galilean frame of reference, the virtual power of the external forces applied to a material point M is equal, at each time, to the virtual power of the quantities of acceleration of M (which is defined in the same way as the virtual power of a force).

The case of a system of n material points $\mathbf{M_1}, \dots, \mathbf{M_n}$

We are now given a system of n material points $M_1, \dots, M_n$. We saw that all the information obtained by applying the fundamental law of dynamics to this system is contained in the following equations:

$$m_i \gamma_i = F_i + \sum_{j=1}^{n} F_{ji},$$

$$F_{ij} = -F_{ji}, F_{ij} \wedge M_i M_j = 0,$$

with the convention: $F_{ii} = 0, i = 1, \dots, n$.

Definition 4.3. *Let S be a system consisting of n material points* $M_1, \dots, M_n$ *and let* $V_1, \dots, V_n$ *be n vectors of* $\mathbb{R}^3$ *called the virtual velocities of* $M_1, \dots, M_n$. *The virtual power of all the forces applied to the system S for this virtual velocity field is the quantity*

$$\sum_{i=1}^{n} \left\{ F_i \cdot V_i + \sum_{j=1,j\neq i}^{n} F_{ji} \cdot V_i \right\} \left(= \sum_{i=1}^{n} m_i \gamma_i V_i \right).$$

Definition 4.4. *We say that the virtual velocity field V_i, $i = 1, \ldots, n$, rigidifies the system S if and only if it is a helicoidal vector field (like the velocity field of a rigid body).*

Theorem 4.2 (Virtual Power Theorem for n Points). *In a Galilean frame of reference and for a rigidifying virtual velocity field, the virtual power of the external forces applied to a system of n material points is, at each time, equal to the virtual power of the quantities of acceleration.*

We will now give the proof of the virtual power theorem and will then make some important remarks.

Proof: We saw that

$$\sum_{i=1}^{n} m_i \gamma_i \cdot V_i = \sum_{i=1}^{n} F_i \cdot V_i + \sum_{i=1}^{n} \sum_{j=1, j \neq i}^{n} F_{ji} \cdot V_i$$

$$= \sum_{i=1}^{n} F_i \cdot V_i + \sum_{1 \leq i < j \leq n} (F_{ji} \cdot V_i + F_{ij} \cdot V_j).$$

From $F_{ij} = -F_{ji}$, it follows that

$$F_{ji} \cdot V_i + F_{ij} \cdot V_j = F_{ji} \cdot (V_i - V_j)$$
$$= F_{ji} \cdot (\omega \wedge M_i M_j)$$

for some vector ω, because the virtual velocity field rigidifies the system. Therefore,

$$F_{ji} \cdot V_i + F_{ij} \cdot V_i = (M_i M_j \wedge F_{ji}) \cdot \omega$$
$$= 0,$$

which yields

$$\sum_{i=1}^{n} m_i \gamma_i \cdot V_i = \sum_{i=1}^{n} F_i \cdot V_i,$$

and the theorem is proven.

Remark 4.1:

1. In the case of n material points, as in the case of a single point, the virtual power theorem is equivalent to the fundamental law of dynamics. We can obtain this result in two ways. A first proof consists of applying the virtual power theorem to each point M_i because this theorem applies to every system of material points. We thus obtain that $m_i \gamma_i \cdot V = F_i \cdot V + \sum_{j \neq i} F_{ji} \cdot V$ for every V and hence the fundamental law for

the system of material points S. A second proof, which is less easy, but has its own interest, consists of applying the virtual power theorem to the system S. We know that

$$\sum_{i=1}^{n} F_i \cdot V_i = \sum_{i=1}^{n} m_i \gamma_i \cdot V_i$$

for every virtual velocity field that rigidifies S. We first take $V_i = V, \forall i$, which gives $\sum_{i=1}^{n} m_i \gamma_i = \sum_{i=1}^{n} F_i$ (theorem of the dynamical resultant or conservation of linear momentum). Then, we take $V_i = \omega \wedge OM_i$; hence,

$$\sum_{i=1}^{n} F_i \cdot (\omega \wedge OM_i) = \sum_{i=1}^{n} m_i \gamma_i \cdot (\omega \wedge OM_i);$$

thus,

$$\omega \cdot \sum_{i=1}^{n} OM_i \wedge F_i = \omega \cdot \sum_{i=1}^{n} OM_i \wedge m_i \gamma_i,$$

for every ω, which yields

$$\sum_{i=1}^{n} OM_i \wedge F_i = \sum_{i=1}^{n} OM_i \wedge m_i \gamma_i,$$

(conservation of angular momentum at O).

2. If the virtual velocity field does not rigidify S, we have

$$\sum_{i=1}^{n} m_i \gamma_i \cdot V_i = \mathcal{P}_{\text{ext}} + \mathcal{P}_{\text{int}}, \tag{4.2}$$

where

$$\mathcal{P}_{\text{ext}} = \sum_{i=1}^{n} F_i \cdot V_i$$

is the virtual power of the external forces to S, and

$$\mathcal{P}_{\text{int}} = \sum_{i=1}^{n} \sum_{j=1, j \neq i}^{n} F_{ji} \cdot V_i$$

is the virtual power of the internal forces to S.

3. The virtual power of the internal forces of a system of n material points is an intrinsic expression independent of the frame of reference (this remark is important, in particular, for thermodynamics). To explain this result, we need the result derived in Chapter 2, Section 2.5 concerning the comparison of velocities and accelerations for two distinct frames

of reference. Furthermore, we need to describe and compare the virtual velocity field with respect to two different frames of reference; we omit the proof here. It is also important to note that, for a rigidifying virtual velocity field, the virtual power of the internal forces vanishes (so far, this has been proven only for a system of material points).

4.2. General material systems: rigidifying velocities

We consider in this section a material system S that fills the domain Ω_t at time t, $t \in I$.

Definition 4.5. *The virtual velocity field of S at time t is a vector field defined on Ω_t : $\{V(x), x \in \Omega_t\}$.*

Remark 4.2:

1. At this point, we make no regularity assumption on the field $\{V(x), x \in \Omega_t\}$. In practice, we will always choose velocity fields that satisfy certain regularity properties.
2. We easily see that the space of virtual velocity fields can be endowed with a vector space structure.

Definition 4.6. *The virtual velocity field $\{V(x), x \in \Omega_t\}$ at time t is rigidifying for S if it is a helicoidal vector field $[\mathcal{V}]$.*

Remark 4.3: The space of the virtual velocities rigidifying S (at time t) can be endowed with a vector subspace structure.

Definition 4.7. *We consider two material systems S and S'. The actions exerted by S' on S are represented by a vector measure $d\varphi_t(x)$ concentrated on Ω_t, the domain filled by S at time t. We define the virtual power of the forces exerted by S' on S for the virtual velocity field $\{V(x), x \in \Omega_t\}$ as the quantity*

$$\int_{\Omega_t} V(x) \, d\varphi_t(x).$$

We easily check that the virtual power is additive with respect to S and S', respectively (we assume that $S \cap S' = \emptyset$), and that it depends linearly on the virtual velocity fields.

It is also easy to see that, if the virtual velocity field is rigidifying, then the virtual power of the forces exerted by S' on S is equal to the scalar product

$$[\mathcal{F}] \cdot [\mathcal{V}],$$

where $[\mathcal{F}]$ is the helicoidal vector field associated with the forces exerted by S' on S, as defined in Chapter 2, Section 2.2.

Definition 4.8. *The virtual power of the quantities of acceleration for the virtual velocity field $\{V(x), x \in \Omega_t\}$ is the quantity*

$$\int_{\Omega_t} \gamma(x, t) \cdot V(x) \, d\mu_t(x).$$

Remark 4.4: When the virtual velocity field is rigidifying for S, the power of the quantities of acceleration is equal to the scalar product of two helicoidal vector fields: that associated with the quantities of acceleration and that associated with the virtual velocity field.

We then immediately obtain the following result.

Theorem 4.3 (Virtual Power Theorem). *For a Galilean frame of reference, for every material system S, at each time t and for every virtual velocity field that rigidifies S, the virtual power of the external forces to S is equal to the virtual power of the quantities of acceleration of S:*

$$[\mathcal{A}] \cdot [\mathcal{V}] = [\mathcal{F}_e] \cdot [\mathcal{V}].$$

Remark 4.5: For a Galilean frame of reference, we easily show that the virtual power theorem above is equivalent to the fundamental law. The proof of this last point reduces to simple calculations on helicoidal vector fields.

Virtual velocity fields rigidifying a partition

The next level of complication is reached after the following definitions. We consider a partition of a material system S into subsystems $S_1, \ldots, S_N$, such that $S_i \neq \emptyset$ for every i, $S_i \cap S_j = \emptyset$ if $i \neq j$ and $\bigcup_{i=1}^{N} S_i = S$.

Definition 4.9. *A virtual velocity field $\{V(x), x \in \Omega_t\}$ in which Ω_t is the domain filled by S at time t rigidifies the partition $\{S_i\}_{i=1,\ldots,N}$ of S if and only if $V_{|S_i}$ rigidifies S_i, $i = 1, \ldots, N$.*

Let $[\mathcal{F}_i]$ be the helicoidal vector field for the forces exerted by the complement of S on S_i, let $[\mathcal{F}_{ji}]$ be the helicoidal vector field for the forces exerted by S_j on S_i, and let $\mathcal{A}_i$ be the helicoidal vector field associated with the quantities of acceleration of S_i. Then, applying the virtual power theorem to S_i, we find

$$[\mathcal{A}_i] \cdot \left[\mathcal{V}^{(i)}\right] = [\mathcal{F}_i] \cdot \left[\mathcal{V}^{(i)}\right] + \sum_{j=1, j \neq i}^{N} [\mathcal{F}_{ji}] \cdot \left[\mathcal{V}^{(i)}\right],$$

where $[\mathcal{V}^{(i)}]$ is the helicoidal vector field associated with $V_{|S_i}$. Consequently,

$$\mathcal{P} = \sum_{i=1}^{N} \mathcal{P}_i = \sum_{i=1}^{N} [\mathcal{F}_i] \cdot [\mathcal{V}^{(i)}] + \sum_{i=1}^{N} \sum_{j=1, j \neq i}^{N} [\mathcal{F}_{ji}] \cdot [\mathcal{V}^{(i)}]$$
$$= \mathcal{P}_{\text{ext}} + \mathcal{P}_{\text{int}}.$$

We can then state the following result:

Theorem 4.4 (Virtual Power Theorem for a Rigidifying Partition). *Let S be a material system, let* $\{S_i, i = 1, \ldots, N\}$ *be a partition of S, and let V be a virtual velocity field on S at time t that rigidifies the partition. Then, the virtual power* $\mathcal{P}_a$ *of the Galilean quantities of acceleration of S is equal to the sum of the virtual power of the external forces to S and of the virtual power of the internal forces to S, which is defined here as the actions of the* S_i *on the* S_j, *namely*

$$\mathcal{P}_a = \mathcal{P}_{\text{ext}} + \mathcal{P}_{\text{int}}, \tag{4.3}$$

$$\mathcal{P}_{\text{ext}} = \sum_{i=1}^{N} [\mathcal{F}_i] \cdot [\mathcal{V}^{(i)}], \quad \mathcal{P}_{\text{int}} = \sum_{i=1}^{N} \sum_{j=1, j \neq i}^{N} [\mathcal{F}_{ji}] \cdot [\mathcal{V}^{(i)}]. \tag{4.4}$$

In Eq. (4.4), it is understood that $[\mathcal{F}_{ii}] = 0$, *and that* $[\mathcal{V}^{(i)}]$ *denotes the helicoidal vector field of the velocity field on* S_i.

Remark 4.6: In general, the real velocity field has no reason to rigidify S. Consequently, the real power of the internal forces has no reason to vanish in general. The computation of this real power (or of the virtual power for a nonrigidifying velocity field) will be studied in Section 4.3 by using the results of Chapter 3 – in particular the notion of stress tensor. We will then reach our ultimate goal for this chapter and give a more general and more satisfying formulation of Theorem 4.4.

4.3. Virtual power of the cohesion forces: the general case

We have just seen that the virtual power of the internal forces of a system S for a virtual velocity field V that rigidifies an arbitrary finite partition $(S_i)_{1 \leq i \leq N}$ of S is given by Eqs. (4.3) and (4.4), where $[\mathcal{V}^{(i)}]$ denotes the helicoidal vector field associated with the virtual velocities on S_i.

Our aim now is to define $\mathcal{P}_{\text{int}}$ for more general virtual velocity fields, to obtain finally a suitable generalization of Eqs. (4.3) and (4.4). To do so, we will give a new expression for the right-hand side of Eq. (4.3). In turn, this expression will naturally lead to the new definition of $\mathcal{P}_{\text{int}}$.

We consider the same partition as above and first assume that $\partial\Omega_i \cap \partial\Omega = \emptyset$; here S_i fills the domain Ω_i, and S fills the domain Ω, and thus $\Omega_1, \ldots, \Omega_N$ is a partition of Ω. Then, we have

$$\sum_{j=1, j\neq i}^{N} [\mathcal{F}_{ji}] \cdot [\mathcal{V}^{(i)}] = \int_{\Omega_i} C(x) \cdot V^{(i)}(x)\, dx,$$

where $C(x)$ is the vector with components $\sigma_{\ell k,k}$. To prove this equality, we start with the expression on the right-hand side. Because $V^{(i)}(x) = V^{(i)}(x_0) + (x_0 - x) \wedge \omega^{(i)}$, we have

$$\int_{\Omega_i} C(x) \cdot V^{(i)}(x)\, dx$$

$$= \int_{\Omega_i} C(x) \cdot \left[V^{(i)}(x_0) + (x_0 - x) \wedge \omega^{(i)} \right] dx$$

$$= V_\ell^{(i)}(x_0) \cdot \int_{\Omega_i} \sigma_{\ell k,k}(x)\, dx + \omega^{(i)} \cdot \left(\int_{\Omega_i} (x - x_0) \wedge C(x)\, dx \right)$$

$$= \text{(Taking } x_0 = 0 \text{ for the sake of simplicity)}$$

$$= V_\ell^{(i)}(0) \cdot \int_{\partial\Omega_i} \sigma_{\ell k} \cdot n_k \, d\Gamma + \omega^{(i)} \cdot \int_{\Omega_i} \begin{pmatrix} -\sigma_{2k,k}x_3 + \sigma_{3k,k}x_2 \\ -\sigma_{1k,k}x_3 + \sigma_{3k,k}x_1 \\ -\sigma_{1k,k}x_2 + \sigma_{2k,k}x_1 \end{pmatrix} dx.$$

Now

$$-\sigma_{2k,k}x_3 + \sigma_{3k,k}x_2 = -(\sigma_{2k}x_3)_{,k} + \sigma_{2k}\delta_{3k} + (\sigma_{3k}x_2)_{,k} - \sigma_{3k}\delta_{2k}.$$

Because $\sigma_{2k}\delta_{3k} - \sigma_{3k}\delta_{2k} = \sigma_{23} - \sigma_{32} = 0$, the two other components of the vector C being treated similarly, it follows that:

$$\int_{\Omega_i} C(x) \cdot V^{(i)}(x)\, dx$$

$$= V^{(i)}(0) \cdot \int_{\partial\Omega_i} T(x, n)\, d\Gamma + \omega^{(i)} \cdot \int_{\partial\Omega_i} \begin{pmatrix} -\sigma_{2k}x_3 n_k + \sigma_{3k}x_2 n_k \\ -\sigma_{1k}x_3 n_k + \sigma_{3k}x_1 n_k \\ -\sigma_{1k}x_2 n_k + \sigma_{2k}x_1 n_k \end{pmatrix} d\Gamma$$

$$= V^{(i)}(0) \cdot \int_{\partial\Omega_i} T(x, n)\, d\Gamma + \omega^{(i)} \cdot \int_{\partial\Omega_i} x \wedge T(x, n)\, d\Gamma;$$

this is exactly the scalar product of $[\mathcal{V}^{(i)}]$ with the helicoidal vector field associated with $\sum_{j=1, j\neq i}^{N} [\mathcal{F}_{ji}]$, which corresponds, in that case, to the actions of $S \backslash S_i$ on S_i. This is precisely the expression on the left-hand side of the equation.

Let us now assume that Ω_i has a common boundary Σ_i' with $\partial\Omega$, and thus $\partial\Omega_i = \Sigma_i' \cup \Sigma_i$, where $\Sigma_i = \cup_{j\neq i}(\partial\Omega_i \cap \partial\Omega_j)$. Then,

$$\sum_{j=1}^{N} [\mathcal{F}_{ji}] \cdot [\mathcal{V}^{(i)}]$$

$$= \text{(the virtual power of the forces exerted by } S\backslash S_i \text{ on } S_i)$$

$$= \int_{\Sigma_i} T(x,n) \cdot V^{(i)}(x_0)\, d\Gamma + \omega^{(i)} \cdot \int_{\Sigma_i} (x - x_0) \wedge T(x,n)\, d\Gamma$$

$$= \int_{\Sigma_i} T(x,n) \cdot \left[V^{(i)}(x_0) + \omega^{(i)} \wedge (x - x_0) \right] d\Gamma$$

$$= \int_{\Sigma_i} T(x,n) \cdot V(x)\, d\Gamma$$

$$= \int_{\partial\Omega_i} T(x,n) \cdot V(x)\, d\Gamma - \int_{\Sigma_i'} T(x,n) \cdot V(x)\, d\Gamma$$

$$= \int_{\Omega_i} C(x) \cdot V(x)\, dx - \int_{\Sigma_i'} T(x,n) \cdot V(x)\, d\Gamma.$$

Finally, summing all these relations with respect to i, we obtain:

Theorem 4.5. *The virtual power of the internal forces to the material system S for a virtual velocity field V that rigidifies a finite partition of S is given at time t by*

$$\mathcal{P}_{\text{int}} = \int_{\Omega_t} \sigma_{ij,j}(x,t)V_i(x)\, dx - \int_{\partial\Omega_t} \sigma_{ij}(x,t)\,n_j(x)V_i(x)\, d\Gamma. \quad (4.5)$$

Because Eq. (4.5) is actually independent of the given partition, however fine it may be, we are led to adopt the expression given in Eq. (4.5) as the definition of the virtual power of the internal forces for every virtual velocity field defined on Ω_t and regular enough for this expression to make sense.

We can also give another expression for $\mathcal{P}_{\text{int}}$. We assume, as usual, that V is of class C^1 with respect to x on Ω_t. Then,

$$\mathcal{P}_{\text{int}} = \int_{\Omega_t} \sigma_{ij,j} V_i\, dx - \int_{\partial\Omega_t} \sigma_{ij} n_j V_i\, d\Gamma$$

$$= \text{(thanks to the Stokes formula)}$$

$$= \int_{\partial\Omega_t} \sigma_{ij} V_i n_j\, d\Gamma - \int_{\Omega_t} \sigma_{ij} V_{i,j}\, dx - \int_{\partial\Omega_t} \sigma_{ij} n_j V_i\, d\Gamma$$

$$= -\int_{\Omega_t} \sigma_{ij} \frac{V_{i,j} + V_{j,i}}{2}\, dx.$$

We finally obtain

$$\mathcal{P}_{\text{int}} = -\int_{\Omega_t} \sigma_{ij}\varepsilon_{ij}(V)\,dx, \tag{4.6}$$

where $\varepsilon_{ij}(V) = \frac{1}{2}(V_{i,j} + V_{j,i})$.

Definition 4.10. *The virtual power of the internal forces of S for the virtual velocity field $V = V(x)$, defined at time t on Ω_t, is the expression given by Eq. (4.5) or (4.6).*

The virtual power theorem

We are now in a position to state the virtual power theorem for a general virtual velocity field that does not necessarily rigidify a partition of S.

We assume, for instance, that the external forces on S are made of

- a volume distribution of forces with volume density f, and
- a surface distribution of forces with surface density $T(x, n) = \sigma \cdot n = F$.

We can then prove the following result.

Theorem 4.6 (The General Virtual Power Theorem). *For every material system S and at each time t, the virtual power of the quantities of acceleration of S with respect to a Galilean frame of reference is the sum of the virtual power of the external forces on S and of the internal forces applied to S:*

$$\boxed{\mathcal{P}_a = \mathcal{P}_{\text{ext}} + \mathcal{P}_{\text{int}}.}$$

Proof: We have (see Chapter 3) that

$$\rho\gamma = f + C,$$

where C is the vector of components $\sigma_{ij,j}$. Consequently, the virtual power of the quantities of acceleration is given by

$$\int_{\Omega_t} \rho\gamma \cdot V\,dx = \int_{\Omega_t} C \cdot V\,dx + \int_{\Omega_t} f \cdot V\,dx.$$

Furthermore, thanks to Eq. (4.5)

$$\int_{\Omega_t} C \cdot V\,dx + \int_{\Omega_t} f \cdot V\,dx = \mathcal{P}_{\text{int}} + \int_{\partial\Omega_t} \sigma_{ij} n_j V_i\,d\Gamma + \int_{\Omega_t} f \cdot V\,dx,$$

and $\int_{\partial\Omega_t} \sigma_{ij} n_j V_i \, d\Gamma + \int_{\Omega_t} f \cdot V \, dx$ is equal to the power $\mathcal{P}_{\text{ext}}$ of the external forces applied to S; that is to say

$$\mathcal{P}_{\text{ext}} = \int_{\partial\Omega_t} F \cdot V \, d\Gamma + \int_{\Omega_t} f \cdot V \, dx$$

because $T_i = \sigma_{ij} n_j$ on Γ_t. This finishes the proof.

Remark 4.7 (Variational Formulations): By using Eqs. (4.6) and (4.7), we find, for every virtual velocity field defined on Ω_t, that is to say for every vector field $\{V(x),\ x \in \Omega_t\}$, that

$$\int_{\Omega_t} \rho\gamma \cdot V \, dx + \int_{\Omega_t} \sigma_{ij}\varepsilon_{ij}(V) \, dx = \int_{\Omega_t} f \cdot V \, dx + \int_{\partial\Omega_t} F \cdot V \, d\Gamma. \quad (4.8)$$

This relation is similar to the variational formulations used in mathematical analysis and numerical analysis for problems stemming from mechanics. These variational formulations are derived by choosing particular vector fields V and by using the stress–strain laws that we will introduce in Chapter 5. We will come back to the variational formulations in Chapter 15, Section 15.5.

Remark 4.8: In all the preceding sections and in what follows, except in Chapters 11 and 12, we implicitly assume that the only internal forces to the system are the cohesion forces introduced in Chapter 3. In the presence of other physical or chemical phenomena, other elements have to be taken into account: for instance, internal electromagnetic forces in magnetohydrodynamics (see some notions in Chapter 11), or internal heat sources (or sinks) due to chemical reactions (see an outline of combustion phenomena in Chapter 12).

4.4. Real power: the kinetic energy theorem

The real power for forces or for the quantities of acceleration can be defined exactly as the virtual power: we just have to replace, at each time t, $V(x)$ by $U(x, t)$.

Having introduced this notion, we can state the kinetic energy theorem that concludes this section.

Theorem 4.7 (The General Kinetic Energy Theorem). *The derivative with respect to time of the kinetic energy of a material system with respect to a Galilean frame of reference is equal, at each time, to the sum of the (real) powers of the external and internal forces applied to S.*

Proof: At this point, it suffices to prove that the real power of the quantities of acceleration is equal to the derivative with respect to time of the kinetic energy. By using the Lagrangian representation of the motion, $x = \Phi(a, t)$, we can write

$$\frac{1}{2}\frac{d}{dt}\int_{\Omega_t} |U(x,t)|^2\, d\mu_t(x) = \frac{1}{2}\frac{d}{dt}\int_{\Omega_0} \left|\frac{\partial\Phi}{\partial t}(a,t)\right|^2 d\mu_0(a)$$

$$= \int_{\Omega_0} \frac{\partial^2\Phi}{\partial t^2}(a,t)\cdot\frac{\partial\Phi}{\partial t}(a,t)\, d\mu_0(a)$$

$$= \int_{\Omega_t} \gamma(x,t)\cdot V(x,t)\, d\mu_t(x),$$

and the result follows.

Remark 4.9: Thanks to Eq. (2.6) and to Eq. (2.14) of Chapter 2 concerning the comparison of the velocities with respect to two distinct frames of reference,[1] we see that the real power of the internal forces is independent of the frame of reference: it is an intrinsic physical (thermodynamical) quantity.[2]

Exercises

1. We consider three horizontal linear springs R_1, R_2, R_3 that are linked consecutively from left to right (we neglect the mass of the springs). The left end A_0 of R_1 and the right end A_3 of R_3 are fixed; material points M_1 and M_2, with mass m_1 and m_2 respectively, are located at the ends A_1 and A_2, common to R_1 and R_2 and to R_2 and R_3 respectively. The spring R_i exerts a force of intensity $-k_i l_i$ proportional to its elongation $l_i (k_i > 0, l_i > 0$ or $l_i < 0)$.
   a) Write the virtual power theorem for the material points M_1 and M_2 separately and then for the system consisting of the two material points.
   b) Same questions as above when the springs are vertical.
2. (*See also Exercise 5, Chapter 13.*) A body in motion fills the domain $\Omega = \Omega_t$. It is fixed on a part Γ_0 of its boundary, where it is submitted to unknown forces; it is submitted to a surface density of force $F = F(a, t)$ on $\Gamma_1 = \partial\Omega \setminus \Gamma_0$. Furthermore, the body is submitted to volume forces

[1] See also Theorem 5.1, Chapter 5.
[2] See also Chapter 6, Section 6.1.

with density $f = f(x, t)$. Write the virtual power theorem (in terms of the Cauchy stress tensor):

a) for an arbitrary virtual velocity field;

b) for a virtual velocity field which vanishes on $\partial\Omega$.

3. See Exercise 1, Part c, Chapter 7.

CHAPTER FIVE

Deformation tensor, deformation rate tensor, constitutive laws

The aim of this chapter is twofold. On the one hand, we introduce various kinematical quantities that will be useful in what follows – especially in Section 5.3. On the other hand, we introduce and address an issue of a totally different nature, more physical and very important: the constitutive laws.

The constitutive laws in mechanics are usually laws that relate the stress tensor to the kinematical quantities; these laws depend on the physics of the material under consideration, contrary to the equations and laws previously written that are valid for all materials. These laws take into account, for instance, the differences of behavior between liquids and solids, or between rubber, wood, steel, and so forth.

We present, without pretending to be exhaustive, a list containing a variety of constitutive laws. These laws are different for solids and fluids; some materials have memory, others do not; some materials have different behaviors depending on certain parameters, whereas others have the same behavior for all regimes, and so forth. The study of the mechanical behavior of materials is called rheology; we also briefly describe some fundamental principles that rheological laws are required to satisfy.

5.1. Further properties of deformations

In Chapter 1, we introduced the following deformation map:

$$\Phi: a \in \Omega_0 \mapsto x \in \Omega_t.$$

We also defined the gradient of Φ (for fixed t) as follows:

$$\mathbf{F} = (\mathbf{F}_{ij}) = \left(\frac{\partial \Phi_i}{\partial a_j}\right) = \nabla \Phi.$$

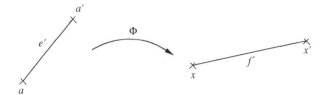

Figure 5.1 Distortion of distances.

In this section, we are going to make precise the role played by $\mathbf{F} = \nabla\Phi$ in the study of the local distortion of distances and angles around one point.

In this section, the time t is fixed and $\Phi(\cdot) = \Phi(\cdot, t, 0)$; the time will vary in Section 5.2.

Distortion of distances

We are given a and a' in Ω_0. We set (see Figure 5.1) $x = \Phi(a)$, $x' = \Phi(a')$, $a' = a + e'$, $|e'| = \lambda$ small, and $x' = x + f'$. Then,

$$x'_i = \Phi_i(a) + \frac{\partial \Phi_i}{\partial a_j}(a) \cdot (a'_j - a_j) + o(|a' - a|),$$

and

$$x' - x = \mathbf{F}e' + o(\lambda);$$

hence, by setting $\bar{e}' = e'/\lambda$,

$$\frac{|x' - x|}{\lambda} = |\mathbf{F}\bar{e}' + o(1)| \simeq |\mathbf{F}\bar{e}'|.$$

Thus,

$$\frac{|f'|^2}{|e'|^2} = \frac{|x' - x|^2}{\lambda^2} \simeq \bar{e}'^T \mathbf{F}^T \mathbf{F}\bar{e}' = \bar{e}'^T C \bar{e}',$$

where $C = (C_{\alpha\beta}) = \mathbf{F}^T \cdot \mathbf{F}$.

This yields, for λ small, the following relation, valid at order $o(\lambda)$:

$$|x' - x| \simeq [(a' - a)^T C (a' - a)]^{1/2}.$$

Distortion of angles

Let a, a', a'' belong to Ω_0. We set (see Figure 5.2) $x = \Phi(a)$, $x' = \Phi(a')$, $x'' = \Phi(a'')$, $e' = a' - a$, $e'' = a'' - a$, $f' = x' - x$, and $f'' = x'' - x$.

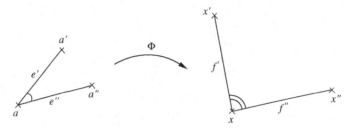

Figure 5.2 Distortion of angles.

Moreover, we assume that $|e'| = |e''| = \lambda$. Then, if $\bar{e}' = e'/\lambda$, $\bar{e}'' = e''/\lambda$,

$$x' - x = \mathbf{F}e' + o(\lambda) = \lambda \mathbf{F}\bar{e}' + o(\lambda),$$
$$x'' - x = \mathbf{F}e'' + o(\lambda) = \lambda \mathbf{F}\bar{e}'' + o(\lambda),$$

and

$$(x'' - x) \cdot (x' - x) = e''^T \mathbf{F}^T \mathbf{F}e' + o(\lambda^2)$$
$$= \lambda^2 \bar{e}''^T C \bar{e}' + o(\lambda^2).$$

Thus,

$$\frac{f'' \cdot f'}{|f''| \cdot |f'|} = \cos[(x'' - x), (x' - x)] = \frac{\bar{e}''^T C \bar{e}'}{(\bar{e}'^T C \bar{e}')^{1/2}(\bar{e}''^T C \bar{e}')^{1/2}} + o(1)$$

$$\simeq \frac{e''^T C e'}{(e''^T C e'')^{1/2} \cdot (e'^T C e')^{1/2}}.$$

More generally, if $|e'| = \lambda(e') \neq |e''| = \lambda(e'')$, then

$$\cos[(x'' - x), (x' - x)] = \frac{(x'' - x) \cdot (x' - x)}{|x'' - x| \cdot |x' - x|}$$

$$\simeq \frac{(a'' - a)^T C(a' - a)}{[(a'' - a)^T C(a'' - a)]^{1/2}[(a' - a)^T C(a' - a)]^{1/2}}.$$

We thus see that the bilinear form and the quadratic form associated with $C = C(x)$ allow us to describe the local distortion of angles and distances near the point x (i.e., it defines the metrics in the tangent space).

Definition 5.1. *The linear operator $C = C(x)$ is called the right Cauchy–Green deformation tensor at x; it is expressed here in the Eulerian variables.*
Similarly the tensor $B = \mathbf{F}\mathbf{F}^T$ is called the left Cauchy–Green deformation tensor.

Remark 5.1: If u denotes the displacement, then $\Phi(a) = u(a) + a$ and

$$C = (\nabla u^T + I) \cdot (\nabla u + I) = I + \nabla u + \nabla u^T + \nabla u^T \cdot \nabla u.$$

When the displacement is rigid

$$|x' - x| = |a' - a|, \quad \forall\, a, a';$$

hence,

$$(x'' - x) \cdot (x' - x) \equiv (a'' - a) \cdot (a' - a),$$

which yields

$$C = I.$$

Conversely, one can prove that, if $C = I$, then the displacement is rigid. The proof, which consists of passing from local information around one point to global information, requires tools and concepts from differential geometry that are beyond our scope here.

Definition 5.2. *The deformation tensor (or Green strain tensor) is the tensor*[1]

$$X = \frac{1}{2}(C - I).$$

We deduce from what precedes that the deformation is rigid if and only if $X = 0$.

To study the small[2] deformations that appear, for instance, in solid mechanics, we introduce the linearized deformation tensor:

Definition 5.3. *The linearized deformation tensor, denoted by $\varepsilon(u)$, is defined by*

$$\varepsilon_{\alpha\beta}(u) = \frac{1}{2}(u_{\alpha,\beta} + u_{\beta,\alpha}), \quad \text{or} \quad \varepsilon(u) = \frac{1}{2}(\nabla u + \nabla u^T),$$

where u denotes the displacement and $u_{\alpha,\beta}(a) = (\partial u_\alpha / \partial a_\beta)(a)$.

Remark 5.2: We have $C = \mathbf{F}^T \mathbf{F}$. Consequently, C is a positive definite matrix (C is definite because $\det(\nabla\Phi) \neq 0$), and it can be written in the form

$$C = W^2,$$

where W is symmetric and positive definite. It then follows that

$$\begin{aligned}
(\mathbf{F}W^{-1})^T (\mathbf{F}W^{-1}) &= W^{-1}\mathbf{F}^T \mathbf{F} W^{-1} \\
&= W^{-1} C W^{-1} \\
&= I,
\end{aligned}$$

---

[1] This tensor is also denoted $\mathbf{E}$ by a number of authors.
[2] The adjective "infinitesimal" is sometimes used in this context. We find it confusing and prefer to avoid it.

because C commutes with W and W^{-1}. We thus deduce that

$$R = \mathbf{F}W^{-1} \text{ is orthogonal,}$$

and $\mathbf{F} = RW$. Furthermore $V = RWR^T$ is symmetric and positive definite as well because

$$(RWR^Tx, x) = (WR^Tx, R^Tx).$$

Then,

$$\mathbf{F} = RW = VR,$$
$$\mathbf{F}\mathbf{F}^T = VRR^TV^T = V^2 = B.$$

Thus, V is the positive square root of $B = \mathbf{F}\mathbf{F}^T$, whereas W is the positive square root of $C = \mathbf{F}^T\mathbf{F}$ ($B \neq C$ of course). Moreover, R is a rotation matrix; W is called the right stretch tensor and R is called the rotation tensor. Similarly, B is the left Cauchy–Green tensor and V is the left stretch tensor.

In the case of small perturbations, we have

$$\mathbf{F} = I + \nabla u$$
$$= I + h,$$

where $h = \nabla u = O(\eta)$. Then,

$$\varepsilon = \frac{1}{2}(h + h^T) \text{ is the symmetric part of } h,$$

$$\varphi = \frac{1}{2}(h - h^T) \text{ is the antisymmetric part of } h.$$

Furthermore, up to order $o(\eta)$,

$$C = \mathbf{F}^T\,\mathbf{F} = (I + h^T)(I + h)$$
$$\simeq I + h^T + h = I + 2\varepsilon,$$
$$\mathbf{F} = I + h = I + \varepsilon + \varphi$$
$$\simeq (I + \varepsilon)(I + \varphi) \simeq (I + \varphi)(I + \varepsilon),$$

where $I + \varepsilon$ is positive definite. Because

$$(I + \varepsilon)^2 \simeq C + o(\eta), \quad (I + \varphi^T)(I + \varphi) \simeq I + o(\eta),$$

we obtain, again up to order $o(\eta)$:

$$W \simeq V \simeq I + \varepsilon,$$
$$R \simeq I + \varphi.$$

5.2. The deformation rate tensor

We now assume that the time varies with initial time t_0. We set

$$x = \Phi(a, t, t_0),$$

$$v(a, t)\, [\equiv U(x, t)] = \frac{\partial \Phi}{\partial t}(a, t, t_0).$$

In the previous section (see Figure 5.2), we saw that, up to order $o(1)$,

$$f''(t) \cdot f'(t) = C_{\alpha\beta}(a, t, t_0)e''_\alpha e'_\beta.$$

Our aim now is to compute the time derivative $(d/dt)[f''(t) \cdot f'(t)]$ at $t = t_0$; that is to say $(\partial/\partial t)[C_{\alpha\beta}(a, t, t_0)]|_{t=t_0}$, or else $(\partial/\partial t)[X_{\alpha\beta}(a, t, t_0)]|_{t=t_0}$.

Considering the deformations between t_0 and $t > t_0$, we write

$$X = \frac{1}{2}(C - I),$$

$$C = \mathbf{F}^T \mathbf{F}.$$

Consequently, we have at the point (a, t, t_0)

$$\frac{1}{2}\frac{\partial C}{\partial t} = \frac{\partial X}{\partial t}$$

$$= \frac{1}{2}\frac{\partial}{\partial t}(\mathbf{F}^T \mathbf{F} - I)$$

$$= \frac{1}{2}\frac{\partial}{\partial t}(\mathbf{F}^T \mathbf{F}).$$

Because $\Phi(a, t_0, t_0) = I$, $\nabla_a \Phi(a, t_0, t_0) = I$; hence, at the point (a, t_0, t_0):

$$\frac{\partial X}{\partial t} = \frac{1}{2}\frac{\partial C}{\partial t} = \frac{1}{2}\left(\frac{\partial \mathbf{F}}{\partial t} + \frac{\partial \mathbf{F}^T}{\partial t}\right).$$

Because $\dfrac{\partial F}{\partial t} = \dfrac{\partial}{\partial t}\nabla_a \Phi = \nabla_a \dfrac{\partial \Phi}{\partial t} = \nabla_a v$, it follows that, at the point (a, t_0, t_0):

$$\frac{\partial X}{\partial t} = \frac{1}{2}(\nabla_a v + \nabla_a v^T).$$

Remark 5.3: In the previous reasoning, a denotes the position at time t_0 taken as the reference or (initial) time. Hereafter (and especially for Definition 5.4), t_0 is an arbitrary time, and we come back to the Eulerian representation; a is then replaced by the actual position x and v by $U(x, t_0)$. With this new notation, $\partial X/\partial t$ is equal, at time t, to $\frac{1}{2}(\nabla_x U + \nabla_x U^T)$.[3]

---

[3] We write here $\nabla_x U$ instead of ∇U to emphasize that the gradient is taken with respect to x.

Definition 5.4. *The tensor* $D = \frac{1}{2}(\nabla U + \nabla U^T)$ *is called the deformation rate tensor (or deformation velocity tensor). The tensor* $\omega = \frac{1}{2}(\nabla U - \nabla U^T)$ *is called the rotation rate tensor. We have*

$$\nabla U = D + \omega.$$

Theorem 5.1. *The velocity field* $U = U(x, t)$ *is, at time* t, *the velocity field of a rigid body motion if and only if, at this time,* $D = 0$ *at every point of the system.*

Proof: Let us assume that the motion is rigid. Then, there exists $\vec{\omega} \in \mathbb{R}^3$, $\vec{\omega} = (\omega_1, \omega_2, \omega_3)$, such that

$$U(x) = U(0) + \vec{\omega} \wedge x = U(0) + \begin{pmatrix} \omega_2 x_3 - \omega_3 x_2 \\ \omega_3 x_1 - \omega_1 x_3 \\ \omega_1 x_2 - \omega_2 x_1 \end{pmatrix}.$$

Therefore,

$$\nabla U = \begin{vmatrix} 0 & -\omega_3 & \omega_2 \\ \omega_3 & 0 & -\omega_1 \\ -\omega_2 & \omega_1 & 0 \end{vmatrix},$$

and hence,

$$\omega = \nabla U \quad \text{and} \quad D = 0.$$

Conversely, let us assume that, at a given time, $D = 0$ at every point. Then, for every i, j, ℓ,

$$\omega_{ij} = \frac{1}{2}(U_{i,j} - U_{j,i}),$$

$$\omega_{ij,\ell} = \frac{1}{2}(U_{i,j\ell} - U_{j,i\ell})$$

$$= \frac{1}{2}(U_{i,j\ell} + U_{\ell,ij} - U_{\ell,ij} - U_{j,\ell i})$$

$$= (D_{i\ell,j} - D_{j\ell,i})$$

$$= 0.$$

Consequently, ω_{ij} is constant in space and so is $\omega = \nabla U$. We deduce that

$$\nabla U(x, t) = \omega = \begin{vmatrix} 0 & -\omega_3 & \omega_2 \\ \omega_3 & 0 & -\omega_1 \\ -\omega_2 & \omega_1 & 0 \end{vmatrix},$$

where the ω_i are constant in space. It follows by integration that

$$U(x) = U(0) + \vec{\omega} \wedge x,$$

where $\vec{\omega}$ is the vector with components $\omega_1, \omega_2, \omega_3$.

Remark 5.4: The vector $\vec{\omega}$ is called the rotation rate vector. It should not be mistaken for the curl vector, curl U, which appears in particular in fluid mechanics. Actually, we easily see that $\vec{\omega} = \frac{1}{2}\text{curl } U$.

Remark 5.5: It is useful to note the similarity between the expressions related to the deformation rates and those related to the linearized deformations under small perturbations. More precisely, we have the following correspondences:

$$\text{Velocity } U = U(x, t) \longleftrightarrow \text{Displacement } u = u(a, t)$$
$$\text{Velocity gradient } \nabla U = \nabla_x U \longleftrightarrow \text{Displacement gradient } \nabla_a u$$
$$\text{Deformation rate tensor } D = D(U) \longleftrightarrow \text{Linearized deformation tensor}$$
$$\epsilon = \epsilon(u)$$
$$\text{Rotation rate tensor } \omega \longleftrightarrow \text{Rotation tensor } \omega$$

5.3. Introduction to rheology: the constitutive laws

As stated in the introduction to this chapter, we now address the question of constitutive laws, also called stress–strain laws, and describe some general principles of rheology.

By simply counting the number of equations and unknown functions, we see that there is need for more equations. Indeed, the unknowns of the problem of mechanics are as follows:

- the displacements in a Lagrangian description, or the velocities in an Eulerian description,
- the stresses, and
- the kinetic state (mass density ρ).

We must add to these unknowns the internal energy e (or the temperature), which we have not yet considered and that will be introduced in Chapter 6.

The fundamental law of dynamics (which is a universal law concerning the displacements) provides the equations

$$\rho\gamma_i = f_i + \sigma_{ij,j}$$

that result in three scalar equations. We thus have 10 unknowns corresponding in the Eulerian description to the density ρ (1 unknown), the velocity vector U (3 components), and the stress tensor (6 components). However, only four equations are available, namely the three previous ones to which we add the continuity equation

$$\frac{\partial\rho}{\partial t} + \text{div}(\rho U) = 0.$$

By a formal comparison between the number of equations and the number of unknowns (and without any discussion of the existence and uniqueness of solutions to these equations), we see that six equations are missing.

To obtain the six missing equations, we will add some equations specific to the material: these are the stress–strain laws relating U (or u) to σ.

The justification and the discovery of these laws constitute an important part of continuum mechanics, which is based on thermodynamics, and called rheology.

In the next section, we briefly describe some fundamental principles of rheology and, in the following sections, we present some important stress–strain laws.

Some principles of rheology

The stress–strain laws for materials must satisfy the following fundamental principles:

1. The laws of thermodynamics and in particular the first and second principles: conservation of energy (see Chapter 6) and the entropy principle.
2. Invariance of the laws by a change of Galilean frame of reference (this principle is also called the material indifference principle).
3. Spatial invariance or localization invariance: the stress–strain law of a material does not depend on its position in space (e.g., invariance by translations in x).
4. Isotropy: once expressed in an orthonormal coordinate system, the law must be invariant (in a sense to be made precise) by a change of orthonormal coordinate system (e.g., a rotation).
5. Causality or determinism: according to this mechanical principle, knowledge of the history of the material up to time t yields knowledge of the state of the stresses at time t.

This last principle must, however, be properly adapted in the case of internal constraints such as, for a fluid, the incompressibility law, div $u = 0$.[4]

We end this section by recalling some basic tools of linear algebra useful, in particular, for point 4 above. We consider a linear operator of $\mathbb{R}^3$, represented by its matrix $A = (A_{ij})$ in an orthonormal basis. We recall that, by the Cayley–Hamilton theorem, A cancels its characteristic polynomial $p = p(\lambda)$, namely,

$$p(A) = 0. \tag{5.1}$$

---

[4] We refer the reader to the books of Germain (1973) and Truesdell (1977) for further information on this point and for a more complete description of the principles of rheology.

We write $p(\lambda)$ in the form

$$p(\lambda) = -\lambda^3 + A_I\lambda^2 - A_{II}\lambda + A_{III}, \qquad (5.2)$$

where

$$
\begin{aligned}
A_I &= \operatorname{tr} A = A_{ii}, \\
A_{II} &= \frac{1}{2}\{(\operatorname{tr} A)^2 - \operatorname{tr} A^2\}, \qquad (5.3) \\
A_{III} &= \det A.
\end{aligned}
$$

These quantities A_I, A_{II}, A_{III} are invariant by a change of orthonormal co-ordinate system (i.e., they depend on the linear operator and not on the chosen basis); these are the only invariants attached to A. In particular, owing to the Cayley–Hamilton theorem,

$$A^3 = A_I A^2 - A_{II} A + A_{III} I,$$

and it is possible to express the trace of A^k, $k \geq 3$, in terms of A_I, A_{II}, and A_{III}. Using the eigenvalues $\lambda_1, \lambda_2, \lambda_3$ (the roots of p), we have

$$A_I = \lambda_1 + \lambda_2 + \lambda_3, \quad A_{II} = \lambda_1\lambda_2 + \lambda_2\lambda_3 + \lambda_3\lambda_1, \quad A_{III} = \lambda_1\lambda_2\lambda_3.$$

The stress–strain laws have different expressions for fluids and solids; we will study these two cases successively.

Main examples in fluid mechanics

In this subsection we do not aim to give a full description of the equations of fluid mechanics. This study will be outlined in the second part of this volume.

Definition 5.5. *For a large class of fluids, the stress–strain law is of the form*

$$\sigma = f(\varepsilon), \qquad (5.4)$$

where $\varepsilon_{ij}(U) = \frac{1}{2}(U_{i,j} + U_{j,i})$ is the deformation rate tensor (ε was previously denoted D).

Remark 5.6: We might consider other laws such as $\sigma = f(\nabla U), \sigma = f(\varepsilon, \omega)$, or $\sigma = f(U, \varepsilon, \omega)$. It can be proven that such laws, where f would explicitly depend on ω (or U), do not satisfy the principles of rheology recalled in the previous subsection. We could consider, for nonhomogeneous media, laws of the form $\sigma = f(\varepsilon, x)$; on the contrary, this generalization raises no difficulty.

Newtonian viscous fluids

A Newtonian viscous fluid is a fluid for which the stress–strain law is linear; that is to say the stress tensor σ is a linear affine function of the strain rate tensor $D = \varepsilon(U)$, namely,

$$\sigma_{ij} = \mu(U_{i,j} + U_{j,i}) + c\, \delta_{ij},$$

where U is the velocity; μ is called the *dynamic viscosity coefficient*. For thermodynamical reasons, c is of the form $\lambda \operatorname{div} U - p$, where p denotes the pressure, and thus

$$\sigma_{ij} = \mu(U_{i,j} + U_{j,i}) + \lambda(\operatorname{div} U)\delta_{ij} - p\delta_{ij}.$$

The second law of thermodynamics yields $\mu \geq 0$. We will also see in Chapter 7 that $2\mu + 3\lambda \geq 0$, again for thermodynamical reasons.

We deduce from this stress–strain law the Navier–Stokes equations that govern the motion of a compressible Newtonian fluid (such as, for instance, water or air)

$$\rho\gamma_i = \rho\left(\frac{\partial U_i}{\partial t} + U_j U_{i,j}\right) = f_i + \mu U_{i,jj} + (\lambda + \mu)(\operatorname{div} U)_{,i} - p_{,i}.$$

For a homogeneous and incompressible fluid $\rho = \rho_0$, $\operatorname{div} U = 0$, and we obtain the incompressible Navier–Stokes equations

$$\rho_0\left(\frac{\partial U_i}{\partial t} + U_j U_{i,j}\right) = f_i + \mu \Delta U_i - p_{,i},$$
$$\operatorname{div} U = 0.$$

As particular cases, we can consider the stationary Navier–Stokes equations, or the evolutionary and stationary Stokes equations derived under the small-motion assumption by suppressing the nonlinear term (see Chapter 7 and the subsequent chapters).

A fluid is called perfect if $\mu = \lambda = 0$. In this case, the stress–strain law becomes

$$\sigma_{ij} = -p\delta_{ij},$$

that is, the stress tensor is spherical. A perfect fluid may or may not be compressible. If it is incompressible and homogeneous, then $\operatorname{div} U = 0$. If it is compressible and barotropic (see Chapter 7), p is a given function of ρ, $p = g(\rho)$. This last relation is the equation of state. Such laws, which are useful in many other cases in fluid mechanics, give a relation between p, ρ, and the temperature T or the internal energy e. We will come back to this point in the forthcoming chapters.

Non-Newtonian fluids

The stress tensor may be a (more or less simple) nonlinear function of the deformation rate tensor. In that case, we say that the fluid is non-Newtonian. Classical examples of non-Newtonian fluids are motor oils, melted plastics, blood, or drying concrete.

Even though this model is not necessarily very realistic, we usually consider, at least for theoretical studies, non-Newtonian fluids whose stress-strain law is of the form (5.4), f being now nonlinear (e.g., the so-called Reiner and Rivlin fluids).

Under the isotropy assumption, one can show that, in the absence of internal constraints, the most general form for Eq. (5.4) then reads

$$\sigma = h_0(\varepsilon_I, \varepsilon_{II}, \varepsilon_{III})I + h_1(\varepsilon_I, \varepsilon_{II}, \varepsilon_{III})\varepsilon + h_2(\varepsilon_I, \varepsilon_{II}, \varepsilon_{III})\varepsilon^2,$$

where ε_I, ε_{II}, and ε_{III} are the invariants of $D = \varepsilon(U)$ (see Eq. (5.3)), namely,

$$\varepsilon_I = \operatorname{tr} \varepsilon = \varepsilon_{ii},$$
$$\varepsilon_{II} = \frac{1}{2}[(\operatorname{tr} \varepsilon)^2 - \operatorname{tr} \varepsilon^2],$$
$$\varepsilon_{III} = \det \varepsilon.$$

We recalled after Eq. (5.3) that any scalar isotropic function of ε is a function of ε_I, ε_{II}, and ε_{III}.

When the fluid is incompressible and homogeneous, the equations become

$$\operatorname{div} U = 0, \tag{5.5}$$
$$\sigma = -pI + h_1(\varepsilon_{II}, \varepsilon_{III})\varepsilon + h_2(\varepsilon_{II}, \varepsilon_{III})\varepsilon^2, \tag{5.6}$$

where the pressure p is undetermined.

Remark 5.7: More generally, h_1 and h_2 depend also on the temperature which will be introduced in Chapter 6.

Main examples in solid mechanics

In solid mechanics, we consider materials for which the Piola-Kirchhoff tensor Π (see Section 3.5) is a function of X, a, and t, where a is the Lagrangian variable, $u = u(a, t)$ is the displacement, and $X = X(u)$ is defined as in section 5.2:

$$\Pi = f(X, a, t); \tag{5.7}$$

here $X_{ij}(u) = \frac{1}{2}(u_{i,j} + u_{j,i} + u_{i,k}u_{k,j})$, and $u_{i,j} = (\partial u_i/\partial a_j)(a, t)$.

In general, we restrict ourselves to small displacements, in which case **F** is, at first order, the identity and $\Pi(a, t)$ is simply the tensor $\sigma(\Phi(a, t), t)$

(see the linearization principle in Chapter 13, Section 13.2); then (5.7) simply becomes

$$\sigma = f(\varepsilon, a, t), \tag{5.8}$$

with $\varepsilon = \varepsilon(u)$, $\varepsilon_{ij}(u) = \frac{1}{2}(u_{i,j} + u_{j,i})$.[5]

Remark 5.8: For the linearized case, we could think, as in Remark 5.6, of stress–strain laws of the form $\sigma = f(u, \nabla_a u)$. However, it can be proven that such laws, where f would explicitly depend on u (or ω, or both), are not consistent with the general principles of rheology described at the beginning of this section.

Elastic media

a) Linear or classical elasticity

Under the small deformations hypothesis, and if $\varepsilon = \varepsilon(u)$ denotes the linearized deformation tensor, a linear elastic medium satisfies the stress–strain law

$$\sigma = f(\varepsilon),$$

where f is a linear function. More precisely, the stress–strain law is of the form

$$\sigma_{ij} = \lambda \varepsilon_{kk} \delta_{ij} + 2\mu \varepsilon_{ij},$$

where $\varepsilon = \varepsilon(u)$, and $u = u(a, t)$ is the displacement; λ and μ are the Lamé coefficients. The second law of thermodynamics requires that $\mu \geq 0$ and $3\lambda + 2\mu \geq 0$.

b) Nonlinear elasticity (hyperelasticity)

We consider the behavior of an elastic medium under the large deformations assumption. A classical example of such a material is rubber.

The stress–strain law reads, at each time,

$$\sigma = \mathcal{G}(\mathbf{F}),$$

where $\mathcal{G}$ characterizes the material and depends on the space variable x in a nonhomogeneous material. Moreover, $\mathbf{F} = RW$, and one can prove that, with another function $\mathcal{G}$, σ is necessarily of the form

$$\sigma = R\mathcal{G}(W)R^T,$$

---

[5] The spatial variable is then denoted by x instead of a for reasons that will be explained in Chapter 13.

which can also be written as

$$\sigma = R\mathcal{H}(C)R^T,$$

where $C = \mathbf{F}^T\mathbf{F} = W^2$.

Additional thermodynamical considerations further lead to require (in large deformations) that the function f in (5.7) be the Fréchet-differential (with respect to X or C) of a scalar function $g = g(X, a, t)$. In the case of small displacements, the function f in (5.8) is the Fréchet-differential (with respect to ε) of a scalar function $g = g(\varepsilon, a, t)$; in the case of linear elasticity, we simply reduce g to its quadratic part with respect to ε (Taylor expansion of g at ε small). The function g is related to the energy; it will reappear under a new name, w, in Chapter 15 for linear elasticity and in Chapter 16 for nonlinear elasticity.

Remark 5.9: We briefly describe, in Chapter 16, several problems of nonlinear elasticity in small deformations. The study of nonlinear elasticity in large deformations raises considerable difficulties, geometrical difficulties adding to mechanical difficulties.

c) Hypoelasticity

In this case, a certain derivative of σ, denoted by $D_J\sigma_{ij}$, is a linear function of D with coefficients depending on σ. The stress–strain law of such materials reads

$$D_J\sigma_{ij} = C_{ijk\ell}D_{k\ell},$$

where the $C_{ijk\ell}$ depend on σ, and where $D_J\sigma_{ij}$ is defined by

$$D_J\sigma_{ij} = \frac{d\sigma_{ij}}{dt} - \omega_{ik}\sigma_{kj} - \omega_{jk}\sigma_{ik}.$$

Viscoelastic materials (materials with memory)

For such media, the stress–strain law contains functions of time derivatives. In classical viscoelastic theory, we set

$$s = \operatorname{tr}\sigma = \sigma_{ii},$$

$$e = \operatorname{tr}\varepsilon = \varepsilon_{ii} = \operatorname{div}u,$$

where $u = u(x, t)$ is the displacement,[6] and thus

$$s_{ij} = \sigma_{ij}^D = \sigma_{ij} - \frac{s}{3}\delta_{ij},$$

$$e_{ij} = \varepsilon_{ij}^D = \varepsilon_{ij} - \frac{e}{3}\delta_{ij}.$$

---

[6] Here also, the spatial (Lagrangian) variable is denoted by x instead of a.

The stress–strain laws then read

$$s(x,t) = 3\kappa(0)e(x,t) + 3 \int_0^t \frac{d\kappa}{d\tau}(\tau)e(x,t-\tau)\,d\tau,$$

$$s_{ij}(x,t) = 2\mu(0)e_{ij}(x,t) + 2 \int_0^t \frac{d\mu}{d\tau}(\tau)e_{ij}(x,t-\tau)\,d\tau,$$

where $\kappa = \kappa(\tau)$ and $\mu = \mu(\tau)$ satisfy $\kappa(0) = \kappa_0$, $\mu(0) = \mu_0$, $\kappa(\infty) = \kappa_\infty$, $\mu(\infty) = \mu_\infty$, the functions κ and μ are monotonically decreasing from their value at $t = 0$ to their value at $t = \infty$, with $\kappa_0, \kappa_\infty > 0$, $\mu_0, \mu_\infty > 0$. If μ and κ are constants, $(d\kappa/d\tau) = (d\psi/d\tau) = 0$, we recover the constitutive law of linear elasticity.

Remark 5.10:

1. To compare with linear elasticity, we can write the constitutive laws of linear elasticity using the functions s, e, s_{ij}, e_{ij}; they read

$$s = \kappa e, \quad s_{ij} = 2\mu e_{ij}, \quad \text{with } \kappa = 3\lambda + 2\mu.$$

2. The constitutive laws of viscoelastic materials can be written in a different form, more similar to linear elasticity. We extend all functions $\kappa(t)$, $\mu(t)$, $e(t)$, ... by 0 for $t < 0$ and consider the Fourier transform (in time) of these functions, namely $\hat{\kappa}$, $\hat{\mu}$, $\hat{e}$, ... Then, the constitutive laws become

$$\hat{s} = 3[\kappa(0) + \widehat{\kappa'}]\hat{e},$$
$$\hat{s}_{ij} = 2[\mu(0) + \widehat{\mu'}]\hat{s}_{ij}.$$

Remark 5.11: One may also consider viscoelastic fluids; their constitutive laws are similar, u denoting now the *velocity*.

Plastic materials

Plastic media are media for which the stresses are compelled to satisfy certain a priori relations; that is, they are restricted to a certain set in the space of tensors. We study these media under the small deformations assumption; as usual, u is the displacement and $\varepsilon = \varepsilon(u)$ is the linearized deformation tensor.

a) Perfectly plastic (or plastic rigid) materials

We consider a homogeneous, isotropic, and incompressible material for which $e = \varepsilon$. We set (this notation is *not* the same as in Eq. (5.3)):

$$\sigma_{II}^D = \sigma_{ij}^D \sigma_{ij}^D = \sigma_{ij}^D \sigma_{ji}^D.$$

Then, the material is compelled to satisfy certain constraints on σ^D, for example $\sigma_{II}^D \leq K^2$. The material has, in this case, a rigid behavior before this bound is reached; that is to say

$$\sigma_{II}^D < K^2;$$

in this region $\varepsilon(u) = 0$, u being then a rigid displacement. Furthermore, ω_{ij} is constant.

In the region where $\sigma_{II}^D = K^2$, we have the relation

$$\sigma_{ij}^D = \frac{K}{\left(\varepsilon_{II}^D\right)^{1/2}} \varepsilon_{ij}^D,$$

which yields, of course,

$$\sigma_{ij}^D \sigma_{ji}^D = K^2 \frac{\varepsilon_{ij}^D \varepsilon_{ji}^D}{\varepsilon_{II}^D} = K^2.$$

b) Elastoplastic materials

As in a), the stresses are subjected to a condition of the type $\sigma_{II}^D \leq K^2$. For such media, when $\sigma_{II}^D < K^2$, the material has a linear elastic behavior, that is, its stress–strain law is of the form

$$\sigma_{ij} = \lambda \varepsilon_{kk} \delta_{ij} + 2\mu \varepsilon_{ij},$$

and in the region where $\sigma_{II}^D = K^2$, the material has the same behavior as in the previous case.

c) Viscoplastic materials

We consider, for simplicity, an incompressible, homogeneous, and isotropic medium. This medium has a rigid behavior if $s_{II} < K^2$ and, if $s_{II} \geq K^2$, we have

$$s_{ij} = \left(2\mu + \frac{K}{(e_{II})^{1/2}}\right) e_{ij},$$

$$\sigma_{ij}^D = \left(2\mu + \frac{K}{\left(\varepsilon_{II}^D\right)^{1/2}}\right) \varepsilon_{ij}^D.$$

The condition $s_{II} \geq K^2$ is equivalent to

$$\sigma_{ij}^D \sigma_{ji}^D = \left(2\mu + \frac{K}{\left(\varepsilon_{II}^D\right)^{1/2}}\right)^2 \varepsilon_{ij}^D \varepsilon_{ij}^D \geq K^2;$$

that is to say,

$$\left(2\mu + \frac{K}{(e_{II})^{1/2}}\right)^2 e_{II} \geq K^2,$$

or

$$2\mu + \frac{K}{(e_{II})^{1/2}} \geq \frac{K}{(e_{II})^{1/2}};$$

hence,

$$\mu \geq 0.$$

One usually calls such stress–strain laws plasticity laws (Hencky's plasticity law for perfectly plastic or plastic rigid materials); these are actually nonlinear elasticity laws with threshold. Actually, plastic phenomena take into account memory phenomena; they are governed by the so-called Prandtl–Reuss law, which is even more involved.

d) Prandtl–Reuss law

The Prandtl-Reuss law involves a convex set C in the space ($\mathbb{R}^6$) of symmetric (stress) tensors; for example, for the models a), b) above, it is the convex $\sigma_{II}^D < K^2$, and C contains the origin.

At each point a of Ω_0 (a the Lagrangian variable, Ω_0 being the initial non-deformed state), at each time t, if $\sigma = \sigma(a, t)$ is in the interior of C, then the law $\sigma = \sigma(\varepsilon)$ is linear as in case b) above; if $\sigma = \sigma(a, t)$ is on the boundary of C, then

$$\varepsilon_{ij}\left(\frac{\partial u}{\partial t}\right) = A_{ijkl} \ as \frac{\partial \sigma_{kl}}{\partial t} + \lambda,$$

where the A_{ijkl} correspond to the inversion of the stress-strain law of linear elasticity (see Section 13.1 for the details), and $\lambda = \lambda(a, t)$ satisfies the following relation (which actually *characterizes* it):

$$\lambda \cdot (\tau - \sigma) \leq 0, \quad \forall \tau \in C.$$

We refer the reader to specialized books for additional details and for the study of these inequations in the context of variational inequalities and convex analysis.

Remark 5.12: As mentioned previously, we refer the reader to the books of Germain (1973, 1986), and of Truesdell (1977), and to the references therein for a more complete study of stress–strain laws.

We will briefly come back to these laws in the forthcoming chapters for Newtonian fluids and elastic solids (linear and nonlinear elasticity).

All the parameters and functions introduced in this chapter such as μ, λ, κ, h_1, and h_2 may also depend on the temperature (introduced in the next chapter), and on other quantities, but, in general, we consider them as absolute constants.

5.4. Appendix. Change of variable in a surface integral

Using the methods and notations of Section 5.1 of this chapter, we prove formula (3.9) of change of variable in a surface integral given in Chapter 3.

With the notations of Section 5.1 and of Section 3.5, assuming that x, x', x'' belong to Σ_t, the vector product

$$(x' - x) \wedge (x'' - x) = f' \wedge f'' (= O(\lambda^2)),$$

approximates, at order λ^3, the vector $n\delta\Gamma_t$, where $\delta\Gamma_t = \delta\Gamma_t(x, t)$ is the area of the parallelepiped constructed over f' and f'' (and where we have written n instead of $n_t(x, t)$).

We recall that the ith component of the vector product $f' \wedge f''$ can be written, using the Einstein summation convention:

$$(f' \wedge f'')_i = \varepsilon_{ijk} f'_j f''_k,$$

where ε_{ijk} is equal to $+1$ or -1 if (i, j, k) is an even or odd permutation of 1, 2, 3 and to 0 in the other cases.

Furthermore

$$f'_j = \frac{\partial \Phi_j}{\partial a_\alpha} e'_\alpha + O(\lambda^2), \tag{5.9}$$

$$n_i \, \delta\Gamma_t = \varepsilon_{ijk} \frac{\partial \Phi_j}{\partial a_\beta} \frac{\partial \Phi_k}{\partial a_\gamma} \, e'_\beta e''_\gamma \, \delta\Gamma + O(\lambda^3), \tag{5.10}$$

$$n_i \frac{\partial \Phi_i}{\partial a_\alpha} \, \delta\Gamma_t = \varepsilon_{ijk} \frac{\partial \Phi_i}{\partial a_\alpha} \frac{\partial \Phi_j}{\partial a_\beta} \frac{\partial \Phi_k}{\partial a_\gamma} \, e'_\beta e''_\gamma + O(\lambda^3). \tag{5.11}$$

Denoting by F_{ij} the components of the matrix $\mathbf{F}$, we easily verify the following formulae of linear algebra:

$$\det \mathbf{F} = \tfrac{1}{6} \varepsilon_{ijk} \, \varepsilon_{\alpha\beta\gamma} \, F_{i\alpha} \, F_{j\beta} \, F_{k\gamma}, \tag{5.12}$$

$$\varepsilon_{\alpha\beta\gamma} \, \det \mathbf{F} = \varepsilon_{ijk} \, F_{i\alpha} \, F_{j\beta} \, F_{k\gamma}. \tag{5.13}$$

Thus, (5.11) yields

$$n_i \frac{\partial \Phi_i}{\partial a_\alpha} \, \delta\Gamma_t = \varepsilon_{\alpha\beta\gamma} \, \det \mathbf{F} \, e'_\beta e''_\gamma + O(\lambda^3).$$

As for $n = n_t(a, t)$, we see that

$$(a' - a) \wedge (a'' - a) = e' \wedge e''(= O(\lambda^2)) = n_0 \; \delta\Gamma_0 + O(\lambda^3),$$

where $n_0 = n_0(a, t)$ and $\delta\Gamma_0 = \delta\Gamma_0(a, t)$ is the area of the parallelepiped constructed over e' and e''; hence

$$n_i \; \frac{\partial\Phi_i}{\partial a_\alpha} \; \delta\Gamma_t = n_{0\alpha} \; \det\mathbf{F} \; \delta\Gamma_0 + O(\lambda^3),$$

and, in vector form,

$$n_0 \; \det\mathbf{F} \; \delta\Gamma_0 = \mathbf{F}^T \cdot n_t \; \delta\Gamma_t + O(\lambda^3). \tag{5.14}$$

Since n_0 has norm 1, this also gives

$$(\det\mathbf{F})^2 \; (\delta\Gamma_0)^2 = n_t^T \cdot \mathbf{F} \cdot \mathbf{F}^T \cdot n_t \; (\delta\Gamma_t)^2 + O(\lambda^5),$$

or, since $B = \mathbf{F} \cdot \mathbf{F}^T$ (see Section 5.1),

$$(\det\mathbf{F})^2 \; (\delta\Gamma_0)^2 = n_t^T \cdot B \cdot n_t \; (\delta\Gamma_t)^2 + O(\lambda^5).$$

In the limit $\lambda \longrightarrow 0$, this yields

$$\left(\frac{d\Gamma_0}{d\Gamma_t}\right)^2 = (\det\mathbf{F})^{-2} \; n_t^T \cdot B \cdot n_t. \tag{5.15}$$

In the limit $\lambda \longrightarrow 0$, (5.14) also gives, with (5.15):

$$n_0 = (n_t^T \cdot B \cdot n_t)^{-1/2} \; \mathbf{F}^T \cdot n_t, \tag{5.16}$$

and, by inversion, using $C = \mathbf{F}^T \cdot \mathbf{F}$:

$$n_t^T \cdot B \cdot n_t = (n_0 \cdot C^{-1} \cdot n_0)^{-1} \tag{5.17}$$

$$\left(\frac{d\Gamma_0}{d\Gamma_t}\right)^2 = \frac{(\det\mathbf{F})^{-2}}{n_0 \cdot C^{-1} \cdot n_0}, \tag{5.18}$$

$$n_t = \frac{1}{(n_0^T \cdot C^{-1} \cdot n_0)^{-1/2}} \; (\mathbf{F}^{-1})^T \cdot n_0, \tag{5.19}$$

$$n_t \; \frac{d\Gamma_t}{d\Gamma_0} = (\mathbf{F}^{-1})^T \cdot n_0(\det\mathbf{F}). \tag{5.20}$$

Equation (3.9) of Chapter 3 follows simply from (5.20).

Exercises

1. The velocity field of a continuum in motion is given by

$$U_1 = -A(x_1^3 + x_1 x_2^2)e^{-kt}, \ U_2 = A(x_1^2 x_2 + x_2^3)e^{-kt}, \ U_3 = 0. \quad (5.21)$$

   Compute the acceleration field, the deformation rate tensor and the rotation rate tensor.

2. We consider, under the small perturbations hypothesis, the displacement field

$$X_1 = kx_2, \quad X_2 = kx_3, \quad X_3 = kx_1, \quad (5.22)$$

   where k is a small constant. Compute the linearized deformation tensor ε, the principal unit elongations (i.e. the eigenvalues of ε), the principal directions of ε (i.e., the eigenvectors of ε), and the rotation vector Ω.

3. We consider a continuum, having a stationary motion with respect to an orthonormal frame $(0x_1 x_2 x_3)$, such that all the components of the deformation rate tensor vanish, except D_{13} and D_{23} which only depend on x_1 and x_2. Give the general form of the velocity field.

4. We consider the planar stationary motion defined by the formula $U = \nabla\varphi$, $\varphi = \varphi(x_1, x_2)$ being the velocity potential.
   a) Show that the rotation rate vector vanishes.
   b) Compute the deformation rate tensor and its principal invariants.

CHAPTER SIX

Energy equations and shock equations

In this last chapter of Part 1, we return to the fundamental concepts of continuum mechanics and develop two new independent subjects.

On the one hand, we introduce some thermodynamical concepts, namely, internal energy, heat, and temperature to express the energy conservation principle, which leads to a new equation.

On the other hand, we study shock waves: contrary to the regularity assumptions consistently made until now, we consider here the case in which some physical and mechanical quantities are piecewise regular, that is, everywhere regular except at the crossing of some surfaces. It is this framework that is used, for instance, in perfect fluid mechanics, to model the shock waves produced by planes flying at transsonic or supersonic speeds.

6.1. Heat and energy

We consider a material system S that fills the domain Ω_t at time t.

Definition 6.1. *For every material system S and at each time t, there exists a measure carried by Ω_t of the form $e(x, t) dx$, where e is nonnegative. By definition*

$$E = \int_{\Omega_t} \rho(x, t) \, e(x, t) \, dx$$

is the internal energy of S at time t, $e(x, t)$ is the mass density of specific internal energy of S at time t, and ρe is the volume density of internal energy.

90

Definition 6.2. *The energy of the system S at time t is the sum of its kinetic energy and of its internal energy:*

$$\mathcal{E} = \int_{\Omega_t} \rho \left(e + \frac{1}{2} U_i U_i \right) dx.$$

The energy $\mathcal{E}$ is sometimes called the total energy *of the system and is thus defined by its volume density*

$$\rho \left(e + \frac{1}{2} U_i U_i \right).$$

Remark 6.1: For fluids, thermodynamics yields relations between ρ, p, and e (p is the pressure). In particular, it postulates the existence of a relation, called the equation of state, of the form $e = g(p, \rho)$. The assumptions on g will be made more precise in the next chapter.

Heat

A material system receives heat at each time:

1. Through its boundary, by contact actions. This corresponds to the heat received by conduction, and we assume that it is defined by a surface density χ. One can prove that χ is necessarily of the form

$$\chi = -q \cdot n,$$

   n being the unit outward normal to $\partial \Omega_t$; q is called the heat current vector. This can be proven exactly as it has been proven in Chapter 3, after having made reasonable physical assumptions, that, for contact actions, $T = \sigma \cdot n$.

   The rate of heat received by S through $\partial \Omega_t$ at time t is thus

$$- \int_{\partial \Omega_t} q \cdot n \, d\Gamma.$$

2. In the volume by distance actions. This corresponds to the heat received by radiation. We assume that it is defined by a volume density $r = r(x, t)$.

   The rate of heat received by radiation by the system S, at time t, is then

$$\int_{\Omega_t} r(x, t) \, dx.$$

Let $Q_S(t)$ be the heat received by S, from time 0 to time t. The rate at which heat is received by S at time t is thus

$$\dot{Q}_S(t) = \frac{d}{dt} Q_S(t) = \int_{\Omega_t} r \, dx - \int_{\partial\Omega_t} q \cdot n \, d\Gamma;$$

that is, upon using Green's formula

$$\dot{Q}_S(t) = \int_{\Omega_t} (r - \operatorname{div} q) \, dx.$$

The evolution of the system is said to be adiabatic if

$$\dot{Q}_S(t) = 0$$

at each time.

The energy conservation principle: the first law of thermodynamics

The energy conservation principle is stated as follows:

At each time t, the derivative with respect to time of the energy $\mathcal{E}$ of a system S is the sum of the power of the external forces applied to the system and of the rate of heat received by the system:

$$\frac{d\mathcal{E}}{dt} = \mathcal{P}_{\text{ext}} + \dot{Q}_S.$$

Remark 6.2: Integrating this last equation between the times t and t', we can also say that the variation of the total energy $\mathcal{E}$ between these times, $\mathcal{E}(t') - \mathcal{E}(t)$, is the sum of the work of the external forces and of the heat received.

Remark 6.3: By comparing the energy conservation principle to the kinetic energy theorem seen in Chapter 4 (and by taking into account the expression for $\mathcal{E}$ above), we can see that the rate of variation of internal energy $E = E(t)$ satisfies

$$\frac{dE}{dt} = \frac{d}{dt} \int_{\Omega_t} \rho e \, dx = \dot{Q}_S - \mathcal{P}_{\text{int}}.$$

The energy conservation equation

Our aim now is to express the energy conservation principle by means of a partial differential equation for the material system S called the energy conservation equation. We give here different forms of this equation.

We assume, as we already have, that the system S is subjected to volume forces with density f and, on its boundary, to forces with surface density F; then

$$\frac{d}{dt}\int_{\Omega_t}\rho\left(e+\frac{1}{2}U_iU_i\right)dx = \int_{\Omega_t}(f\cdot U + r)\,dx + \int_{\partial\Omega_t}(F\cdot U + \chi)\,d\Gamma.$$

We apply the energy conservation principle to an arbitrary subsystem $S' \subset S$ filling the subdomain $\Omega'_t \subset \Omega_t$ at time t. We transform the integral on $\partial\Omega'_t$ into a volume integral; hence, because $F = \sigma\cdot n$:

$$\int_{\partial\Omega'_t}F\cdot U\,d\Gamma = \int_{\partial\Omega'_t}\sigma_{ij}n_iU_j\,d\Gamma = \int_{\Omega'_t}(\sigma_{ij}U_j)_{,i}\,dx,$$

$$\int_{\partial\Omega'_t}\chi\,d\Gamma = -\int_{\partial\Omega'_t}q\cdot n\,d\Gamma = -\int_{\Omega'_t}q_{i,i}\,dx,$$

and we obtain

$$\int_{\Omega'_t}\left\{\frac{\partial}{\partial t}\left[\rho\left(e+\frac{1}{2}U_iU_i\right)\right] + \left[\rho U_i\left(e+\frac{1}{2}U_jU_j\right) - U_j\sigma_{ij} + q_i\right]_{,i}\right\}$$

$$= \int_{\Omega'_t}(f_iU_i + r)\,dx.$$

This relation is valid for every $\Omega'_t \subset \Omega_t$ and yields, as usual, the equation

$$\frac{\partial}{\partial t}\left[\rho\left(e+\frac{1}{2}U_iU_i\right)\right] + \left[\rho U_i\left(e+\frac{1}{2}U_jU_j\right) - U_j\sigma_{ij} + q_i\right]_{,i} = f_iU_i + r. \tag{6.1}$$

Because

$$\rho\gamma\cdot U = \rho U_i\left(\frac{\partial U_i}{\partial t} + U_jU_{i,j}\right),$$

it follows that

$$\frac{\partial}{\partial t}\left(\rho\frac{|U|^2}{2}\right) + \left(\rho\frac{|U|^2}{2}U_j\right)_{,j}$$

$$= \underline{\frac{|U|^2}{2}\left(\frac{\partial\rho}{\partial t} + \mathrm{div}(\rho U)\right)} + \rho\frac{\partial}{\partial t}\left(\frac{|U|^2}{2}\right) + \rho U_j\left(\frac{|U|^2}{2}\right)_{,j}.$$

The underlined term vanishes thanks to the continuity equation, and thus there remains

$$\frac{\partial}{\partial t}\left(\rho\frac{|U|^2}{2}\right) + \left(\rho\frac{|U|^2}{2}U_j\right)_{,j} = \rho U_i\frac{\partial U_i}{\partial t} + \rho U_jU_iU_{i,j} = \rho\gamma\cdot U. \tag{6.2}$$

Similarly,

$$\frac{\partial}{\partial t}(\rho e) + \operatorname{div}(\rho U e) = e\left(\frac{\partial \rho}{\partial t} + \operatorname{div}(\rho U)\right) + \rho \frac{\partial e}{\partial t} + \rho U \cdot (\nabla e)$$

$$= \rho \frac{\partial e}{\partial t} + \rho (U \cdot \nabla)e;$$

that is,

$$\frac{\partial}{\partial t}(\rho e) + \operatorname{div}(\rho U e) = \rho \frac{De}{Dt}. \tag{6.3}$$

We deduce from Eq. (6.2) that

$$\frac{\partial}{\partial t}\left(\rho \frac{|U|^2}{2}\right) + (U \cdot \nabla)\rho \ U_j\left(\frac{|U|^2}{2}\right)_{,j} - \sigma_{ij,j} U_i = f_i U_i. \tag{6.4}$$

Therefore, thanks to Eqs. (6.3) and (6.4), Eq. (6.1) becomes

$$\boxed{\rho \frac{De}{Dt} + q_{i,i} - \sigma_{ij} U_{i,j} = r.} \tag{6.5}$$

Equation (6.5) is called the energy equation. It indicates that the variation of specific internal energy

$$\frac{De}{Dt} = \frac{1}{\rho}\sigma_{ij} U_{ij} + \frac{1}{\rho}(r - q_{i,i})$$

is due to the power of the internal forces and the exchanges of heat.

Particular cases

For certain materials (for instance for perfect gases), $e = C_V \theta$, where θ is the temperature, $q = -\kappa \operatorname{grad} \theta$, C_V, $\kappa > 0$, and the energy equation becomes

$$\boxed{\rho C_V \frac{\partial \theta}{\partial t} + \rho C_V (U \cdot \nabla)\theta - \kappa \Delta \theta = r + \sigma_{ij} U_{i,j}.} \tag{6.6}$$

When the medium is a perfect incompressible gas, $\sigma_{ij} = -p\delta_{ij}$ and $\operatorname{div} U = 0$, which yields

$$\sigma_{ij} U_{i,j} = -p\delta_{ij} U_{i,j} = -p \operatorname{div} U = 0,$$

and the heat equation becomes

$$\boxed{\rho_0 C_V \left(\frac{\partial \theta}{\partial t} + U \cdot \nabla \theta\right) - \kappa \Delta \theta = r.} \tag{6.7}$$

For other media, the term $\sigma_{ij} U_{i,j}$ often vanishes or is negligible.

6.2. Shocks and the Rankine–Hugoniot relations

Although we did not emphasize it, we have already implicitly encountered
the possibility of having discontinuous quantities at interfaces: for example, if
two different materials are in contact, then the density may be discontinuous
at the interface (e.g., air and water), or the stress tensor may be discontinuous
because the constitutive laws are different.

We now address a more fundamental type of discontinuity that occurs
already at the kinematic level (i.e., for velocities) and typically inside a given
medium: the shock wave.

In the modeling of shock waves, the velocities and other physical quantities
may be discontinuous at the crossing of certain surfaces. We develop here the
mathematical framework for the study of such discontinuities. This section is
essentially independent of Section 6.1.

The central axioms of continuum mechanics are modified as follows. We
still assume that the mapping Φ, which maps the position at time 0 to the
position at time t, is one-to-one. We again denote by $x = \Phi(a) = \Phi(a, t, t_0)$
the corresponding mapping.

Furthermore, we consider a moving surface Σ_t with equation

$$x = \mathcal{E}(a, t), \quad a \in \Sigma_{t_0}.$$

When $\mathcal{E} = \Phi$, Σ_t moves with the flow, but this will not be the case in general;
hence, in general $\Phi^{-1}(\Sigma_t) \neq \Sigma_{t_0}$ and $\Phi(\Sigma_{t_0}) \neq \Sigma_t$ (see Figure 6.1). As usual
$(\partial\Phi/\partial t)(a, t)$ is the velocity of the fluid (again denoted by $U(x, t)$ in the
Eulerian description); similarly $(\partial\mathcal{E}/\partial t)(a, t)$ is the velocity of the surface Σ_t
(also denoted by $W(x, t)$ in the Eulerian description). We finally assume that
Σ_t separates Ω_t into two domains Ω_t^1 and Ω_t^2, and that $\Phi_{|\Phi^{-1}(\Omega_t^1)}$ and $\Phi_{|\Phi^{-1}(\Omega_t^2)}$

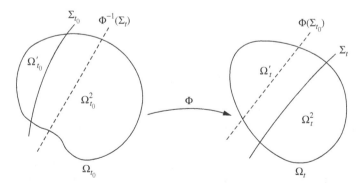

Figure 6.1 The map Φ.

are two mappings from $\Phi^{-1}(\Omega_t^i)$ into Ω_t^i, for $i = 1, 2$, of class C^1, and their inverse are of class C^1 also.

Principle of the study

Our first aim is to compute quantities of the form

$$\frac{d}{dt} \int_{\Omega_t} C \, dx,$$

where C has discontinuities on Σ_t. We have seen that such derivatives occur in the derivation of the conservation equations.

We first notice that, thanks to Remark 1.7, the formula

$$\frac{d}{dt} \int_{\Omega_t} C \, dx = \int_{\Omega_t} \frac{\partial C}{\partial t} \, dx + \int_{\partial \Omega_t} CU \cdot n \, d\Gamma \qquad (6.8)$$

applies to each Ω_t^i, $i = 1, 2$, in the form

$$\frac{d}{dt} \int_{\Omega_t^i} C \, dx = \int_{\Omega_t^i} \frac{\partial C}{\partial t} \, dx + \int_{\partial \Omega_t^i \cap \partial \Omega_t} CU \cdot n \, d\Gamma + \int_{\Sigma_t} CW \cdot n \, d\Gamma, \quad (6.9)$$

provided we interpret U in Eq. (6.8) as the velocity of the boundary; hence, $U = W$ on Σ_t. If N denotes the normal to Σ_t, from Ω_t^1 to Ω_t^2, then $n = N$ for $i = 1$ and $n = -N$ for $i = 2$.

Summing the relations (6.9) for $i = 1, 2$, we obtain

$$\frac{d}{dt} \int_{\Omega_t} C \, dx = \int_{\Omega_t} \frac{\partial C}{\partial t} \, dx + \int_{\partial \Omega_t} CU \cdot n \, d\Gamma - \int_{\Sigma_t} [CW] \cdot N \, d\Gamma,$$

where $X_i = X_{|\Omega_t^i}$, $i = 1, 2$, and $[X] = X_2 - X_1$ is the jump of X through Σ_t. Consequently,

$$\frac{d}{dt} \int_{\Omega_t} C \, dx = \int_{\Omega_t} \frac{\partial C}{\partial t} \, dx + \int_{\partial \Omega_t^1 \cap \partial \Omega_t} CU \cdot n \, d\Gamma + \int_{\partial \Omega_t^2 \cap \partial \Omega_t} CU \cdot n \, d\Gamma$$

$$- \int_{\Sigma_t} [CU] \cdot N \, d\Gamma + \int_{\Sigma_t} [CV] \cdot N \, d\Gamma$$

$$= \int_{\Omega_t^1 \cup \Omega_t^2} \left[\frac{\partial C}{\partial t} + \operatorname{div}(CU) \right] dx + \int_{\Sigma_t} [CV] \cdot N \, d\Gamma,$$

where $V = U - W$ is the relative velocity of the fluid with respect to Σ_t, and $\{\partial C / \partial t + \operatorname{div}(CU)\}$ denotes the piecewise regular function equal to $\partial C / \partial t + \operatorname{div}(CU)$ in Ω_t^1 and in Ω_t^2.

Consequence: the shock conditions

If we know that

$$\frac{d}{dt} \int_{\Omega'_t} C \, dx = \int_{\Omega'_t} f \, dx, \quad \forall \, \Omega'_t \subset \Omega_t,$$

then

$$\frac{\partial C}{\partial t} + \operatorname{div}(CU) = f \quad \text{in } \Omega^i_t, \quad i = 1, 2,$$

and

$$\int_{\Sigma'_t} [CV] \cdot N \, d\Gamma = 0, \quad \forall \, \Sigma'_t \subset \Sigma_t;$$

hence,

$$[CV] \cdot N = 0 \quad \text{on } \Sigma_t.$$

When C is a scalar, setting $v = V \cdot N$, we rewrite the last relation as

$$(Cv)_2 - (Cv)_1 = 0. \tag{6.10}$$

Application to conservation laws: the Rankine–Hugoniot relations

a) Conservation of mass

In this case $C = \rho$, and Eq. (6.10) becomes

$$\rho_2 v_2 = \rho_1 v_1,$$

where v is the normal component of V, $v = V \cdot N$. Two situations may occur:

1. The case of a contact discontinuity, that is, $v_2 = v_1$. In this case, one has $\rho_2 = \rho_1$. Furthermore $U_2 \cdot n = U_1 \cdot n$ and $W \cdot n = 0$. The velocity of Σ_t is tangent to the stream lines.
2. In the other case, we have a shock wave. In this case, $\rho_2 \neq \rho_1$, and the discontinuity wave surface moves with respect to the medium. Its velocity has a component normal to the streamlines as well as a component tangential to them.

b) Conservation of momentum

We take here $C = \rho U_i, i = 1, 2, 3$, which gives

$$\frac{d}{dt} \int_{\Omega_t} \rho U_i \, dx = \int_{\Omega_t} (f_i + \sigma_{ij,j}) \, dx.$$

Owing to the presence of the term $\sigma_{ij,j}$, where σ is discontinuous as well on Σ_t, we need to modify the previous proof and extend it to the case in which C is a vector or a tensor. A similar reasoning, not developed here, gives

$$[\rho v U] = [\sigma \cdot N],$$

and, setting $\rho_1 v_1 = \rho_2 v_2 = m$, we find

$$m[U] = [\sigma \cdot N] = [T].$$

In the case of a perfect fluid, $\sigma_{ij} = -p\delta_{ij}$, and hence,

$$\sigma_{ij} N_j = -pN_i,$$

which gives

$$m[U] + [pN] = 0,$$

or else, by projecting onto N and onto the tangent plane to Σ_t:

$$(U_T)_1 = (U_T)_2,$$

and

$$p_1 + \rho_1 v_1^2 = p_2 + \rho_2 v_2^2$$

because $W \cdot N$ is continuous, U_T being the tangential velocity on the shock surface.

Conservation of energy

We obtain, as a consequence of the conservation of energy, the condition

$$\left[\rho v \left(e + \frac{1}{2} U^2 \right) - T \cdot U + q \cdot N \right] = 0,$$

with $T = \sigma \cdot N$. Writing $U = V + W$, we see that

$$[mU^2] = m[V^2] + 2m[V] \cdot W,$$

$$[T \cdot U] = [T \cdot V] + [T \cdot W]$$

$$= [T \cdot V] + [T]W$$

$$= [T \cdot V] + m[V] \cdot W,$$

which yields

$$\left[\rho v \left(e + \frac{1}{2} V^2 \right) \right] - [TV] + [q \cdot N] = 0.$$

Remark 6.4: Actually, q must be continuous (conduction law), and there is therefore no discontinuity for q. An interesting case to study is the adiabatic case in which $q = 0$.

Remark 6.5: We do not pursue the study of shock waves any further because we would then need more results from thermodynamics, introducing in particular entropy and the second law of thermodynamics. This is a natural continuation of Section 6.1 that we refrain from developing.

To go further:

In the bibliography of Part One, we give several references, in French and English, which allow to go further into several aspects related to the general principles: more detailed analysis of the hypotheses, proof of some of the results that we have assumed, additional developments (rheology, thermodynamics). In French, the books by Duvaut, Germain, and Salençon are relevant to this part; the point of view is sometimes different, less mathematical. In English, the books by Gurtin and Spencer (very concise) are very useful, as well as the book by Segel. The books by Goldstein and Ziegler emphasize by analytical mechanics and mechanics of rigid bodies; the book by Truesdell makes in particular a detailed analysis of the "axioms" of mechanics. The book by Zemanski is one of the classical references on thermodynamics.

Exercises

1. The second principle of thermodynamics states that there exists a funtion s, called entropy (per unit mass), such that the variation of entropy, in an elementary transformation, is the sum of the variation of entropy due to external supply and of the entropy produced inside the system. For a closed system, the second principle is often written in the form

$$\delta Q = T ds - \delta f, \quad \delta f \geq 0 \qquad \text{(Jouget relation)}, \qquad (6.11)$$

where T is the temperature, and δf is called the non-compensated work. Furthermore, it follows from the first and second principles that

$$de = T ds - p \, d\frac{1}{\rho}, \qquad (6.12)$$

where s is the entropy per unit mass and p is the pressure.

a) Show that the enthalpy per unit of mass $h = e + p/\rho$ satisfies

$$dh = T ds + \frac{1}{\rho} dp. \qquad (6.13)$$

b) We consider a perfect gas. In that case, we have

$$e = \frac{c_v}{\gamma}\rho^{\gamma-1}e^{s/c_v}, \quad c_v = \frac{r}{\gamma-1}. \tag{6.14}$$

Show that $p/\rho = (\gamma - 1)e = rT$ and that $e = c_v T$.

2. Let h and s be defined as in Exercise 1. We call free energy and free enthalpy the functions $F = e - Ts$ and $G = h - Ts$ respectively. Show that

$$dF = -sdT - pd\frac{1}{\rho}, \tag{6.15}$$

$$dG = -sdT + \frac{1}{\rho}dp. \tag{6.16}$$

3. Let h and s be defined as in Exercise 1. We consider a two-dimensional shock wave in a perfect fluid. We assume that the discontinuity is a stationary vertical straight line. We further assume that the external forces are negligible and that the evolution is adiabatic. Finally, we assume that the velocity vector is horizontal. As usual, we will use the index 1 for the quantities before the shock and the index 2 after the shock.
   a) Write the conservation of mass, momentum, energy, and enthalpy.
   b) Same question as in a) for a perfect gas for which the state equation is $p = \rho rT$ and the enthalpy is $h = c_p T$.
   c) Show that $(p_2/p_1) - 1 = -\gamma M_1^2(\rho_1/\rho_2 - 1)$ (here, $M = v/c$, $c^2 = \partial p/\partial \rho$, is the Mach number and, for a perfect gas, $p = \rho^\gamma e^{s/c_v}$).

4. We consider here the same situation as in 3.a).
   a) Show that $h_2 - h_1 = \frac{1}{2}v_1v_2(p_2/p_1 - \rho_1/\rho_2)$ and that $p_2 - p_1 = v_1v_2(\rho_2 - \rho_1)$.
   b) Deduce that $h_2 - h_1 = \frac{1}{2}(p_2 - p_1)(1/\rho_1 + 1/\rho_2)$ and that $e_2 - e_1 = \frac{1}{2}(p_1 + p_2)(1/\rho_1 - 1/\rho_2)$.

5. Same questions as in 3.a) and b) for a straight line making an angle ε with the horizontal. Furthermore, we assume that the velocity v_1 before the shock is horizontal and that the velocity v_2 after the shock makes an angle β with the shock.

PART II

PHYSICS OF FLUIDS

CHAPTER SEVEN

General properties of Newtonian fluids

7.1. General equations of fluid mechanics

Our aim in this section is to return to, and study in more detail, the equations of motion of a Newtonian fluid as well as the boundary conditions that are associated with them.

Throughout this chapter, the fluid is represented by a material system S that fills the domain Ω_0 at time $t = 0$ and the domain Ω_t at time t. We may have $\Omega_t = \Omega_0$, which corresponds, for instance, to the case in which the fluid in motion entirely fills a container of constant shape.

In the Eulerian representation, we will denote the density by $\rho = \rho(x, t)$, the volume forces by $f = f(x, t)$, and the surface forces applied to the boundary $\partial\Omega_t$ of Ω_t by $F = F(x, t)$. We change our notation and, from now on, we denote the velocity in the Eulerian representation by $u = u(x, t)$ instead of $U(x, t)$; $u(x, t)$ is then the velocity of the fluid particle occupying the position $x \in \Omega_t$ at time t.

The equations

The general equations governing the motion of a fluid are the continuity equation (or mass conservation equation)

$$\frac{\partial\rho}{\partial t} + (\rho u_i)_{,i} = 0, \quad \text{in } \Omega_t, \tag{7.1}$$

and the conservation of momentum equation, which is a consequence of the fundamental law of dynamics, namely

$$\rho\gamma_i = \rho\left(\frac{\partial u_i}{\partial t} + u_j\frac{\partial u_i}{\partial x_j}\right) = f_i + \sigma_{ij,j}, \quad \text{in } \Omega_t.$$

For a Newtonian fluid, we have

$$\sigma_{ij} = 2\mu\varepsilon_{ij} + \lambda\varepsilon_{kk}\delta_{ij} - p\delta_{ij},$$

where $p = p(x, t)$ is the pressure and $\varepsilon_{ij} = \varepsilon_{ij}(u) = \frac{1}{2}(u_{i,j} + u_{j,i})$. Consequently,

$$\sigma_{ij} = \mu\left(u_{i,j} + u_{j,i}\right) + \lambda(\operatorname{div} u)\delta_{ij} - p\delta_{ij},$$

which yields

$$\begin{aligned}\sigma_{ij,j} &= \mu u_{i,jj} + \mu u_{j,ji} + \lambda(\operatorname{div} u)_{,i} - p_{,i}\\ &= \mu\Delta u_i + (\lambda + \mu)(\operatorname{div} u)_{,i} - p_{,i},\end{aligned}$$

and thus, the conservation of momentum equation can be rewritten as

$$\rho\left(\frac{\partial u_i}{\partial t} + u_j\frac{\partial u_i}{\partial x_j}\right) + p_{,i} - \mu\Delta u_i - (\lambda + \mu)(\operatorname{div} u)_{,i} = f_i, \quad \text{in } \Omega_t. \quad (7.2)$$

In vector form, this reads

$$\rho\left[\frac{\partial u}{\partial t} + (u \cdot \nabla)u\right] + \operatorname{grad} p - \mu\Delta u - (\lambda + \mu)\operatorname{grad} \operatorname{div} u = f, \quad \text{in } \Omega_t. \quad (7.2')$$

Equations (7.1) and (7.2) (or (7.2')) are called the Navier–Stokes equations for compressible fluids.[1]

By comparing the components, one easily checks that

$$u_j\frac{\partial u}{\partial x_j} = \operatorname{grad}\frac{|u|^2}{2} + (\operatorname{curl} u) \wedge u,$$

and Eq. (7.2) also reads, in vector form, as follows:

$$\rho\left(\frac{\partial u}{\partial t} + \operatorname{grad}\frac{|u|^2}{2} + (\operatorname{curl} u) \wedge u\right)$$
$$+ \operatorname{grad} p - \mu\Delta u - (\lambda + \mu)\operatorname{grad} \operatorname{div} u = f, \quad \text{in } \Omega_t. \quad (7.2'')$$

In Eqs. (7.2), (7.2'), and (7.2''), the constants λ and μ satisfy $\mu \geq 0$ and $3\lambda + 2\mu \geq 0$ (hence, $(\lambda + \mu) \geq 0$). These positivity conditions are imposed by thermodynamic considerations; in Section 7.3, we will come back to the condition $3\lambda + 2\mu \geq 0$.

---

[1] Some authors prefer to use the name Navier–Stokes equations for incompressible fluids only. We will use this name without distinguishing between the two cases.

Inviscid fluid

A fluid for which $\lambda = \mu = 0$ is called an inviscid, or perfect, or nonviscous fluid. For such a fluid, $\sigma_{ij} = -p\,\delta_{ij}$ (i.e., the stress tensor is spherical) and Eqs. (7.2), (7.2'), and (7.2'') become

$$\rho\left(\frac{\partial u_i}{\partial t} + u_j\frac{\partial u_i}{\partial x_j}\right) + p_{,i} = f_i, \tag{7.3}$$

$$\rho\left[\frac{\partial u}{\partial t} + (u \cdot \nabla)u\right] + \operatorname{grad} p = f, \tag{7.3'}$$

$$\rho\left[\frac{\partial u}{\partial t} + \operatorname{grad}\frac{|u|^2}{2} + (\operatorname{curl} u) \wedge u\right] + \operatorname{grad} p = f, \tag{7.3''}$$

where $x \in \Omega_t, t \in I$. Equations (7.3) are called the Euler equations of inviscid fluids.

Incompressible fluid

A fluid is said to be incompressible if the volume of any quantity of fluid remains constant during its motion. For such a fluid, by using Eq. (1.8) of Chapter 1 (with $C = 1$), we see that

$$\operatorname{div} u = 0. \tag{7.4}$$

If we further assume that the initial state is homogeneous, which means

$$\rho(x, 0) = \rho_0 = \text{Const.},$$

then

$$\rho(x, t) = \rho_0, \quad \forall x \in \Omega_t, \quad \forall t \in I, \tag{7.5}$$

and Eq. (7.4) is equivalent to the continuity equation.

When the fluid is incompressible and homogeneous, Eqs. (7.2), (7.2'), and (7.2'') become, respectively,

$$\frac{\partial u_i}{\partial t} + u_j\frac{\partial u_i}{\partial x_j} + \frac{1}{\rho}\,p_{,i} = \nu\Delta u_i + f_i', \tag{7.6}$$

$$\frac{\partial u}{\partial t} + (u \cdot \nabla)u + \frac{1}{\rho}\operatorname{grad} p = \nu\Delta u + f', \tag{7.6'}$$

$$\frac{\partial u}{\partial t} + \operatorname{curl} u \wedge u + \operatorname{grad}\left(\frac{|u|^2}{2} + \frac{p}{\rho}\right) = \nu\Delta u + f', \tag{7.6''}$$

where $x \in \Omega_t$ and $t \in I$; $f' = f/\rho$ and $\nu = \mu/\rho$ is called the kinematic viscosity coefficient. Equations (7.4) and (7.6) constitute the Navier–Stokes equations for incompressible homogeneous fluids.

Remark 7.1 (Nonhomogeneous Incompressible Fluid): Equations (7.4) and (7.6) remain valid for a nonhomogeneous incompressible fluid (for which Eq. (7.5) is not valid). Equation (7.4) is, however, no longer equivalent to the continuity equation, but the latter can be simplified and becomes

$$\frac{\partial \rho}{\partial t} + (u \cdot \nabla)\rho = 0, \quad \text{in } \Omega_t. \tag{7.1'}$$

The vorticity equation

The vorticity is the vector $\omega = \text{curl } u$.

We assume that the fluid is incompressible and homogeneous. Setting $\omega = \text{curl } u$ and taking the curl of (7.6″), we find

$$\boxed{\frac{\partial \omega}{\partial t} + \text{curl}(\omega \wedge u) - \nu\Delta\omega = \text{curl } f', \quad \text{in } \Omega_t.} \tag{7.7}$$

We can give an alternate formulation of this equation, which is not the same in space dimension 2 or 3. Let us recall that, in space dimension 3, ω is a vector function, whereas in space dimension 2, $\omega = \omega_3 e_3$ has only one nonzero component. We will set in this case $\omega = \omega_3$, and we will say that the curl is a scalar function.

The term $\text{curl}(\omega \wedge u)$ is different, depending on whether the flow is two-dimensional or three-dimensional. A straightforward calculation gives

$$\text{curl}(\omega \wedge u) = (u \cdot \nabla)\omega, \quad \text{if } n = 2,$$
$$\text{curl}(\omega \wedge u) = (u \cdot \nabla)\omega - (\omega \cdot \nabla)u, \quad \text{if } n = 3.$$

Indeed, if $n = 2$, the only nonvanishing component of $\text{curl}(\omega \wedge u)$ is the third one, which is equal to

$$D_1(\omega_3 u_1) + D_2(\omega_3 u_2) = u_1(D_1\omega_3) + u_2(D_2\omega_3) = (u \cdot \nabla)\omega$$

because $\text{div } u = 0$ ($u_3 = 0$, $\omega_3 = D_1 u_2 - D_2 u_1$, $D_i = \partial/\partial x_i$). If $n = 3$, we can derive the relation above by a simple comparison of the components. For instance, the third component of $\text{curl}(\omega \wedge u)$ is

$$D_1(\omega_3 u_1 - \omega_1 u_3) - D_2(\omega_2 u_3 - \omega_3 u_2)$$
$$= (\text{div } \omega)u_3 + (u \cdot \nabla)\omega_3 - (\omega \cdot \nabla)u_3$$
$$= (u \cdot \nabla)\omega_3 - (\omega \cdot \nabla)u_3$$

because div $\omega = 0$. The curl equation then becomes

$$\frac{\partial \omega}{\partial t} + (u \cdot \nabla)\omega - \nu\Delta\omega = \text{curl } f', \quad \text{in space dimension 2}, \qquad (7.7')$$

$$\frac{\partial \omega}{\partial t} + (u \cdot \nabla)\omega - \nu\Delta\omega = (\omega \cdot \nabla)u + \text{curl } f', \quad \text{in space dimension 3}. \qquad (7.7'')$$

If the fluid is inviscid, one has to take $\nu = 0$ in these equations; for instance, in space dimension 2, Eq. (7.7') becomes

$$\frac{\partial \omega}{\partial t} + (u \cdot \nabla)\omega = \text{curl } f'. \qquad (7.7''')$$

Remark 7.2: If f is the gradient (with respect to x) of a function $V = V(x, t)$ (one says in mechanics that "the forces derive from a potential"), then curl $f = 0$. This is true, for instance, for gravitational forces.

Equation of state: barotropic fluid

For a compressible fluid, Eqs. (7.1) and (7.2) are not sufficient to describe the motion fully. Indeed, they provide four equations for the five unknowns ρ, p, and u. The missing equation is provided by the equation of state of the fluid (see Chapter 5).

The equation of state of a fluid expresses p as a prescribed function of ρ (the density) and e (the internal energy). In the case of a barotropic fluid, p only depends on ρ as described by

$$p = g(\rho).$$

For thermodynamic reasons, g must satisfy

$$g > 0, \quad \frac{dg}{d\rho} > 0 \quad \text{and} \quad \frac{d^2 g}{d\rho^2} \geq 0.$$

We then set $c^2 = dg/d\rho$, c being the *local sound velocity* in the fluid.

Examples
1. If the gas is perfect with constant specific heat in isothermal evolution, then $p = k\rho, k > 0, c = \sqrt{k}$.
2. If the evolution is adiabatic, then $p = k\rho^\gamma, k > 0, \gamma > 1$ ($\gamma = 1.4$ for the air); hence, $c^2 = k\gamma\rho^{\gamma-1} = k^{1/\gamma}\gamma p^{(\gamma-1)/\gamma}$.

Boundary conditions

To describe the motion of a fluid completely we need to know, in addition to the general equations described above, the behavior of the physical quantities at the boundary of the domain filled by the fluid: these are the *boundary conditions*.

a) Case of a rigid wall

If the boundary is materialized and, say, fixed in a Galilean frame of reference, we write first that the fluid cannot cross the boundary, which can be expressed by the kinematic condition

$$u \cdot n = 0, \quad \forall x \in \partial\Omega_t, \quad \forall t \in I. \tag{7.8}$$

When the fluid is viscous and the boundary is materialized, the fluid does not slip along the boundary; hence, the velocity of the fluid particle occupying position x at time t is the same as the velocity $g(x, t)$ of the material boundary, that is,

$$u(x, t) = g(x, t), \quad \forall x \in \partial\Omega_t, \quad \forall t \in I. \tag{7.8'}$$

b) Case of a nonmaterialized boundary (open boundary)

We consider here a nonmaterialized boundary Σ_t (called open boundary), as in Figure 7.1.(a); we can also consider the interface Σ_t' between two fluids as in Figure 7.1.(b).

Kinematic conditions. For a perfect fluid, we write that $u \cdot n$ is continuous and that $u \cdot n = v \cdot n$ on Σ_t, where v is the velocity of Σ_t, which may vanish;

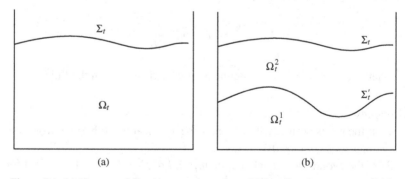

(a)   (b)

Figure 7.1 (a) Nonmaterialized or open boundary; (b) Interface between two fluids.

that is,

$$u \cdot n = v \cdot n \quad \text{on } \Sigma_t, \quad \text{case (a)},$$
$$u^1 \cdot n = u^2 \cdot n = v \cdot n \quad \text{on } \Sigma'_t, \quad \text{case (b)}. \tag{7.9}$$

In the case of viscous fluids, the conditions of Eq. (7.9) are still valid; in case (b), we also write that the velocity u is continuous:

$$u^1 = u^2 \quad \text{on } \Sigma'_t.$$

Fluid Conditions. We have noted in Chapter 3 (Remark 3.3) that the constraint vectors are continuous, which gives

$$\sigma \cdot n = 0 \quad \text{on } \Sigma_t, \quad \text{case (a)},$$
$$\sigma^1 \cdot n = \sigma^2 \cdot n \quad \text{on } \Sigma'_t, \quad \text{case (b)}.$$

In the case of a perfect fluid, $\sigma_{ij} = -p\,\delta_{ij}$, and these conditions reduce to

$$p = 0 \quad \text{on } \Sigma_t, \quad \text{case (a)},$$
$$p^1 = p^2 \quad \text{on } \Sigma'_t, \quad \text{case (b)}.$$

Other cases

When the temperature is not constant, we need to introduce other thermodynamic quantities (the internal energy e, the enthalpy h, the entropy s, etc.) and the equations of thermodynamics, an overview of which was given in Part 1. Some elementary aspects of thermodynamics will appear in Chapter 9. Similarly, electromagnetic phenomena may be involved: one then speaks of magnetohydrodynamics, some aspects of which will be mentioned in Chapter 11. Finally, in the context of combustion, chemical phenomena are involved; see Chapter 12.

7.2. Statics of fluids

We consider here a fluid at rest. Consequently $u(x, t) \equiv 0$, which yields $\Omega_t \equiv \Omega_0$ for every t. The equations studied previously become

$$f = \text{grad } p \tag{7.10}$$

for every Newtonian fluid, whether this fluid is viscous or not, compressible or not, because $\sigma_{ij} = -p\delta_{ij}$. We can then state the following result.

Theorem 7.1. *If a fluid is at rest, the volume forces derive from a potential.*

Next, we prove Archimedes' principle for a body (fully) immersed in a fluid.

Theorem 7.2. *(Archimedes' Principle). The pressure forces exerted by a fluid at rest on a completely immersed body define a helicoidal vector field that is the opposite of that associated with the gravity forces acting on a volume of fluid identical to that occupied by the body.*

Proof: Let Ω be the domain occupied by the immersed body. Because the fluid is at rest, we have, in the fluid,

$$\sigma_{ij} = -p\delta_{ij},$$

and, with e_3 pointing upwards in the vertical direction,

$$\text{grad } p = -\rho g e_3.$$

The resultant of the pressure forces exerted by the fluid on the body is thus equal to

$$-\int_{\partial\Omega} pn \, d\Gamma = -\int_{\Omega} \text{grad } p \, dx,$$

$$= \int_{\Omega} \rho g \vec{e}_3 \, dx;$$

this resultant is opposite to that of the gravity forces on an identical volume of fluid equal to

$$-\int_{\Omega} \rho g \vec{e}_3 \, dx.$$

Furthermore, the momentum at O of the pressure forces is equal to $-\int_{\partial\Omega} x \wedge np \, d\Gamma$ (if it is assumed that O is the center of mass of Ω). The first component of the momentum at O is thus

$$\int_{\partial\Omega} (n_2 x_3 - n_3 x_2)p \, d\Gamma = \text{(by Stokes formula)} \int_{\Omega} [(x_3 p)_{,2} - (x_2 p)_{,3}] \, dx$$

$$= \int_{\Omega} (x_3 p_{,2} - x_2 p_{,3}) \, dx$$

$$= -\int_{\Omega} (x \wedge \text{grad } p)_1 \, dx \quad \text{(1st component)}$$

$$= \int_{\Omega} (x \wedge \rho g e_3)_1 \, dx = 0 \quad (O \text{ center of mass of } \Omega).$$

Proceeding similarly with the other two components, we see that the momentum at O of the pressure forces is indeed opposite to the momentum at O of the gravity forces of the considered volume, namely,

$$-\int_{\partial\Omega} x \wedge np \, d\Gamma = \int_{\Omega} x \wedge \rho g e_3 \, dx = 0.$$

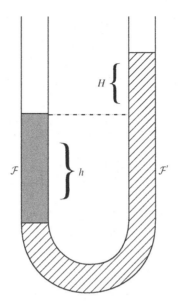

Figure 7.2 Contact between two fluids.

Statics of incompressible fluids

For an incompressible homogeneous fluid, we have

$$f = \operatorname{grad} p \quad \text{and} \quad f = -\rho_0 g \vec{e}_3.$$

Therefore,

$$p = -\rho_0 g(x_3 - h).$$

We now give some applications of these results.

a) Contact between two fluids

We consider two fluids $\mathcal{F}$ and $\mathcal{F}'$ with their respective densities and pressures (ρ, p) and (ρ', p'); they are put in contact in a U-shaped tube, as shown in Figure 7.2. We then obtain the following relation determining the position of the fluids at equilibrium:

$$\rho g h = \rho' g(h + H).$$

b) The Venturi device

The devices represented in Figures 7.3 and 7.4 were very important measuring devices used in wind tunnels. Nowadays, they are generally replaced by laser velocimetry devices but, for our purpose, it is interesting to see their principle.

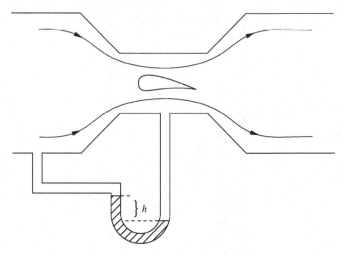

Figure 7.3 The Venturi device.

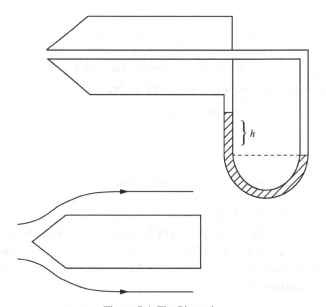

Figure 7.4 The Pitot tube.

In the Venturi device (Figure 7.3), because the tube has a small cross section, we can assume that the fluid that it contains is at rest. Then, the difference between pressures at the ends of the tube is $\rho g h$. Thanks to the Bernoulli theorem that we will encounter in the next chapter, this difference of pressure gives us the velocity of the fluid in the collar of the wind tunnel.

c) The Pitot tube

The Pitot tube, which is similar to the Venturi device, measures the difference of pressure between the front and the rear of an obstacle (e.g., a rocket). With the help of the Bernoulli theorems, which we will see in Chapter 8, it allows in some cases the calculation of the difference of velocities.

Statics of compressible barotropic fluids: a first approximation in meteorology

We assume that the atmosphere is a compressible fluid at rest (in Chapter 10, we will consider less-simplifying assumptions). Then,

$$f = -\rho g \vec{e}_3 = \text{grad } p;$$

hence,

$$\frac{\partial p}{\partial x_1} = \frac{\partial p}{\partial x_2} = 0 \quad \text{and} \quad \frac{\partial p}{\partial x_3} = -\rho g.$$

Furthermore, the air being a barotropic fluid, the pressure is related to the density by a relationship of the form[2]

$$p = \varphi(\rho),$$

which yields

$$\frac{\partial \rho}{\partial x_1} = \frac{\partial \rho}{\partial x_2} = 0 \quad \text{and} \quad \frac{\partial \rho}{\partial x_3} = -\frac{\rho g}{\varphi'(\rho)}.$$

When $p = k\rho$ (which is true for a perfect fluid with constant specific heat), it follows that

$$\frac{\partial \rho}{\partial x_3} = \frac{d\rho}{dx_3} = -\frac{\rho g}{k},$$

which yields

$$\rho = \rho_0 \exp\left(-\frac{gx_3}{k}\right).$$

Because $k = p_0/\rho_0$, the pressure reads

$$\boxed{p = p_0 \exp\left(-\frac{g\rho_0 x_3}{p_0}\right),}$$

where $x_3 = 0$ denotes the ground level, and p_0 and ρ_0 are, respectively, the pressure and the density at the ground level.

---

[2] The equation of state was written $p = g(\rho)$ before, but here g is used for the gravitational acceleration.

7.3. Remark on the energy of a fluid

Our aim in this section is to show that the coefficients λ and μ appearing in Eq. (7.2) satisfy the relations $\mu \geq 0$ and $3\lambda + 2\mu \geq 0$. To do so, we consider the power of the internal forces. We have

$$\mathcal{P}_{\text{int}} = -\int_{\Omega_t} \sigma_{ij} u_{i,j} \, dx = -\int_{\Omega_t} \sigma_{ij} \varepsilon_{ij} \, dx.$$

Now

$$\sigma_{ij} u_{i,j} = -p u_{i,i} + \lambda (\text{div } u)^2 + 2\mu \varepsilon_{ij} \varepsilon_{ij}$$
$$= -p \, \text{div } u + \varphi(\varepsilon).$$

Let us set

$$\mathcal{P}^{(1)} = -\int_{\Omega_t} p \, \text{div } u \, dx \quad \text{and} \quad \mathcal{P}^{(2)} = -\int_{\Omega_t} \varphi[\varepsilon(u)] \, dx.$$

For an incompressible fluid, we will have $\mathcal{P}^{(1)} = 0$. In all cases (i.e., whether the fluid is compressible or not), $\mathcal{P}^{(2)}$ is equal to the power due to friction, and the laws of thermodynamics require that $\mathcal{P}^{(2)} \leq 0$; that is to say that φ is a semidefinite positive quadratic form. But

$$\varphi = \lambda(\varepsilon_{11} + \varepsilon_{22} + \varepsilon_{33})^2 + 2\mu \left(\varepsilon_{11}^2 + \varepsilon_{22}^2 + \varepsilon_{33}^2 \right)$$
$$+ 4\mu \left(\varepsilon_{12}^2 + \varepsilon_{13}^2 + \varepsilon_{23}^2 \right).$$

Therefore, φ is positive definite if and only if the quadratic form $q = \lambda(x + y + z)^2 + 2\mu(x^2 + y^2 + z^2)$ is positive definite. By homogeneity, it is necessary and sufficient that q be positive when $x + y + z = 1$; that is to say $\mu \geq 0$ and q is positive for $x + y + z = 1$.

Thus, there should be no (x, y, z) satisfying

$$x^2 + y^2 + z^2 < -\frac{\lambda}{2\mu} \quad \text{and} \quad x + y + z = 1,$$

which implies that the distance from O to the plane R with equation $x + y + z = 1$ is such that

$$[d(O, R)]^2 \geq -\frac{\lambda}{2\mu}.$$

Because $d(O, R) = 1/\sqrt{3}$, it follows that

$$3\lambda + 2\mu \geq 0.$$

Exercises

1. A viscous, homogeneous, incompressible and stationary fluid fills the volume $\Omega \subset \mathbb{R}^3$. It is submitted to volume forces with volume density $f = f(x)$ and surface forces with surface density $F = F(x)$ on the boundary $\Sigma = \partial\Omega$.
   a) Write the equations of motion in terms of the stress tensor σ, the velocity u, and f.
   b) The constitutive law of such a fluid reads

$$\sigma^D = 2\mu\varepsilon^D, \tag{7.11}$$

$$\sigma_{ii} = (3\lambda + 2\mu)\varepsilon_{ii} - 3p, \tag{7.12}$$

   where p is the pressure and ε the deformation rate tensor. Write the equations of motion in terms of u, p, and f.
   c) Write the virtual power theorem for an arbitrary virtual velocity field $v = v(x)$.

2. We denote by p, ρ, and T the pressure, density, and temperature of the atmosphere and we assume that the quantities are related by the perfect gas law $p = R\rho T$. We assume that the atmosphere is at rest and that T is a known function of the altitude x_3. Compute the variation of p with respect to x_3.

3. A perfect, compressible fluid moves in a rectilinear tube with very small constant section, parallel to the $(0x)$ axis with unit vector $\vec{x}$ (the volume forces are neglected). Let $\rho(x, t)$, $p(x, t)$, and $\vec{U}(x, t) = u(x, t)\vec{x}$ be the density, pressure, and velocity of the fluid at x, at time t. We assume that the fluid satisfies the equation of state $p = k\rho$. Show that u satisfies the equation

$$\frac{\partial^2 u}{\partial t^2} + \frac{\partial}{\partial x}\left(2u\frac{\partial u}{\partial t} + u^2\frac{\partial u}{\partial x}\right) = k\frac{\partial^2 u}{\partial x^2}. \tag{7.13}$$

4. Rewrite Archimedes' principle for a partially immersed body, assuming that the fluid and the body are at rest (thus, we assume that the surface of the fluid is planar and horizontal).

CHAPTER EIGHT

Flows of inviscid fluids

This chapter is devoted to the study of flows of nonviscous (perfect) Newtonian fluids. We give some general results and then consider several specific flows.

8.1. General theorems

We establish in this section several simple general theorems concerning flows of perfect fluids; these are known as the Bernoulli and Laplace theorems.

We will only consider incompressible fluids (for which $(1/\rho)\,\text{grad}\,p$ is a gradient). We note, however, that the following analysis also applies to perfect barotropic fluids for which the term $(1/\rho)\,\text{grad}\,p$ is also a gradient, for, when $p = g(\rho)$, it can be written as

$$\frac{1}{\rho}\,\text{grad}\,p = \frac{1}{\rho}\,g'(\rho)\,\text{grad}\,\rho = \text{grad}\,G(\rho),$$

where $G(\rho)$ is an antiderivative of $g'(\rho)/\rho$.

The Bernoulli, Kelvin, and Lagrange theorems

We further assume that the mass density of forces derives from a potential, that is,

$$\frac{f}{\rho} = -\text{grad}\,V.$$

We then deduce from the fundamental law of dynamics that the acceleration γ satisfies

$$\gamma = -\text{grad}\left(\frac{p}{\rho} + V\right).$$

116

We set $\mathcal{H} = (p/\rho) + V + \frac{1}{2}|u|^2$, and we obtain the following equation (see Chapter 7):

$$\frac{\partial u}{\partial t} + \text{curl } u \wedge u + \text{grad } \mathcal{H} = 0. \tag{8.1}$$

This yields the following theorem.

Theorem 8.1 (Stationary Flows). *If the fluid is incompressible and inviscid, if the mass density of forces derives from a potential and the flow is stationary, then $\mathcal{H}$ remains constant along each trajectory.*

Proof: Because the flow is stationary, Eq. (8.1) becomes

$$(\text{curl } u) \wedge u + \text{grad } \mathcal{H} = 0. \tag{8.2}$$

Taking then the scalar product of Eq. (8.2) with u, we obtain

$$u \cdot \text{grad } \mathcal{H} = 0. \tag{8.3}$$

Along a trajectory, u is parallel to the unit tangent vector and thus, owing to Eq. (8.3), the derivative of $\mathcal{H}$ with respect to the curvilinear coordinate vanishes; consequently, $\mathcal{H}$ remains constant.

Remark 8.1: If the mass density of forces vanishes (that is to say $V = 0$), then

$$\frac{p}{\rho} + \frac{1}{2}|u|^2 = \text{const.} = \frac{p_0}{\rho_0} + \frac{1}{2}|u_0|^2,$$

along the trajectories, where p_0, ρ_0, and u_0 are the values of p, ρ, and u at some point of the trajectory. Therefore, the determination of p reduces to that of kinematic quantities or, conversely, $|u|$ is known once p is given (as well as p_0, ρ_0, and $|u_0|$).

Definition 8.1. *A flow is irrotational if $\omega = \text{curl } u$ vanishes everywhere.*

For irrotational flows, we have the following result:

Theorem 8.2 (Irrotational flows). *Assume that an incompressible inviscid fluid fills the domain Ω_t at time t. If, moreover, the flow is irrotational, then there exists a velocity potential, denoted by $\Phi(x, t)$ such that, on any simply connected part of Ω_t, we have*

$$\frac{\partial \Phi}{\partial t} + \mathcal{H} = C(t),$$

where C depends only on time.

Proof: Because curl $u = 0$, u can be written locally in the form

$$u = \text{grad } \Phi.$$

We then deduce from Eq. (8.1) that

$$\text{grad}\left(\frac{\partial \Phi}{\partial t} + \mathcal{H}\right) = 0,$$

which yields, locally,

$$\frac{\partial \Phi}{\partial t} + \mathcal{H} = C(t),$$

where C depends on the time t but not on x. This last relation is actually valid with the same constant $C(t)$ in any simply connected component of Ω_t: the potential Φ is then one-to-one. In addition, by replacing Φ by $\Phi - C^*$, where C^* is an antiderivative of C, we can assume that $C = 0$. In a connected but not simply connected component of Ω_t, the functions are not one-to-one and one has to remove lines or surfaces from Ω_t to make it simply connected.

A straightforward consequence of Theorems 8.1 and 8.2 is the following result.

Corollary 8.1. *We assume, in addition to the hypotheses of Theorem 8.2, that the flow is stationary. Then, $\mathcal{H}$ is constant on every simply connected part of the domain filled by the fluid.*

Theorem 8.3 (Bernoulli's Theorem). *For an irrotational flow of an inviscid incompressible fluid, the velocity potential Φ is a harmonic function:*

$$\Delta \Phi = 0.$$

Proof: Because the fluid is irrotational,

$$u = \text{grad } \Phi,$$

locally, for some function Φ. The fluid being moreover incompressible

$$0 = \text{div } u = \text{div grad } \Phi$$

$$= \Delta \Phi.$$

We will now give some consequences of the vorticity equation introduced in Chapter 7.

Theorem 8.4 (Kelvin's Theorem). *We consider an inviscid incompressible fluid for which the mass (or volume) density of forces derives from a potential. Then, the flux of the vorticity vector through a surface moving with the flow remains constant. Also, the circulation of the velocity vector along a closed curve that moves with the flow remains constant.*

Remark 8.2: Let us recall that "a set that moves with the flow" is a set whose preimage in the domain filled by the fluid at the initial time remains constant.

Proof: We are given a surface Σ_t that moves with the flow. We admit here the following equation, similar to Eqs. (1.7) and (1.8) of Chapter 1, in which n is a unit normal to Σ_t:

$$\frac{d}{dt} \int_{\Sigma_t} \omega \cdot n \, d\Gamma = \int_{\Sigma_t} \left\{ \frac{\partial \omega}{\partial t} + \mathrm{curl}(\omega \wedge u) \right\} \cdot n \, d\Gamma.$$

Using the curl equation (Eq. (7.7) of Chapter 7) with $\nu = 0$ and $f' = 0$, we see that the time derivative of the integral $\int_{\Sigma_t} \omega \cdot n \, d\Gamma$ vanishes, and the first part of the theorem follows. Then, we recall without proof the following formula from vector analysis: if Σ_t is a surface with boundary Γ_t, then

$$\oint_{\Gamma_t} u \cdot \tau \, d\ell = \int_{\Sigma_t} \mathrm{curl}\, u \cdot n \, d\Gamma = \int_{\Sigma_t} \omega \cdot n \, d\Gamma,$$

τ being the unit tangent vector to $\Gamma_t = \partial \Sigma_t$ such that (τ, n) is direct. Hence, if Γ_t and Σ_t move with the flow, the integrals above remain constant.

A first consequence of Theorem 8.3 is that the curl vector is a vector carried by the flow (we need differential geometry to define this notion rigorously). Nevertheless, it is now easy to prove the following result.

Theorem 8.5 (Lagrange's Theorem). *We consider the flow of an inviscid incompressible fluid. If, at a given time, the flow is irrotational, then the flow remains irrotational for all times.*

Proof: According to Theorem 8.4, if Σ_t is a surface moving with the flow,

$$\int_{\Sigma_t} \omega \cdot n \, d\Gamma = \int_{\Sigma_{t'}} \omega \cdot n \, d\Gamma,$$

for every t and t'. Therefore, if $\omega = 0$ at time t', then

$$\int_{\Sigma_t} \omega \cdot n \, d\Gamma = 0, \quad \forall \, \Sigma_t \subset \Omega_t, \quad \forall \, t \in I.$$

As a result, $\omega = 0$ in the whole domain Ω_t for every time t.

Remark 8.3: Theorem 8.4 applies only to perfect fluids. In the case of a viscous fluid, there will be a spontaneous generation of vorticity, even if the flow is initially irrotational: this is due to the turbulence phenomenon, which we briefly address in Chapter 9, Section 9.5.

8.2. Plane irrotational flows

Our aim in this section is to study the plane irrotational and stationary flows of inviscid incompressible fluids. We begin with several remarks of more general interest.

Definition 8.2. *A flow is a plane flow parallel to the plane π if every velocity vector is parallel to π and if the velocity field is invariant by any translation orthogonal to π.*

Remark 8.4: In practice, the plane π will be chosen to be parallel to (Ox_1x_2) so that every velocity vector will be of the form

$$\vec{u} = \begin{pmatrix} u_1(x_1, x_2, 0) \\ u_2(x_1, x_2, 0) \\ 0 \end{pmatrix}.$$

Velocity potential and stream function

We are given a plane incompressible flow. In what follows, the velocity vector will be denoted by $\vec{u} = (u, v)$, and we will write x, y instead of x_1, x_2:

$$\vec{u} = [u(x, y), v(x, y), 0].$$

The flow being incompressible,

$$\frac{\partial u}{\partial x} + \frac{\partial v}{\partial y} = 0,$$

and thus, there exists a stream function Ψ, defined locally, such that

$$u = \frac{\partial \Psi}{\partial y}, \quad v = -\frac{\partial \Psi}{\partial x}.$$

Furthermore, the flow is irrotational; thus, curl $\vec{u} = 0$, and there also exists a velocity potential Φ, defined locally, such that

$$\vec{u} = \text{grad } \Phi;$$

that is to say

$$u = \frac{\partial \Phi}{\partial x}, \quad v = \frac{\partial \Phi}{\partial y}.$$

Because div $\vec{u} = 0$, we see that $\Delta\Phi = 0$; that is, Φ is a harmonic function. Similarly, because curl $\vec{u} = 0$, we see that $\Delta\Psi = 0$.

It is easy to see that the functions Φ and Ψ satisfy the Cauchy–Riemann equations

$$\frac{\partial \Phi}{\partial x} = \frac{\partial \Psi}{\partial y}, \quad \frac{\partial \Phi}{\partial y} = -\frac{\partial \Psi}{\partial x},$$

and, consequently,

$$f(z, t) = (\Phi + i\Psi)(z, t), \quad z = x + iy,$$

is an analytic function of z. The function f is called the complex potential of the flow. We note that it is not necessarily a one-to-one function in the whole domain filled by the fluid. Furthermore,

$$\frac{df}{dz} = u - iv = \overline{u + iv}.$$

a) Boundary conditions

Let us now specify the boundary conditions. On the boundary of a fixed obstacle, we write the nonpenetration condition

$$\vec{u} \cdot n = 0,$$

which is locally equivalent to

$$\frac{\partial \Phi}{\partial n} = 0, \quad \text{or} \quad \frac{\partial \Psi}{\partial \tau} = 0.$$

Remark 8.5: The preceding remarks can be extended to compressible barotropic fluids for which $(1/\rho)\nabla p$ is a gradient.

b) Stream lines

The stream lines are the curves $\Psi = \text{const}$. Indeed, on a stream line, the normal is parallel to $(\Psi_{,x}, \Psi_{,y})$, and the tangent is parallel to $(\Phi_{,x}, \Phi_{,y}) = (u, v)$.

c) Equipotential lines

The equipotential lines are the curves $\Phi = \text{const}$.

Remark 8.6: The stream lines and the equipotential lines are orthogonal at every point.

Elementary stationary plane flows

Our aim is now to study elementary flows corresponding to simple functions f. Apart from its intrinsic interest, this study will be useful, as we will see later, in the consideration of more complex real flows such as the flow around an

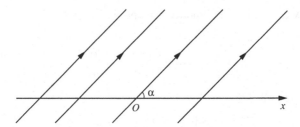

Figure 8.1 Streamlines of a uniform flow.

airfoil in two-dimensional space. This corresponds to the classical approach of airfoil theory in the early times of aeronautics, which was based on the theory of analytic functions. At present, we tend to use numerical methods instead (see the end of Section 8.2) for studying the flows around airfoils, but these "historical" developments are still of interest.

Let us notice that every stream line that does not encounter a singularity point of the velocity vector (see below) can be assumed to be a solid (material) line as far as the boundary conditions are concerned. We thus obtain a perfect fluid flow in each of the regions limited by the curve.

a) Uniform flows

For a uniform flow, we have $f(z) = V_0 z \exp(-i\alpha)$. In this case, $\Phi = V_0(x\cos\alpha + y\sin\alpha)$, $\Psi = V_0(y\cos\alpha - x\sin\alpha)$, $u = V_0\cos\alpha$, $v = V_0\sin\alpha$, and $(df/dz) = V_0 e^{-i\alpha}$.

This is thus a uniform flow of velocity V_0 in a direction at angle α with Ox. The stream lines are the straight lines making an angle α with Ox, and the equipotential lines are the straight lines making an angle α with the Oy axis (see Figure 8.1).

b) Sources and sinks

We consider the case in which $f(z) = (D/2\pi)\log z = (D/2\pi)(\log r + i\theta)$, $z = re^{i\theta}$, D being a real constant. Thus,

$$\Phi = \frac{D}{2\pi}\log r, \quad \Psi = \frac{D}{2\pi}\theta, \quad \frac{df}{dz} = \frac{D}{2\pi z} = \frac{D}{2\pi r}e^{-i\theta},$$

$$u = \frac{D}{2\pi r}\cos\theta, \quad v = \frac{D}{2\pi r}\sin\theta.$$

The stream lines are the straight lines containing O, the equipotential lines are the circles centered at O, and O is a source or a sink (see Figure 8.2).

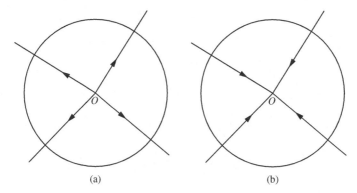

Figure 8.2 Streamlines of a source (a), and a sink (b).

By definition, the flowrate of the source is described by

$$\int_C \vec{u} \cdot n \, d\Gamma,$$

where C is any circle centered at O; hence, the flowrate is given by

$$\int_0^{2\pi} \frac{D}{2\pi r_0}(\cos^2 \theta + \sin^2 \theta) r_0 \, d\theta = D.$$

When $D > 0$, we say that O is a source, and when $D < 0$, we say that O is a sink.

c) Singular point vortex

We set $f(z) = (\Gamma/2\pi i) \log z = -(i\Gamma/2\pi)(\log r + i\theta)$, $\Gamma \in \mathbb{R}$. Hence, $\Phi = (\Gamma/2\pi)\theta$, $\Psi = -(\Gamma/2\pi)\log r$, $(df/dz) = -(i\Gamma/2\pi z) = -(i\Gamma/2\pi r)e^{-i\theta}$, $u = -(\Gamma/2\pi r)\sin\theta$, and $v = (\Gamma/2\pi r)\cos\theta$.

The equipotential lines are the straight lines containing O, and the stream lines are the circles centered at O. We notice that the flow is invariant by rotation around O (see Figure 8.3).

d) Doublet

We set $f(z) = -(K/2\pi z) = -(K/2\pi r)e^{-i\theta}$, $K \in \mathbb{R}$. Then, $(df/dz) = (K/2\pi z^2)$, $\Psi = (K/2\pi r)\sin\theta$, $\Phi = -(K/2\pi r)\cos\theta$. Therefore, the stream lines are the circles centered on the Oy axis and tangent to the Ox axis. Similarly, the equipotential lines are the circles centered on the Ox axis and tangent to the Oy axis (see Figure 8.4).

We say that we have a doublet located at O, of axis Ox, and with intensity or momentum K. More generally, we can define a doublet located at a point M_0 with complex coordinate z_0 and with a direction making an angle α with

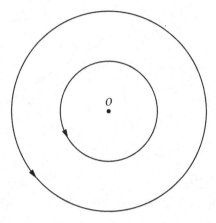

Figure 8.3 Streamlines of a singular point vortex.

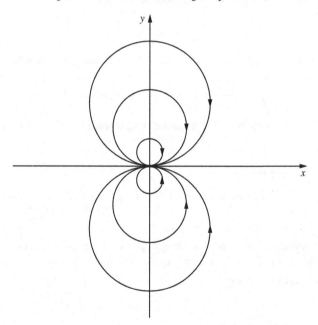

Figure 8.4 Streamlines of a doublet.

the Ox axis by setting $f(z) = -Ke^{i\alpha}/[2\pi(z - z_0)]$ (to do so, make the change of variables $z = z_0 + Ze^{i\alpha}$).

e) *Flow in an angle or around an angle*

For such flows, $f(z) = az^2 = ar^2 e^{2i\theta}$, $a \in \mathbb{R}$. Then, $\Psi = ar^2 \sin 2\theta = (a/2)xy$, $\Phi = ar^2 \cos 2\theta = a(x^2 - y^2)$, and $(df/dz) = 2az$. We then conclude that the stream lines are the hyperbolas with asymptotes Ox, Oy (and the straight lines $x = 0$ and $y = 0$). Also, the equipotential lines are the hyperbolas whose asymptotes are the bisectors.

f) Flow with circulation around a circle

This flow brings us closer to the flow around an airfoil; it is obtained by superposing a doublet in O, a uniform flow, and a singular point vortex.

The complex potential thus reads

$$f(z) = V_0 \left(z + \frac{R^2}{z} \right) - \frac{i\Gamma}{2\pi} \log z;$$

in polar coordinates, the equation of the stream lines reads

$$V_0 \left(r - \frac{R^2}{r} \right) \sin \theta - \frac{\Gamma}{2\pi} \log r = C,$$

and, if $C = -(\Gamma/2\pi) \log R$, we find a circle centered at O and of radius R. This function f can thus describe the flow around the circle centered at O of radius R.

We can also consider the flow around a circle with incidence α, which corresponds to the complex potential

$$f(z) = V_0 \left(z e^{-i\alpha} + \frac{R^2}{z} e^{i\alpha} \right) - \frac{i\Gamma}{2\pi} \log z. \tag{8.4}$$

We will not describe these flows, which can be rather complex depending on the values of α, Γ, V_0, and R; however, let us mention that the circulation of the velocity vector around the circle is precisely equal to Γ. As we will see, the circulation is important because it is related to the lift (the vertical component of the resultant of the pressure forces exerted by the fluid on the airfoil).

Computation of the forces on a wall: Blasius formulas

We indicate here how the resultant of the pressure forces exerted by the fluid on a closed curve C may be computed using the analytic function f, which corresponds to the complex potential in one of the regions limited by C (and such that C is a stream line). The corresponding formulas are called the Blasius formulas. Of course, for a three-dimensional flow, these are the forces per unit length in the orthogonal direction to the plane.

The density of surface forces is $T = \sigma \cdot n = -pn$ and, thanks to Bernoulli's theorem, $p + \frac{1}{2}\rho |\vec{u}|^2 = p_1$, where the constant p_1 is the pressure at the rest point along the stream line (i.e., the point of vanishing velocity or the point of maximal pressure). Thus,

$$p = p_1 - \frac{1}{2}\rho \left| \frac{df}{dz} \right|^2.$$

We set $n = n_1 - in_2$, $T = T_1 - iT_2$. Therefore,

$$T = -pn$$
$$= -p \left(\frac{dy + idx}{ds} \right) = -ip\frac{d\bar{z}}{ds}$$
$$= -ip_1\frac{d\bar{z}}{ds} + \frac{i}{2}\rho\frac{df}{dz} \cdot \frac{d\bar{f}}{ds}.$$

On $\mathcal{C}$,

$$\frac{df}{ds} = \frac{d\Phi}{ds} + i\frac{d\Psi}{ds}, \quad \frac{d\Psi}{ds} = 0, \quad \text{and} \quad df = d\bar{f},$$

and thus, the general resultant of the pressure forces is equal to

$$\mathcal{F} = -\int_{\mathcal{C}} ip_1\frac{d\bar{z}}{ds} + \frac{i\rho}{2}\int_{\mathcal{C}} \left(\frac{df}{dz} \right)^2 dz;$$

that is to say

$$\boxed{\mathcal{F} = \frac{i\rho}{2}\int_{\mathcal{C}} \left(\frac{df}{dz} \right)^2 dz.} \tag{8.5}$$

Similarly, the momentum at O of the pressure forces is given by

$$\mathcal{M} = \int_{\mathcal{C}} (xT_2 - yT_1)\,ds = \int_{\mathcal{C}} \text{Re}(izT)\,ds$$
$$= \text{Re}\int_{\mathcal{C}} zp_1\frac{d\bar{z}}{ds}\,ds - \frac{\rho}{2}\text{Re}\int_{\mathcal{C}} \frac{df}{dz}\frac{d\bar{f}}{ds}z\,ds$$
$$= p_1\text{Re}\int_{\mathcal{C}} \frac{d}{ds}\left(\frac{|z|^2}{2} \right)ds - \frac{\rho}{2}\text{Re}\int_{\mathcal{C}} z\left(\frac{df}{dz} \right)^2 dz;$$

hence,

$$\boxed{\mathcal{M} = -\frac{\rho}{2}\text{Re}\int_{\mathcal{C}} z\left(\frac{df}{dz} \right)^2 dz.} \tag{8.6}$$

Equations (8.5) and (8.6) are called the Blasius formulas.

Remark 8.7: The integrals appearing in the expressions of $\mathcal{F}$ and $\mathcal{M}$ can be computed, for instance, by the residue method well known in complex analysis. In particular, for the calculation of these integrals, it is possible to deform the curve $\mathcal{C}$ continuously as long as it does not encounter a singularity point.

Airfoil theory: the analytic approach

a) The Kutta–Joukowski condition

As we said, in the classical approach, the computation of airfoils is based on the theory of analytic functions. The aim is to determine the closed curve $\mathcal{C}$ and the analytic function f defined in the infinite open set delimited by $\mathcal{C}$; $\mathcal{C}$ must be a stream line and must satisfy certain desirable properties – in particular, it should produce a satisfactory lift (vertical component of $\mathcal{F}$).

Such functions f are obtained by adding (superposing) simple functions f as above and by using conformal mappings, i.e., setting,

$$\tilde{z} = \tilde{h}(z), \quad z = h(\tilde{z}),$$

where h and $\tilde{h}$ are analytic functions in some suitable regions of $\mathbb{C}$. The interested reader can consult the specialized literature for more details. We only indicate here a few essential aspects.

An important particular case is the one in which we start with the flow around a circle as above (see Eq. (8.4)); then, we consider a conformal mapping having a singular point on the circle. This gives a flow around a profile having an angle $\delta\pi$, $0 < \delta < 1$, as in Figure 8.5. The function h is then of the form

$$z = z_F + a(\tilde{z} - \tilde{z}_{\tilde{F}})^q + \text{(terms of order} > q),$$

where

$$(2 - \delta)\pi = q\pi, \quad q = 2 - \delta.$$

We have

$$\frac{dz}{d\tilde{z}} = a(2 - \delta)(\tilde{z} - \tilde{z}_{\tilde{F}})^{1-\delta} + \text{(terms of order} > 1 - \delta),$$

$$\frac{df}{dz} = \frac{d\tilde{f}}{d\tilde{z}} \cdot \frac{d\tilde{z}}{dz} \quad [\tilde{f}(\tilde{z}) = f(h(\tilde{z}))];$$

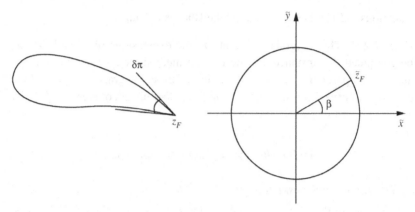

Figure 8.5 Trailing edge.

then (df/dz) and the velocities are infinite at z_F, which is unacceptable unless

$$\frac{d\tilde{f}}{d\tilde{z}}(\tilde{z}_F) = 0, \tag{8.7}$$

that is, with $\tilde{z}_F = Re^{i\beta}$,

$$V_0\left(e^{-i\alpha} - e^{i(\alpha-2\beta)}\right) - \frac{i\Gamma}{2\pi R}e^{-i\beta} = 0. \tag{8.7'}$$

The condition defined by Eqs. (8.7) and (8.7′), which is called the Kutta–Joukowski condition, allows the determination of the circulation Γ; it is important in aeronautical engineering.

By applying the Blasius formulas, a simple calculation using analytic function theory shows that $\mathcal{M} = 0$ and $\mathcal{F} = i\rho\Gamma V_0 e^{i\alpha}$; thus, the drag (i.e., the component of $\mathcal{F}$ on the velocity at infinity $V_0 e^{i\alpha}$) vanishes, whereas the lift, which is the component of $\mathcal{F}$ in an orthogonal direction, has an algebraic measure on this axis equal to $-\rho\Gamma V_0$; the lift is proportional to Γ, which explains the practical importance of the circulation Γ and of the Kutta–Joukowski formula in aeronautical engineering.

b) Numerical methods

As mentioned above, analytic methods tend to be replaced at present by numerical methods that allow more flexibility in the choice of the line $\mathcal{C}$. Also, besides solving the fluid mechanics problem, they open the way to the solution of optimization problems of an economical nature: for instance, we look for a profile $\mathcal{C}$ that maximizes or minimizes certain parameters of the flow (e.g., that maximizes the lift, but other problems could be important as well).

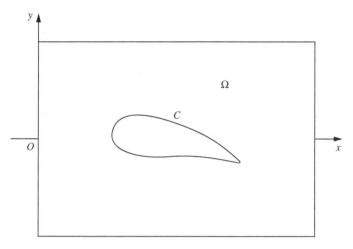

Figure 8.6 Numerical methods.

Numerical methods do not allow us, in general, to make calculations in the whole space; therefore, we usually replace the space $\mathbb{R}^2$ by a large enough rectangle, $0 < x < \ell$, $|y| \leq L$, and we then solve a boundary value problem for Φ or Ψ in the open set consisting of the part of the rectangle outside of $\mathcal{C}$ (see Figure 8.6).

For instance, for Ψ, we write that

$$\Delta \Psi = 0, \quad \text{in } \Omega,$$

$$\Psi = \gamma = \text{const.}, \quad \text{on } \mathcal{C};$$

the constant γ is unknown, but the Kutta–Joukowski condition gives an algebraic condition that "compensates" for this lack of information:

$$\int_{\mathcal{C}} \frac{\partial \Psi}{\partial n} d\ell = \Gamma.$$

Finally, the boundary conditions on the boundary of the rectangle are fixed empirically by assuming that the rectangle is large enough so that we can neglect, on its edges, the perturbations induced by the profile. Thus, for instance, we can reasonably assume that

$$\begin{cases} \Psi = V_0 y, & \text{at } x = 0, \quad |y| \leq L, \\ \Psi = V_0 y, & \text{or } \dfrac{\partial \Psi}{\partial n} = \dfrac{\partial \Psi}{\partial x} = 0, \quad \text{at } x = \ell, \quad |y| \leq L, \\ \Psi = \pm V_0 L, & \text{for } y = \pm L, \quad 0 < x < \ell. \end{cases}$$

The preceding equations constitute a mathematically well-posed problem whose numerical solution is relatively easy with today's computers.

8.3. Transsonic flows

In this section, we consider plane, irrotational, and stationary flows of *compressible* inviscid fluids. As in Section 8.2, the velocity field will be denoted by

$$\vec{u} = \begin{pmatrix} u \\ v \\ 0 \end{pmatrix} (x, y).$$

The continuity equation then becomes

$$\operatorname{div}(\rho \vec{u}) = 0,$$

that is to say

$$\frac{\partial}{\partial x}(\rho u) + \frac{\partial}{\partial y}(\rho v) = 0,$$

and the irrotationality condition is

$$\frac{\partial u}{\partial y} = \frac{\partial v}{\partial x}.$$

Furthermore, the momentum equation reduces to

$$\rho \operatorname{grad}\left(\frac{1}{2}|\vec{u}|^2\right) + \operatorname{grad} p = f;$$

it allows us to compute p once ρ and $|\vec{u}|$ are known (provided $\rho \operatorname{grad}\left(\frac{1}{2}|\vec{u}|^2\right)$ is indeed a gradient).

In what follows, we will assume that $f = 0$ and that the fluid is barotropic. The momentum equation then becomes (see Chapter 7)

$$\nabla\left(\frac{1}{2}|\vec{u}|^2\right) + \frac{1}{\rho}g'(\rho)\nabla\rho = 0, \tag{8.8}$$

which implies (the integral denoting a primitive):

$$\int \frac{g'(\rho)}{\rho}d\rho + \frac{1}{2}|\vec{u}|^2 = \text{const.} \tag{8.9}$$

The density ρ is a function of the magnitude $|\vec{u}|$ of $\vec{u}$ according to Eq. (8.9). The quantity $(dp/d\rho) = c^2 = g'(\rho)$ is called (represents) the local sound velocity. Finally, we set $[dp(\rho)]/\rho = dh(\rho)$ with h denoting the specific enthalpy. Then, Eq. (8.9) can be rewritten as

$$h + \frac{1}{2}|\vec{u}|^2 = \text{const.} \tag{8.10}$$

On the other hand, owing to the irrotationality, a velocity potential Φ exists such that, locally,

$$u = \frac{\partial \Phi}{\partial x}, \quad v = \frac{\partial \Phi}{\partial y}.$$

Similarly, there exists a funtion Ψ satisfying

$$\rho u = \frac{\partial \Psi}{\partial y}, \quad \rho v = -\frac{\partial \Psi}{\partial x}.$$

The functions Φ and Ψ then satisfy the following equations:

$$\frac{\partial}{\partial x}\left(\rho \frac{\partial \Phi}{\partial x}\right) + \frac{\partial}{\partial y}\left(\rho \frac{\partial \Phi}{\partial y}\right) = 0, \tag{8.11}$$

$$\frac{\partial}{\partial x}\left(\frac{1}{\rho} \frac{\partial \Psi}{\partial x}\right) + \frac{\partial}{\partial y}\left(\frac{1}{\rho} \frac{\partial \Psi}{\partial y}\right) = 0. \tag{8.12}$$

According to Eq. (8.9), ρ is a function of $|\vec{u}|^2$, and Eqs. (8.11) and (8.12) are nonlinear equations for Φ and Ψ, respectively. Despite their apparent simplicity, these equations are actually very complicated, and their mathematical theory (addressing such questions as existence and uniqueness of solutions and their approximation) is far from being complete at this time.

In the flow, the functions h, p, ρ, and c are decreasing functions of the variable $q = |\vec{u}|$. Indeed, $h = \text{const.} - q^2/2$, which implies that $h(q)$ is decreasing. Because $(dp/dh) = \rho > 0$, the function $p(h)$ is increasing, and thus $q \mapsto p(q)$ is decreasing. Furthermore, $p = g(\rho)$, with $g'(\rho) > 0$. Therefore, $p(\rho)$ is increasing, which implies that $\rho(p)$ is increasing and $\rho(q)$ is decreasing. Finally, $c^2 = (dg/d\rho)$ with $(d^2 g/d\rho^2) \geq 0$; hence, $(dc^2/d\rho) \geq 0$ and $c^2(\rho)$ is increasing. We then deduce that $c = c(q)$ is a decreasing function.

Definition 8.3. *The local Mach number of the flow is the number $M = (q/c)$. The Mach number is an increasing function of q. At a point where $M > 1$, the flow is said to be supersonic, and it is said to be subsonic if $M < 1$. The line $M = 1$ is the sonic line.*

We are now interested in the quantity ρq:

$$\frac{d(\rho q)}{\rho q} = \frac{dq}{q} + \frac{d\rho}{\rho} = \frac{dq}{q} + \frac{dh}{c^2}$$

$$= \frac{dq}{q} - \frac{q}{c^2}dq = \frac{dq}{q}(1 - M^2).$$

Thus, ρq, as a function of q, reaches a maximum for $M = 1$.

All the scalar functions are expressed (or can be expressed) in terms of q; it remains to express $\vec{u}$ in terms of q. We have

$$\frac{d\rho}{\rho} = \frac{dp}{c^2\rho} = \frac{dh}{c^2} = -q\frac{dq}{c^2},$$

$$\frac{1}{\rho}\nabla\rho = -\frac{1}{2c^2}\nabla(q^2).$$

Therefore, the equation satisfied by Φ becomes

$$\rho\Phi_{,xx} + \rho\Phi_{,yy} + \nabla\rho \cdot \nabla\Phi = 0;$$

that is to say

$$\Phi_{,xx} + \Phi_{,yy} + \frac{1}{\rho}\nabla\rho \cdot \nabla\Phi = 0.$$

Thus,

$$\Phi_{,xx} + \Phi_{,yy} - \frac{1}{c^2}\left(\Phi_{,x}^2\Phi_{,xx} + 2\Phi_{,x}\Phi_{,y}\Phi_{,xy} + \Phi_{,y}^2\Phi_{,yy}\right) = 0.$$

After simplification, we obtain the following equation for Φ:

$$\left(1 - \frac{\Phi_{,x}^2}{c^2}\right)\Phi_{,xx} - \frac{2}{c^2}\Phi_{,x}\Phi_{,y}\Phi_{,xy} + \left(1 - \frac{\Phi_{,y}^2}{c^2}\right)\Phi_{,yy} = 0. \quad (8.13)$$

We consider the coefficients of $\Phi_{,xx}$, $\Phi_{,xy}$, $\Phi_{,yy}$ and set

$$\frac{\delta}{4} = \frac{\Phi_{,x}^2\Phi_{,y}^2}{c^4} - \left(1 - \frac{\Phi_{,x}^2}{c^2}\right)\left(1 - \frac{\Phi_{,y}^2}{c^2}\right)$$

$$= \frac{\Phi_{,x}^2 + \Phi_{,y}^2}{c^2} - 1$$

$$= M^2 - 1.$$

According to the theory of partial differential equations (see Section 8.5), when the flow is supersonic, $M > 1$, Eq. (8.13) is hyperbolic and there is

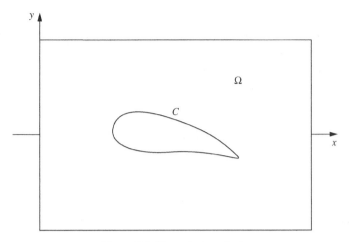

Figure 8.7 Numerical methods.

wave propagation (shock waves). In the region where the flow is subsonic, $M < 1$, Eq. (8.13) is elliptic. As we have already said, Eq. (8.13) is at this time out of reach from the theoretical point of view. However, it can be solved numerically fairly well. To do so, as in the incompressible case, we restrict the spatial domain to a large rectangle containing $\mathcal{C}$, $0 < x < \ell$, $|y| \leq L$ (see Figure 8.7). The rectangle is sufficiently large for the perturbations induced by $\mathcal{C}$ to be small (negligible) on the outer boundary of the rectangle.

Then, Eq. (8.13) is solved in the domain Ω consisting of the part of the rectangle outside of $\mathcal{C}$. This equation is supplemented by the boundary conditions

$$\frac{\partial \Phi}{\partial n} = 0, \quad \text{on } \mathcal{C},$$

and, for instance (this is an approximation), the following:

$$\begin{cases} \pm\dfrac{\partial \Phi}{\partial n} = -\dfrac{\partial \Phi}{\partial x} = -V_0, & \text{for } x = 0 \text{ or } \ell, \quad |y| \leq L, \\ \pm\dfrac{\partial \Phi}{\partial n} = \dfrac{\partial \Phi}{\partial y} = 0, & \text{for } y = \pm L, \quad 0 < x < \ell. \end{cases}$$

Remark 8.8:

1. We can derive a similar boundary value problem for the function Ψ. However, this equation is less interesting than in the plane irrotational case where the density is a function uniquely defined by the value of

$|\nabla\Phi|$ as we just saw. However, for a given value of $|\nabla\Psi|$, there exist two possible values of the density which correspond, respectively, to subsonic and supersonic flows.

2. The next step, in aerodynamics, is to abandon the irrotationality condition and to introduce the viscosity.

8.4. Linear accoustics

Sound propagates in air or water by infinitesimal variations of the density and pressure, for these fluids are compressible. The equations of linear acoustics are deduced from the equations of compressible fluids by assuming that the motions of the fluid (air, water) are very small and by using asymptotic expansions.

Indeed, because of the complexity of the problems to be solved, we want to take advantage of the fact that certain parameters are small (or large), and we are then led to make linearizations that lead to simplified models. The principle of linearization is very simple: we have a small parameter η, and we make assumptions on the scale order of the different physical quantities with respect to η. This last point is the most delicate one, and it necessitates, in the absence of rigorous mathematical proofs, a good understanding of the physics of the problem. As examples of linearized flows, we can mention the linearized models for the study of wave propagation in fluids or gases, or the Stokes equations. Other wave phenomena will be studied in the fourth part of this book by using much more involved asymptotic expansions.

Our aim in this section is to study small motions of a perfect compressible fluid, and this will lead to the equations of linear acoustics. We assume that the fluid is barotropic, near rest, and that it is not subjected to any external force. We will denote by p_0, ρ_0, and c_0 the state of the fluid at rest; p_0, ρ_0, and c_0 denote the pressure, the volume density, and the sound velocity, respectively. We finally make the following assumption.

"Infinitesimal" motion assumption

We assume that the velocity u, the pressure p, and the density ρ are of the form

$$u = \eta\bar{u} + o(\eta), \quad p = p_0 + \eta\bar{p} + o(\eta), \quad \rho = \rho_0 + \eta\bar{\rho} + o(\eta),$$

the asymptotic expansions above being understood in $C^k(\bar{\Omega} \times [0, T])$, $k = 2$ or 3 in general.

We then make formal asymptotic expansions of the equations. The continuity equation (at order $o(\eta)$) thus becomes

$$\eta \frac{\partial \bar{\rho}}{\partial t} + \mathrm{div}(\rho_0 \eta \bar{u}) = 0;$$

hence,

$$\frac{\partial \bar{\rho}}{\partial t} + \rho_0 \,\mathrm{div}\, \bar{u} = 0. \tag{8.14}$$

In the momentum equation, we similarly neglect the terms of order η^2 or higher. There remains

$$\rho_0 \frac{\partial \bar{u}}{\partial t} + \nabla \bar{p} = 0. \tag{8.15}$$

Because $\bar{p} = \frac{dg}{d\rho}(\rho_0) \cdot \bar{\rho} + o(1)$, we find, neglecting the terms in η of order greater than 2,

$$\bar{p} = c_0^2 \bar{\rho}. \tag{8.16}$$

Then, by elimination of $\bar{u}$, we finally obtain

$$\frac{\partial^2 \bar{\rho}}{\partial t^2} - c_0^2 \Delta \bar{\rho} = 0. \tag{8.17}$$

This last equation is called the wave equation. It is hyperbolic (see the appendix to this book), and it models wave propagation phenomena (e.g., here, propagation of sound); c_0 is precisely the velocity of propagation of the waves (see Chapter 18).

Remark 8.9: The prototype of the wave equation is $\Box \rho = 0$, where $\Box$ denotes the Dalembertian, and corresponds to $c_0 = 1$ in Eq. (8.17).

By proper substitution, we can also obtain similar wave equations for $\bar{p}$ and $\bar{u}$:

$$\frac{\partial^2 \bar{p}}{\partial t^2} - c_0^2 \Delta \bar{p} = 0,$$

$$\frac{\partial^2 \bar{u}}{\partial t^2} - c_0^2 \,\mathrm{grad}\,\mathrm{div}\, \bar{u} = 0.$$

Exercises

1. We consider the motion of a stationary, perfect, incompressible fluid with specific mass ρ submitted to gravity and for which the velocity

U is given by

$$U_1 = -\frac{x_2}{r}\,\Phi(r), \quad U_2 = \frac{x_1}{r}\,\Phi(r), \quad U_3 = \Psi(r), \quad r = \sqrt{x_1^2 + x_2^2}. \quad (8.18)$$

a) Show that the motion is compatible with the dynamics of perfect, in-compressible fluids.
b) Compute the streamlines.
c) Compute Φ and Ψ for the motion to be irrotational and then give the pressure law.

2. We consider the planar motion of a stationary, incompressible, irrota-tional fluid defined by the complex potential $f(z) = m \log\left(z - \frac{1}{z}\right)$, $m > 0$.
   a) Compute the stream function ψ and the velocity potential ϕ.
   b) Show that the coordinate axes and the unit circle are streamlines.

3. We consider a planar, non-stationary flow defined at each time t by the velocity vector $U = (u, v)$,

$$u = \alpha c e^{ky} \sin[k(x - ct)],$$
$$v = -\alpha c e^{ky} \cos[k(x - ct)],$$

   with $\alpha, k, c > 0$. We denote by $0z$ the axis orthogonal to $(0xy)$. Show that the fluid is incompressible and irrotational and compute the streamlines at $t = 0$.

4. We consider a perfect gas having an adiabatic, stationary, and irrotational motion. We assume that the velocity potential is of the form $\Phi(r)$, r being the distance of the fluid particle to the origin.
   a) Show that the velocity $\vec{U}$ and the density ρ only depend on r.
   b) Show that $r^2 \rho U = $ const (where $\vec{U} = U(r)\vec{e_r}$).
   c) Compute r^2 in terms of the Mach number M and show that the motion is impossible inside a sphere (S) to be determined (we assume that $U^2 = c_0^2 M^2/[1 + (\gamma - 1)M^2/2]$ and $\rho = \rho_0(1 + \frac{\gamma - 1}{2} M^2)^{-\frac{1}{1-\gamma}}$, c_0 being the sound velocity and γ the adiabatic index).

5. An inviscid fluid of density ρ flows by a small opening of radius r out of the base of a large container. We further assume that the pressure in the opening is constant and equal to that at the free surface in the container and that the equilibrium of the fluid in the container is not affected (i.e., the flow remains stationary). Compute the flow rate.

CHAPTER NINE

Viscous fluids and thermohydraulics

This chapter is devoted to the study of some problems of viscous fluid mechanics and of thermohydraulics. We describe several simple classical problems and introduce through these elementary examples very complex phenomena of viscous fluids – in particular turbulence and boundary layers.

9.1. Equations of viscous incompressible fluids

To start with, we briefly recall the Navier–Stokes equations governing the motion of a viscous incompressible homogeneous fluid (these equations have been described in Chapter 7). The velocity $u = (u_1, u_2, u_3)$ and the pressure p, functions of x and t, satisfy the equations

$$\rho_0 \left(\frac{\partial u}{\partial t} + \sum_{i=1}^{3} u_i \frac{\partial u}{\partial x_i} \right) + \operatorname{grad} p = \mu \, \Delta u + f, \tag{9.1}$$

$$\operatorname{div} u = 0, \tag{9.2}$$

which we supplement with the no-slip boundary condition

$$u = \varphi \quad \text{at the boundary}, \tag{9.3}$$

where φ is a sufficiently regular given function in the case of a rigid boundary; in the case of a free (open) boundary, the boundary conditions are

$$u \cdot n = 0, \quad (\sigma \cdot n)_\tau = 0, \quad \text{at the boundary}, \tag{9.3'}$$

where $(v)_\tau$ denotes the tangential component of a vector v.

9.2. Simple flows of viscous incompressible fluids

We now describe several classical examples of viscous incompressible stationary flows.

Poiseuille flow between two parallel planes

a) The two-dimensional case

We consider stationary two-dimensional flows between two fixed parallel planes with equations $x_2 = \pm h$ and without external forces. We look for velocity fields $u = (u_1, u_2)$ such that $u_2 = 0$ and $u_1 = u_1(x_2)$ so that div $u = 0$. The Navier–Stokes equations then reduce to

$$-\mu \frac{\partial^2 u_1}{\partial x_2^2} + \frac{\partial p}{\partial x_1} = 0,$$

$$\frac{\partial p}{\partial x_2} = 0,$$

which yields $p = p(x_1)$. Moreover,

$$\frac{dp}{dx_1}(x_1) = \mu \frac{d^2 u_1}{dx_2^2}(x_2),$$

and this common value is necessarily a constant denoted by $-K$; K is the gradient of the pressure drop (linear rate of pressure drop) and is equal to the pressure drop by unit length. Then,

$$p = -Kx_1,$$

$$u_1 = -\frac{K}{2\mu}x_2^2 + \alpha x_2 + \beta;$$

the no-slip boundary conditions then yield

$$u_1 = \frac{K}{2\mu}\left(h^2 - x_2^2\right).$$

We observe that the velocity profile is parabolic (Figure 9.1).

If we assume that the upper plane $x_2 = h$ moves with velocity $U_0 \, \vec{e}_1$ and that the lower plane is fixed (in this case, we assume for simplicity that its equation is $x_2 = 0$), we still have

$$u_1 = -\frac{K}{2\mu}x_2^2 + \alpha x_2 + \beta,$$

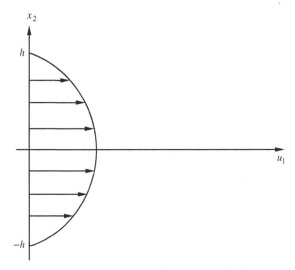

Figure 9.1 Velocity profile for the Poiseuille flow.

and, taking into account the boundary conditions:

$$u_1 = -\frac{K}{2\mu}x_2(x_2 - h) + U_0\frac{x_2}{h}.$$

If $K = 0$, u_1 is linear, and the corresponding flow is called the Couette flow.

b) The Three-Dimensional Case

We consider again stationary flows between two parallel planes of equations $x_3 = \pm h$ in the absence of external forces, and we look for a velocity field u such that $u_2 = u_3 = 0$ and $u_1 = u_1(x_3)$. As above, it follows that

$$p = -Kx_1,$$
$$u_1 = \frac{K}{2\mu}\left(h^2 - x_3^2\right), \qquad (9.4)$$

when the planes are fixed, and

$$u_1 = -\frac{K}{2\mu}x_3(x_3 - h) + U_0\frac{x_3}{h},$$

when the upper plane moves with velocity $U_0\vec{e}_1$ and the lower plane (of equation $x_3 = 0$) is fixed.

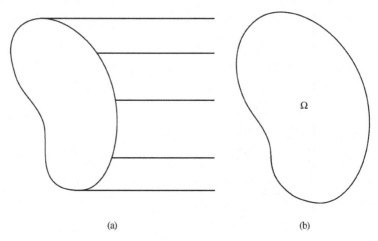

(a)   (b)

Figure 9.2 (a) Flow in a cylindrical tube; (b) Section Ω of the tube.

Poiseuille flow in a cylindrical tube

We now consider a cylindrical tube with axis Ox_1 and cross section Ω, and we look for a flow such that the stream lines are parallel to Ox_1 so that $u_2 = u_3 = 0$ and $u_1 = u_1(x_2, x_3)$ (Figure 9.2). It then follows that

$$-\mu\,\Delta u_1(x_2, x_3) + \frac{\partial p}{\partial x_1} = 0,$$

$$\frac{\partial p}{\partial x_2} = \frac{\partial p}{\partial x_3} = 0.$$

Consequently, p depends only on x_1, and we infer, as above, the existence of a constant K such that

$$\frac{dp}{dx_1} = \mu\,\Delta u_1 = -K.$$

The flow is entirely determined by the function $u_1(x_1, x_2)$, which is a solution of the following boundary value problem in Ω (which is called the Dirichlet boundary value problem):

$$\begin{cases} -\mu\,\Delta u_1 = K & \text{in } \Omega, \\ u_1 = 0 & \text{on } \partial\Omega. \end{cases}$$

Let us study the particular case in which Ω consists of the disc with polar equation $r \le R$. We recall that, in cylindrical coordinates,

$$\Delta u_1 = \frac{\partial^2 u_1}{\partial r^2} + \frac{1}{r}\frac{\partial u_1}{\partial r} + \frac{1}{r^2}\frac{\partial^2 u_1}{\partial \theta^2},$$

and we look for a solution u_1 such that $\partial u_1/\partial\theta = 0$; hence,

$$\frac{1}{r}\frac{\partial}{\partial r}\left(r\frac{\partial u_1}{\partial r}\right) = -\frac{K}{\mu},$$

and

$$u_1 = -\frac{K}{4\mu}r^2 + \alpha\log r + \beta.$$

The continuity of u at 0 implies $\alpha = 0$ and, taking into account the boundary conditions ($u_1 = 0$ at $r = R$), we finally obtain

$$u_1 = \frac{K}{4\mu}(R^2 - r^2). \tag{9.5}$$

Remark 9.1: This result is at the very heart of empirical (engineering) calculations for flows in pipes. The flowrate in the channel is given by

$$D = \int_\Omega u_1(x_2, x_3)\,dx_2\,dx_3 = 2\pi\int_0^R u_1(r)r\,dr.$$

Hence,

$$D = \frac{K\pi R^4}{8\mu}.$$

The flowrate is thus proportional to the linear pressure drop K and to the fourth power of the radius. We recover the experimental laws of Poiseuille.

Remark 9.2: In practice, these results are valid for small radii and moderate pressure gradients. If these conditions are not satisfied, the flow may be turbulent, and the solution may become time dependent or stationary but different from this one (we will come back to this in more detail in Section 9.5, which introduces turbulence).

Flows between two coaxial cylinders (Couette–Taylor flows)

We start by writing the Navier–Stokes equations in cylindrical coordinates in the absence of external forces; we will not present the calculations (change of coordinate system) leading to these equations. Let (r, θ, z) be the system of cylindrical coordinates. We set $u = u_r e_r + u_\theta e_\theta + u_z e_z$, where (e_1, e_2, e_3), as usual, denotes the canonical basis of $\mathbb{R}^3$ and $(e_r, e_\theta, e_z = e_3)$ is the local basis in cylindrical coordinates:

$$\frac{\partial u_r}{\partial t} - \nu\left(\Delta u_r - \frac{u_r}{r^2} - \frac{2}{r^2}\frac{\partial u_\theta}{\partial \theta}\right) + u \cdot \nabla u_r - \frac{u_\theta^2}{r} + \frac{1}{\rho}\frac{\partial p}{\partial r} = 0, \quad (9.6)$$

$$\frac{\partial u_\theta}{\partial t} - \nu\left(\Delta u_\theta + \frac{2}{r^2}\frac{\partial u_r}{\partial \theta} - \frac{u_\theta}{r^2}\right) + u \cdot \nabla u_\theta + \frac{u_r u_\theta}{r} + \frac{1}{\rho r}\frac{\partial p}{\partial \theta} = 0, \quad (9.7)$$

$$\frac{\partial u_z}{\partial t} - \nu\Delta u_z + u \cdot \nabla u_z + \frac{1}{\rho}\frac{\partial p}{\partial z} = 0; \quad (9.8)$$

we have, for a scalar function v:

$$\nabla v = \frac{\partial v}{\partial r}e_r + \frac{1}{r}\frac{\partial v}{\partial \theta}e_\theta + \frac{\partial v}{\partial z}e_z,$$

$$\Delta v = \frac{1}{r}\frac{\partial}{\partial r}\left(r\frac{\partial v}{\partial r}\right) + \frac{1}{r^2}\frac{\partial^2 v}{\partial \theta^2} + \frac{\partial^2 v}{\partial z^2}.$$

We now study the stationary flow between two concentric cylinders of axis Oz, of radii R_1 and R_2, with $0 < R_1 < R_2$. We assume that these cylinders rotate with angular velocities ω_1 and ω_2, respectively. We look for radial solutions, that is, solutions such that $u_z = 0$, and u_r, u_θ, and p depend only on r. Thus,

$$\text{div } u = \frac{1}{r}\frac{\partial}{\partial r}(ru_r) + \frac{1}{r}\frac{\partial u_\theta}{\partial \theta} + \frac{\partial u_z}{\partial z}$$

$$= \frac{1}{r}\frac{\partial}{\partial r}(ru_r) = 0,$$

which yields

$$ru_r = c_0 \text{ (const.)};$$

hence,

$$u_r = \frac{c_0}{r},$$

and, according to the boundary conditions ($u_r = 0$, at $r = R_1, R_2$), $c_0 = 0$ and $u_r = 0$. It now remains to determine u_θ and p. The equations become, after some simplifications (and also setting $\rho = 1$)

$$-\frac{u_\theta^2}{r} + \frac{\partial p}{\partial r} = 0, \quad (9.9)$$

$$-\nu\left[\frac{1}{r}\frac{\partial}{\partial r}\left(r\frac{\partial u_\theta}{\partial r}\right) - \frac{u_\theta}{r^2}\right] = 0. \quad (9.10)$$

Equation (9.10) is equivalent to

$$\frac{\partial^2}{\partial r^2}(ru_\theta) - \frac{1}{r}\frac{\partial}{\partial r}(ru_\theta) = 0.$$

We set $y = (ru_\theta)'$, $(\partial/\partial r =')$, and it follows that

$$y' - \frac{1}{r}y = 0;$$

hence,

$$y = Cr,$$

which leads to

$$\frac{\partial}{\partial r}(ru_\theta) = Cr.$$

Finally,

$$u_\theta = Cr + \frac{D}{r},$$

where C and D are constants. These constants are determined by the following boundary conditions:

$$u_\theta = R_i\omega_i \quad \text{for } r = R_i, \quad i = 1, 2.$$

Hence, by straightforward calculations

$$C = \frac{R_1^2\omega_1 - R_2^2\omega_2}{R_1^2 - R_2^2},$$

$$D = -\frac{R_1^2 R_2^2(\omega_1 - \omega_2)}{R_1^2 - R_2^2}.$$

Finally,

$$u_\theta = \frac{R_1^2\omega_1 - R_2^2\omega_2}{R_1^2 - R_2^2}r - \frac{R_1^2 R_2^2(\omega_1 - \omega_2)}{r\left(R_1^2 - R_2^2\right)}. \tag{9.11}$$

Substituting Eq. (9.11) in Eq. (9.9), we obtain

$$\frac{\partial p}{\partial r} = C^2 r + \frac{D^2}{r^3} + 2\frac{CD}{r};$$

hence,

$$p = C^2\frac{r^2}{2} - \frac{D^2}{2r^2} + 2CD\log r,$$

where C and D are as defined in the equations above.

Remark 9.3: The expressions of u_θ and p above allow, by an elementary calculation, the determination of the torque (per unit of length) that must be applied to the rotating cylinders to maintain their uniform rotations.

9.3. Thermohydraulics

The general equations of Newtonian fluids described in Chapter 7 were derived under the assumption that the fluid does not conduct heat or that the temperature remains constant. We now present the Boussinesq equations. These equations govern the evolution of slightly compressible fluids when thermal phenomena are taken into account.

The local temperature $T = T(x, t)$ is governed by Eq. (6.6) of Chapter 6, which is equivalent to the energy equation. With the change of notations undertaken in Chapter 6 (θ, U, κ replaced by T, u, $\tilde{\kappa}$, respectively), assuming that $C_V = 1$ and, as is customary, neglecting the term $\sigma_{ij} U_{i,j}$, we find

$$\rho_0 \left(\frac{\partial T}{\partial t} + (u \cdot \nabla)T \right) - \tilde{\kappa} \Delta T = r; \tag{9.12}$$

in the absence of heat sources, $r = 0$, and, setting $\kappa = \tilde{\kappa}/\rho_0$, we obtain

$$\frac{\partial T}{\partial t} + (u \cdot \nabla)T - \kappa \Delta T = 0; \tag{9.13}$$

κ is called the thermal diffusion coefficient.

In the Boussinesq approximation for slightly compressible homogeneous fluids, Eqs. (7.1) and (7.2) of Chapter 7, which express the mass and momentum conservation laws, are simplified as follows:

ρ is assumed to be constant ($= \rho_0$) everywhere in the equations, except in the gravity force term.

For the gravity forces, ρ is replaced by a linear function of T (linear state equation for the fluid) as follows:

$$\rho = \rho_0 - \alpha(T - T_0).$$

Thus, Eq. (7.1) of Chapter 7 reduces to

$$\text{div } u = 0, \tag{9.14}$$

and Eq. (9.2) becomes

$$\frac{\partial u}{\partial t} + (u \cdot \nabla)u + \nabla \frac{p}{\rho_0} - \nu \Delta u = \frac{\alpha g}{\rho_0} \mathbf{k}(T - T_0), \tag{9.15}$$

where **k** is the unit upward vertical vector. Equations (9.13), (9.14), and (9.15) constitute the Boussinesq equations of slightly compressible fluids.

As usual, we supplement these equations with boundary conditions: for u, they are those described in Section 1 above or in Chapter 7. For T, the boundary conditions are of the Dirichlet (conductive material at the boundary, prescribed temperature) or the Neumann type (prescribed heat rate on $\partial\Omega$). Boundary conditions of mixed type combine these two situations.

The Bénard problem in thermohydraulics consists of studying the heating, from the bottom, of a horizontal layer of fluid confined between the planes $x_2 = 0$ and $x_2 = L_2$.

We assume that the fluid is confined in the domain $\Omega = (0, L_1) \times (0, L_2)$ in space dimension two, and $\Omega = (0, L_1) \times (0, L_2) \times (0, L_3)$ in space dimension three. If (e_1, e_2, e_3) denotes the canonical basis of $\mathbb{R}^3$, e_2 vertical pointing upward, then the equations of motion are (9.13), (9.14), and (9.15) with $\mathbf{k} = e_2$. Let us assume, for instance, that the boundaries of Ω are materialized and fixed, that the lower and upper walls conduct heat and are heated at the temperatures T_1 and T_2, respectively, and $T_1 > T_2$; we finally assume that the lateral walls do not conduct heat. Under these hypotheses, Eqs. (9.13), (9.14), and (9.15) are supplemented with the following boundary conditions:

$$u = 0, \quad \text{on } \partial\Omega, \tag{9.16}$$

$$\left.\begin{array}{l} T = T_1, \quad \text{for } x_2 = 0, \quad \text{and } T = T_2, \quad \text{for } x_2 = L_2 \\ \dfrac{\partial T}{\partial n} = 0, \quad \text{on the lateral boundary of } \Omega; \end{array}\right\} \tag{9.17}$$

$\partial T/\partial n = n \cdot \operatorname{grad} T$ is the normal derivative of T, that is, the scalar product of $\operatorname{grad} T$ with the unit outward normal vector n.

It is easy to see that the following relations define a solution of Eqs. (9.14) and (9.15) that corresponds to the equilibrium of the fluid, the temperature being diffused by conduction:

$$\left.\begin{array}{l} u_s = 0, \quad T_s = T_1 + \dfrac{x_2}{L_2}(T_2 - T_1), \\[2mm] p_s = \alpha g \left[(T_1 - T_0)x_2 + \dfrac{x_2^2}{2L_2}(T_2 - T_1) \right], \\[2mm] \rho_s = \rho_0 - \alpha(T_1 - T_0) - \dfrac{\alpha x_2}{L_2}(T_2 - T_1). \end{array}\right\} \tag{9.18}$$

This is a stationary solution.

Setting $\theta = T - T_s$ and $q = p - p_s$, we then obtain the following equations:

$$\frac{\partial u}{\partial t} - \nu \Delta u + (u \cdot \nabla)u + \frac{1}{\rho_0}\nabla q = e_2 \frac{g\alpha}{\rho_0}\theta, \qquad (9.19)$$

$$\operatorname{div} u = 0, \qquad (9.20)$$

$$\frac{\partial \theta}{\partial t} - \kappa \Delta \theta + (u \cdot \nabla)\theta = -\frac{T_2 - T_1}{L_2}u_2. \qquad (9.21)$$

The boundary conditions are again Eq. (9.16) for u and, for θ:

$$\left.\begin{array}{ll} \theta = 0, & \text{for } x_2 = 0 \text{ and } x_2 = L_2, \\ \partial\theta/\partial n = 0, & \text{on the lateral boundary of } \Omega. \end{array}\right\} \qquad (9.22)$$

Remark 9.4: The boundary conditions of Eqs. (9.16) and (9.17) are an example of boundary conditions suitable for this problem. Other suitable boundary conditions, corresponding to other physical situations, may be considered; for instance, a free (open) boundary at the upper boundary of the fluid, $x_2 = L_2$.

Remark 9.5: We will see in Section 9.5 how the stationary solutions of the type defined by Eqs. (9.4), (9.5), and (9.11) for fluid flows or of the type defined by Eq. (9.18) for thermohydraulics can actually appear and persist. This issue is related to the notions of stability and turbulence.

Before describing these concepts, it is convenient to introduce the nondimensional form of these equations and the related concept of similarity.

9.4. Equations in nondimensional form: similarities

It is common in physics to write the equations describing a phenomenon in nondimensional form independent of the system of units: the different quantities (length, time, etc.) are then measured by reference to chosen quantities that are characteristic of the phenomenon. This produces the following advantages: Nondimensional numbers appear that regroup several physical quantities. These numbers are characteristic of the phenomenon; consequently, similarities between different phenomena appear, and the nondimensionalization procedure highlights small or large parameters whose magnitude is essential for an understanding of the physical phenomenon. As an example, we will write the nondimensional form of the Navier–Stokes and Boussinesq equations.

We first consider the Navier–Stokes equations of incompressible fluids:

$$\frac{\partial u}{\partial t} + (u \cdot \nabla)u + \frac{1}{\rho_0}\nabla p = \nu \Delta u + \frac{1}{\rho_0}f,$$

$$\operatorname{div} u = 0.$$

We set

$$x = L_c x', \quad u = U_c u', \quad t = T_c t',$$
$$p = P_c p', \quad f = F_c f',$$

where L_c, U_c, T_c, P_c, and F_c are, respectively, the characteristic length, velocity, time, pressure, and force for the flow. We will choose, for example, for L_c the diameter of the volume filled by the fluid for a confined flow or the diameter of the body bathing in the fluid for the flow around a body; U_c will be a characteristic velocity of the fluid such as the mean velocity of the boundaries in a confined flow or the velocity at infinity in a nonconfined flow. Having chosen these two quantities, we will then be able to choose the other quantities by their natural expressions corresponding to their physical units (but this is not compulsory and not always desirable). This choice yields the following characteristic time, pressure, and force:

$$T_c = L_c U_c, \quad P_c = \rho_0 U_c^2, \quad F_c = \rho_0 L_c / T_c^2.$$

The Navier–Stokes equations then become

$$\frac{\partial u'}{\partial t'} + (u' \cdot \nabla')u' + \nabla' p' = \frac{1}{R}\Delta' u' + f', \tag{9.23}$$

$$\operatorname{div}' u' = 0; \tag{9.24}$$

here $R = U_c L_c / \nu$ is called the *Reynolds number* of the flow.

We deduce from Eqs. (9.23) and (9.24) that the flow now only depends on one single parameter: the Reynolds number. For instance, if we know the stationary flow around a sphere of radius $2a$ placed in a fluid whose velocity at infinity is $U_\infty/2$, the flow around a sphere of radius a placed in a fluid of the same nature whose velocity at infinity is U_∞ is fully determined thanks to the similarity relations: both Reynolds numbers are equal, and we pass from one solution to the other by the proper scaling. This corresponds to the flow similarity phenomenon: such similarity relations between flows are at the very heart of the scale-model method in aerodynamical engineering by which, for example, the flow around an airfoil (or an entire plane) is experimentally simulated in a wind tunnel by the flow around a smaller model.

Remark 9.6: A fundamental difficulty of fluid mechanics arises because the Reynolds number takes very large values for the most interesting flows – for instance from 10^6 to 10^8 in aerodynamics or in industrial fluid mechanics (pumps, etc.). This number is already quite large for a flow as common as that produced by a spoon that one stirs in a cup of coffee. Reynolds numbers of the order of unity correspond, in fluid mechanics, to very slow flows or to fluids as viscous as honey. Reynolds numbers even larger than the values indicated above appear in meteorology in which other complicated features have to be taken into consideration.

In a similar way we can obtain the nondimensional form of the Boussinesq equations (9.19) to (9.21) corresponding to the Bénard problem. The nondimensionalization is performed as above for the variables x, u, t, and p with $L_c = L_2$, $T_c = (L_2/g)^{1/2}$. For the temperature, we set $\theta = (T_1 - T_2)\theta'$, and it then follows that

$$\frac{\partial u'}{\partial t'} + (u' \cdot \nabla')u' - v'\Delta'u' + \nabla'p' = e_2\alpha'\theta', \qquad (9.25)$$

$$\mathrm{div}'\, u' = 0, \qquad (9.26)$$

$$\frac{\partial \theta'}{\partial t'} + (u'\nabla')\theta' - \kappa'\Delta'\theta' = u'_2. \qquad (9.27)$$

We have set

$$\kappa' = \frac{\kappa L_2}{T_1 - T_2}, \quad v' = \frac{v}{g^{1/2}L_2^{3/2}}, \quad \alpha' = \frac{\alpha}{\rho_0}(T_1 - T_2).$$

The flow now only depends on *two nondimensional numbers*, namely κ' and v'. It is common in thermohydraulics to consider two numbers among the following:

$$\left(\frac{1}{v'}\right)^2 = Gr = \text{Grashof number,}$$

$$\frac{v'}{\kappa'} = Pr = \text{Prandtl number,}$$

$$\frac{1}{v'\kappa'} = Ra = \text{Rayleigh number.}$$

Remark 9.7: For simplicity of notation, we will drop the primes in the nondimensional equations (9.25) to (9.27), and we will eventually refer to Eqs. (9.25) to (9.27) written without the primes.

9.5. Notions of stability and turbulence

This section and the next one are brief introductions to extremely complex phenomena of fluid mechanics that are currently the object of important research work in experimental laboratories as well as from the computational and theoretical perspectives: stability, turbulence, and boundary layers. We thought that we could not fully ignore such important phenomena and, although our presentation will be very superficial, we will introduce them through very simple models.

The stability issue (and then the turbulence issue) appears, for instance, when the following question is asked: Under what conditions can one of the stationary flows described by Eqs. (9.4), (9.5), (9.11), or (9.18) occur and persist when the corresponding forces and boundary conditions are applied. Hereafter, we have rather arbitrarily chosen to discuss the case of thermohydraulics, but the phenomena and discussions are quite similar to what follows in the other cases presented before and in many more similar situations that we will not depict here.

Returning to the notations of Section 9.3, we perform the following mathematical operation whose interest will be apparent later: We consider Eqs. (9.25), (9.26), and (9.27), which are the nondimensionalized versions of Eqs. (9.19), (9.20), and (9.21), but we now omit the primes; we multiply Eq. (9.27) by θ and integrate the equation thus obtained, in x, over Ω. We obtain

$$\frac{1}{2}\frac{d}{dt}\int_\Omega \theta^2\,dx + \int_\Omega [(u\cdot\nabla)\theta]\theta\,dx - \kappa\int_\Omega \Delta\theta\cdot\theta\,dx = \int_\Omega u_2\theta\,dx. \quad (9.28)$$

Using the Green–Stokes formula and the boundary conditions of Eqs. (9.16) and (9.22) transposed to the nondimensional variables, we find

$$2\int_\Omega [(u\cdot\nabla)\theta]\theta\,dx = \sum_{i=1}^{3}\int_\Omega u_i\frac{\partial\theta^2}{\partial x_i}\,dx$$

$$= \int_{\partial\Omega} u\cdot n\,\theta^2 d\Gamma - \int_\Omega (\operatorname{div} u)\theta^2\,dx = 0,$$

$$-\int_\Omega \Delta\theta\cdot\theta\,dx = -\int_{\partial\Omega}\frac{\partial\theta}{\partial n}\theta\,d\Gamma + \int_\Omega |\operatorname{grad}\theta|^2\,dx = \int_\Omega |\operatorname{grad}\theta|^2\,dx.$$

We admit here the following mathematical inequality, called the Poincaré inequality; for every function θ satisfying Eq. (9.22),

$$\int_\Omega \theta^2\,dx \le \int_\Omega |\operatorname{grad}\theta|^2\,dx.$$

We then deduce from Eq. (9.28):

$$\frac{d}{dt}\int_\Omega \theta^2\,dx + 2\kappa\int_\Omega \theta^2\,dx \le 2\int_\Omega u_2\theta\,dx. \quad (9.29)$$

A similar relation can be obtained in the same way for u, starting from Eq. (9.25):

$$\frac{d}{dt} \int_\Omega |u|^2\, dx + 2\nu \int_\Omega |u|^2\, dx \leq 2\alpha' \int_\Omega u_2 \theta\, dx. \qquad (9.30)$$

Adding Eq. (9.30) to Eq. (9.29), we obtain

$$\frac{d}{dt} \int_\Omega (\theta^2 + |u|^2)\, dx + 2 \int_\Omega (\kappa\theta^2 + \nu|u|^2)\, dx \leq 2(1 + \alpha') \int_\Omega \theta u_2\, dx. \qquad (9.31)$$

Consequently, if

$$\kappa\nu > (1 + \alpha')^2, \qquad (9.32)$$

there exists $\delta > 0$ such that

$$\kappa\theta^2 + \nu|u|^2 - 2(1 + \alpha')\theta u_2 \geq \delta(\theta^2 + |u|^2);$$

then,

$$\frac{d}{dt} \int_\Omega (\theta^2 + |u|^2)\, dx + \delta \int_\Omega (\theta^2 + |u|^2)\, dx \leq 0.$$

Hence, by integration,

$$\int_\Omega (\theta^2 + |u|^2)(x, t)\, dx \leq \exp(-\delta t) \int_\Omega (\theta^2 + |u|^2)(x, 0)\, dx. \qquad (9.33)$$

This last inequality means that, *whatever the initial distribution (at time $t = 0$) of the velocities and temperatures in the Bénard experiment, after a certain time, the stationary regime described in Eq. (9.18) is established*: the fluid is at rest, and the temperature is a linear function of x_2. Inequality (9.33) means that the stationary regime is reached as $t \to \infty$; in practice, because (for example) $\exp(-10)$ is an extremely small number, this regime is reached, say, for $t = 10/\delta$.

Using the Rayleigh number introduced in Section 9.4, condition (9.32) becomes

$$Ra < (1 + \alpha')^{-2}. \qquad (9.34)$$

A natural question now is: what happens if this condition is not satisfied? Before answering this question, we notice that if T_1 alone varies, with the other characteristic quantities of the experiment remaining unchanged ($T_1 \geq T_2$), then for $T_1 = T_2$ or $T_1 - T_2$ small, the Rayleigh number is very small and the condition of inequality (9.34) is satisfied. On the other hand, it is no longer satisfied if Ra (or T_1) becomes larger.

As we said, for Ra small, the flow actually observed after a transient regime is the fluid at rest with a linear distribution of the temperatures in the z

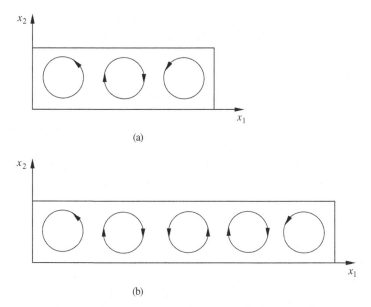

Figure 9.3 The Bénard rolls. (a) Three rolls; (b) Five rolls.

direction. When Ra grows, the liquid at the bottom, which is warmer than the liquid on top, tends to move upward, and the fluid starts moving. The stationary solution of Eq. (9.18), which still mathematically exists, becomes physically unstable. Typically, other stationary solutions that are stable appear: We say that we have a *bifurcation* of stationary solutions. If L_2/L_1 and L_2/L_3 are small enough, the so-called Bénard rolls appear, far enough from the boundaries $x_1 = 0$ or L_1; the number of rolls appearing depends on the conditions of the experiment; Figure 9.3 indicates the corresponding stream lines of the flow.

If Ra (or T_1) is even larger, the flow never becomes stationary, and we observe rolls with blurred and moving shapes. Finally, if Ra (or T_1) is very large, the flow does not have an apparent structure at all anymore. Numerous very small nonstationary vortices appear. The flow observed is *fully turbulent*.

If we perform a Fourier analysis in time of an observed signal of the flow, such as the velocity at a point of the fluid measured by laser velocimetry, then the signal is flat in the stationary regime; in intermediate nonstationary regimes several frequencies appear, corresponding to frequencies that are or are not rationally independent (periodic or quasi-periodic flows). In the purely turbulent regime, we observe a continuous spectrum of frequencies.

Remark 9.8: All that has just been said for thermohydraulics extends to the case of the fluid flows described in Sections 9.1 and 9.2, the important parameter being the *Reynolds number.*

9.6. Notion of boundary layer

Finally, we describe another important phenomenon of fluid mechanics whose existence is related to the fact that the Reynolds number is very large: the boundary layer. In the next subsection we describe the phenomenon in the context of a physically unrealistic but nevertheless instructive experiment; in the final subsection, we give a mathematical model.

A plane wall instantaneously set-in motion

We consider the following experiment: a viscous incompressible fluid fills the whole half-space $x_3 > 0$; it is at rest and, at time $t = 0$, we suddenly set in motion the plane $x_3 = 0$ with the constant velocity $V\vec{e}_1$. This experiment is unrealistic because it would require an infinite energy and an infinite quantity of fluid for its implementation; it is nevertheless enlightening. We then look for a flow of the form $(u, p) = (v(x_3, t), 0, 0, \text{const.})$ (it can be shown that no other solution actually exists). Consequently, v is solution of the equation

$$\frac{\partial v}{\partial t} - v\frac{\partial^2 v}{\partial x_3^2} = 0. \tag{9.35}$$

If we set $x_3' = ax_3$, $t' = a^2 t$, we find that this equation is invariant (i.e., $v(x_3, t) = v(ax_3, a^2 t)$). Hence, it is natural to look for solutions that depend only on $\theta = x_3/\sqrt{2vt}$; that is,

$$v(x_3, t) = f(\theta), \quad \theta = \frac{x_3}{\sqrt{2vt}}.$$

Therefore,

$$\frac{\partial v}{\partial t} = -\frac{\theta}{2t}f'(\theta),$$

$$\frac{\partial v}{\partial x_3} = \frac{1}{\sqrt{2vt}}f'(\theta),$$

$$\frac{\partial^2 v}{\partial x_3^2} = \frac{1}{2vt}f''(\theta),$$

and thus we are led, for the determination of f and of the flow, to the resolution of the ordinary differential equation

$$f''(\theta) + \theta f'(\theta) = 0;$$

hence

$$f(\theta) = A + B \int_0^\theta e^{-s^2/2}\, ds.$$

We find A and B by using the boundary and initial conditions, and we obtain

$$v(x_3, t) = V\left[1 - 2\,\mathrm{erf}\left(\frac{x_3}{\sqrt{2\nu t}}\right)\right],$$

$$\mathrm{erf}\, x = \frac{1}{\sqrt{2\pi}} \int_0^x \exp\left(-\frac{\xi^2}{2}\right) d\xi.$$

Because uniqueness has been assumed, we have found the solution to our problem (see Figure 9.4). Let us comment on this solution:

1. Because $\mathrm{erf}\, x \le \mathrm{erf}(+\infty) = 1/2$, $v(x_3, t) > 0$ as soon as $t > 0$. Thus, the whole fluid is set in motion instantaneously. The propagation velocity is thus infinite, owing to the viscosity, contrary to hyperbolic equations (the wave equations, for instance) for which the propagation velocity is finite (see Chapter 17 concerning the propagation of waves at finite speed).

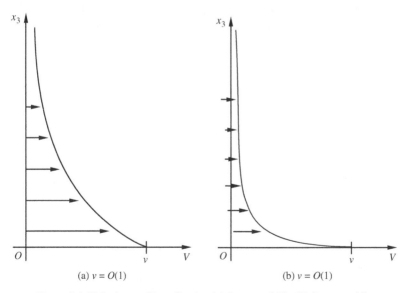

(a) $v = O(1)$ (b) $v = O(1)$

Figure 9.4 Velocity profile at fixed t: (a) for $\nu = O(1)$, (b) for $\nu = o(1)$.

2. For $x_3 > 0$ fixed, we have $\lim_{t \to +\infty} v(x_3, t) = V$. Therefore, the whole fluid tends to acquire the velocity V.

3. For fixed x_3 and t, if $\nu \to 0$, then $\theta \to +\infty$ and $v \to 0$: the velocity at every interior point of the fluid becomes small (tends to zero). We notice that for a (nonviscous) inviscid fluid submitted to the same experiment, the fluid remains at rest. Thus, the solution $v = v_\nu(x_3, t)$ tends to the inviscid fluid solution at every point of the fluid $x_3 > 0$ but not on the axis $x_3 = 0$.

   For small ν, the velocity varies very rapidly in a small region near $x_3 = 0$; we will comment on this in the next paragraph.

4. Boundary layer

   Let $0 < k < 1$ be fixed. We look for the set of points where $v > kV$ (and thus $kV < v < V$). Because erf is an increasing function, this is equivalent to a condition of the form

$$\theta < \theta_0(k).$$

   Therefore (and the value of k, $k = 0(1)$ does not matter here), we will have $v > kV$ in a strip of thickness, in the direction of x_3, of order $\sqrt{\nu}$. This strip is called the boundary layer (its thickness being thus of order $\sqrt{\nu}$).

More generally, for large Reynolds numbers, there exists in the neighborhood of the boundaries a layer of thickness of order $1/\sqrt{R}$, the boundary layer of the flow, where large velocity gradients of order $\sqrt{R}$ occur. This also happens for nonturbulent flows, but the boundary layer phenomena are much more complex, and very little is known in the turbulent case.

Mathematical model

There are simple mathematical models that display some of the many diverse features of the boundary layer phenomenon (we also speak of singular perturbation phenomenon). The following simple example is one of the first mathematical models explaining the Gulf Stream marine currents near the east coast of America.

We consider the ordinary differential equation

$$\varepsilon\, u''(x) + u'(x) + a = 0, \quad 0 < x < 1,$$
$$u(0) = 0, \quad u(1) = 0,$$

where $0 < a < 1$ and $\varepsilon > 0$ is small. It is easy to check that

$$u(x) = u(x, \varepsilon) = a\,\frac{1 - \exp(-x/\varepsilon)}{1 - \exp(-1/\varepsilon)} - ax.$$

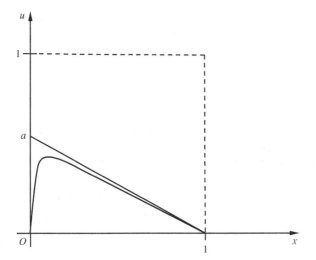

Figure 9.5 A simple mathematical model of boundary layer. Graphs of $u(x, \varepsilon)$ and of $u_0(x)$.

Consequently

$$\lim_{\varepsilon \to 0, \ x>0 \text{ fixed}} u(x, \varepsilon) = u_0(x) = a(1 - x),$$

and

$$\lim_{\varepsilon \to 0} u(0, \varepsilon) = 0 \neq u_0(0) = a.$$

We notice the loss of one of the boundary conditions, at the wall $x = 0$, and, for ε small, a boundary layer occurs near this wall. Furthermore, $u(x, \varepsilon)$ converges uniformly on $[\alpha, 1]$, $\forall \alpha > 0$, to its limit $u_0(x)$, but it does not and cannot converge uniformly to its limit on $[0, 1]$ (see Figure 9.5).

Remark 9.9: Methods that allow the detection and study of the boundary layer phenomenon starting only from the equations and the boundary conditions have been developed: this is the subject of asymptotic analysis that is possibly helped and supplemented by numerical methods.

Exercises

1. We consider in the domain $\Omega_\infty = \{(x_1, x_2), 0 < x_2 < L\}$ the flow of a viscous, homogeneous, incompressible fluid submitted to gravity and

which conducts heat. In the Boussinesq approximation, the equations read

$$\rho_0(\frac{\partial v}{\partial t} + (v \cdot \nabla)v) - \mu\Delta v + \nabla p = -\rho_0 g(T - T_*)e_2, \qquad (9.36)$$

$$\text{div } v = 0, \qquad (9.37)$$

$$c_v(\frac{\partial T}{\partial t} + (v \cdot \nabla)T) - v_0\Delta T = 0, \qquad (9.38)$$

where $x \in \Omega, 0 < t < T_{max}$, e_2 is the unit vertical vector pointing down-wards, g is the gravity constant, $c_v > 0$ and T_* is the reference temperature.

a) We assume that the temperature is maintained at $T = T_0$ on $x_2 = 0$ and at $T = T_1$ on $x_2 = L$. Compute the stationary solution satisfying $v = 0$ and $T = \psi(x_2)$ (we call $p_0(x_1, x_2)$ the pressure).

From now on, we set $T = \psi(x_2) + \theta$, $p = p_0 + P$. Rewrite (1)-(3) for the new variables v, P, θ. We will call (1')-(3') the corresponding equations.

b) We assume that the solutions (v, P, θ) of (1')-(3') are periodic with period 2π in x_1 and that v vanishes at $x_2 = 0$ and $x_2 = L$. We set $\Omega = (0, 2\pi) \times (0, L)$.
- (i) Show that $\int_\Omega [(v \cdot \nabla)v]v dx = 0$.
- (ii) Show that $\int_\Omega [(v \cdot \nabla)\theta]\theta dx = 0$.
- (iii) Show that $\int_\Omega \nabla p \cdot v dx = 0$.

c) Give an expression of

$$\frac{1}{2}\frac{d}{dt}\int_\Omega (\rho_0|v|^2 + c_v\theta^2)\, dx, \qquad (9.39)$$

assuming that the boundary conditions on v, P, and θ are homogeneous.

d) Show that (9.39) is negative when $g = 0$ and $T_1 = T_0$.

2. Consider the incompressible Navier-Stokes equations with periodic boundary conditions in the cube $\Omega = (0, L)^3$ (such boundary conditions are useful in the theory of turbulence):

$$\frac{\partial u}{\partial t} - \nu\Delta u + (u \cdot \nabla)u + \nabla p = f,$$

$$\text{div } u = 0,$$

$$u \text{ is periodic in } x_1, x_2, \text{ and } x_3,$$

where, for simplicity, the density is set to 1. Consider the Fourier expansions of u, p, and f:

$$u = \sum_{k\in\mathbb{Z}^M} u_k e^{\frac{2i\pi k \cdot x}{L}}, p = \sum_{k\in\mathbb{Z}^M} p_k e^{\frac{2i\pi k \cdot x}{L}}, f_k = \sum_{k\in\mathbb{Z}^M} f_k e^{\frac{2i\pi k \cdot x}{L}}.$$

a) Write the equations for u_k and p_k.

b) Solve the stationary, linear Navier-Stokes equations

$$-\nu\Delta u + \nabla p = f,$$
$$\operatorname{div} u = 0,$$
$$u \text{ is periodic in } x_1, x_2, \text{ and } x_3,$$

using Fourier series.

CHAPTER TEN

Magnetohydrodynamics and inertial
confinement of plasmas

The purpose of this chapter and of the next one is to present complex problems of fluid mechanics involving other physical phenomena such as electromagnetism (for this chapter) and chemistry (for the next one).

In this chapter, we consider flows of fluids conducting electricity in the presence of electric currents and of electromagnetic fields; the purpose of magnetohydrodynamics (MHD) is to study such flows. The MHD equations consist of the Maxwell equations, which govern the electromagnetic quantities, and of the equations of fluid mechanics in which we include the electromagnetic forces.

In this chapter, we will limit ourselves to a single fluid flow: in certain circumstances (e.g., very high temperatures, intense electromagnetic fields), the conductive fluid is ionized and becomes a plasma; it then becomes desirable to study the flows of the positive and negative particles separately and, in such a case, one has to deal with a multifluid flow, which is a situation similar to the flow of the mixture of two miscible fluids. That case does not present any particular difficulty; multifluid flows are considered in the next chapter.

The modeling of the electromagnetic phenomena leads to the Maxwell equations, and it can be conducted essentially like the modeling of the mechanical phenomena in Chapters 1 to 5: describing the physical quantities and writing the physical conservation laws, partial differential equations, and boundary conditions. It would be too long, and not particularly instructive from the modeling point of view, to conduct this study here; we will be content with describing the Maxwell equations and will refer the reader to the specialized literature for their physical justification.

Section 10.1 presents the Maxwell equations, and Section 10.2 addresses those of magnetohydrodynamics. Finally, in Section 10.3, we present one of the important applications of MHD: the equilibrium of a plasma confined in a Tokamak machine. The Tokamak consists of a fluid (plasma) subjected to intense pressures, temperatures, and magnetic fields; it is one of the experimental machines with which physicists hope, in the long term, to reach controlled thermonuclear fusion as a source of nonpolluting energy.

10.1. The Maxwell equations and electromagnetism

We start this section by introducing new physical quantities, namely, the electromagnetic quantities. These quantities are the electric charge density, represented by a scalar q, the electric current density J, the electric induction D, the magnetic induction B, the magnetic field H, and the electric field E that are vectors in $\mathbb{R}^3$. These quantities are related by the conservation laws that we will write without justification, but one may derive these equations by the same methods as those used for obtaining the mass and momentum conservation laws studied in the first part of this book, starting from the necessary physical assumptions.

Conservation of the electric charge

The conservation of the electric charge can be mathematically written as:

$$\frac{d}{dt} \int_\Omega q \, dx = - \int_{\partial\Omega} J \cdot n \, d\Gamma + \int_\Omega g \, dx, \qquad (10.1)$$

where Ω is a bounded regular domain of $\mathbb{R}^3$ with boundary $\partial\Omega$, and g denotes the source of electric charges per unit volume and time. Because Eq. (10.1) is satisfied for every open set Ω, it follows that, pointwise,

$$\frac{\partial q}{\partial t} + \operatorname{div} J = g. \qquad (10.2)$$

We then introduce the vector D, called electric induction, and a vector G such that

$$\operatorname{div} D = q,$$
$$\operatorname{div} G = g. \qquad (10.3)$$

Because

$$\operatorname{div}\left(\frac{\partial D}{\partial t} + J - G\right) = 0, \qquad (10.4)$$

we deduce the existence of a vector field H called magnetic field such that

$$\frac{\partial D}{\partial t} + J - \operatorname{curl} H = G. \qquad (10.5)$$

Faraday's law

The Faraday law, relating the magnetic induction B to the electric field E, is stated as follows:

> *For every fixed surface Σ with boundary $\partial\Sigma$, the derivative with respect to time of the flux of magnetic induction B through Σ is equal in magnitude and opposed to the circulation of the electric field E along $\partial\Sigma$.*

This is expressed by the equation

$$\frac{d}{dt}\int_{\Sigma} B \cdot n \, d\Sigma + \int_{\partial\Sigma} E \cdot d\ell = 0. \tag{10.6}$$

Now, a classical vector analysis formula, which we recall without proof, gives

$$\int_{\partial\Sigma} E \cdot d\ell = \int_{\Sigma} \operatorname{curl} E \cdot n \, d\Sigma, \tag{10.7}$$

where the orientations of n and of $\partial\Sigma$ are consistent. It then follows from Eqs. (10.6) and (10.7) that

$$\frac{d}{dt}\int_{\Sigma} B \cdot n \, d\Sigma + \int_{\Sigma} \operatorname{curl} E \cdot n \, d\Sigma = \int_{\Sigma} \left(\frac{\partial B}{\partial t} + \operatorname{curl} E\right) \cdot n \, d\Sigma = 0,$$

and because Σ is an arbitrary fixed surface, we obtain the equation

$$\frac{\partial B}{\partial t} + \operatorname{curl} E = 0. \tag{10.8}$$

Finally, taking the divergence of Eq. (10.8), we see that

$$\frac{\partial}{\partial t}\operatorname{div} B = 0.$$

If $\operatorname{div} B = 0$ at initial time,[1] we find the equation

$$\operatorname{div} B = 0, \tag{10.9}$$

valid at all times.

Equations (10.2), (10.3), (10.5), (10.8), and (10.9) constitute the Maxwell equations, the fundamental general equations of electromagnetism. We need to supplement them with a set of interface and constitutive laws.

---

[1] This assumption is related to the question, in physics, of the existence of the magnetic monopole.

Interface laws

Let us now consider the case in which an open set Ω of $\mathbb{R}^3$ is divided into two subdomains Ω^1 and Ω^2 by a smooth surface Σ and in which one or several of the quantities D, H, B, and E are discontinuous at the crossing of Σ.

Using the conservation laws written in integral form in Eqs. (10.1) and (10.6) and the equations already derived, namely (10.2), (10.3), (10.5), (10.8) and (10.9), we can obtain the interface conditions on Σ.[2] For instance, we write, for D,

$$\int_{\partial\Omega^1} D \cdot n \, d\Gamma = \int_{\Omega^1} q \, dx,$$

$$\int_{\partial\Omega^2} D \cdot n \, d\Gamma = \int_{\Omega^2} q \, dx,$$

$$\int_{\partial\Omega} D \cdot n \, d\Gamma = \int_{\Omega} q \, dx,$$

where n is the unit outward normal (to Ω^1, Ω^2, Ω). Hence, on Σ, with n pointing then from Ω^1 to Ω^2,

$$(D^2 - D^1) \cdot n = q \quad \text{on } \Sigma, \tag{10.10}$$

where $D^i = D|_{\Omega^i}$, $i = 1, 2$. Similarly, if H, B, and E are discontinuous at the crossing of Σ, we deduce from the relation

$$\int_{\Omega} \text{curl } H \, dx = \int_{\partial\Omega} n \wedge H \, d\Gamma,$$

from the Faraday law, and from Eqs. (10.5), (10.8), and (10.9) that

$$n \wedge (H^2 - H^1) = J, \tag{10.11}$$

$$(B^2 - B^1) \cdot n = 0, \tag{10.12}$$

$$n \wedge (E^2 - E^1) = 0. \tag{10.13}$$

Constitutive laws

As in the case of fluids and solids (see Chapters 5 and 7), the Maxwell equations are not sufficient to describe the electromagnetic evolution of the medium because they provide 7 independent scalar equations ((10.3), (10.5), and (10.8)) for 16 unknowns, namely q, D, J, B, H, and E. They must be supplemented with constitutive laws that express the difference of behavior of the medium (for instance, a medium more or less conductive of electricity). These laws relate D to E on the one hand and B to H on the other hand. We

---

[2] The situation is similar to that of Chapter 6, Section 6.2, but simpler because Ω is fixed here.

will only retain the most common form of these laws. For other less classical forms of these laws, and for a discussion of their physical validity, the reader is refered to the books mentioned in the reference list. We will thus assume hereafter that the fields and inductions are proportional, which is expressed by

$$
\begin{aligned}
D &= \varepsilon E, \\
B &= \mu H,
\end{aligned}
\tag{10.14}
$$

where ε and μ are called, respectively, the dielectric constant of the medium and the magnetic permeability of the medium. In the simplest cases, and in particular in air, they are assumed to be independent of the electromagnetic quantities and even constant.

A second law, called Ohm's law, relates the current density to the electric field by the relation

$$
J = \sigma E,
\tag{10.14'}
$$

where σ is the conductivity, which is also independent of the electromagnetic quantities. The media for which Ohm's law is valid will be called stable media.

The three relations written above do provide the nine equations that were needed to supplement the Maxwell equations.

Remark 10.1: The laws previously described are linear laws. We can note that, contrary to the Maxwell equations, these laws do not satisfy the Galilean invariance, and thus, in general, we will only be able to apply them in one specified Galilean frame (this unnatural restriction gives rise to the relativity theory).

Electromagnetism in a stable medium

To describe the electromagnetic evolution of a medium Ω, we need to supplement the Maxwell equations and the constitutive laws with boundary and initial conditions, that is, some equations at the boundary of Ω and the values of the electromagnetic quantities at the initial time.

Taking into account the constitutive laws defined by Eqs. (10.14) and (10.14') above, the unknowns of the problem, reduced to B, D, and J, satisfy the equations

$$
\frac{\partial D}{\partial t} + J - \operatorname{curl}\left(\frac{1}{\mu}B\right) = G, \quad \text{in } \Omega,
\tag{10.15}
$$

$$
\frac{\partial B}{\partial t} + \operatorname{curl}\left(\frac{1}{\varepsilon}D\right) = 0, \quad \text{in } \Omega.
\tag{10.16}
$$

We supplement Eqs. (10.15) and (10.16) with the initial data, at $t = 0$, as follows:

$$B(x, 0) = B_0(x), \quad D(x, 0) = D_0(x). \tag{10.17}$$

If the boundary $\partial \Omega$ of Ω is assumed to be superconducting, the boundary conditions that we impose (also satisfied by B_0 and D_0) read

$$B \cdot n = 0, \quad \text{on } \partial \Omega, \tag{10.18}$$

$$D \wedge n = 0, \quad \text{on } \partial \Omega. \tag{10.19}$$

Remark 10.2: The quantities ε, μ, and σ may depend on x and t, but they remain bounded, strictly positive for ε and μ, and positive for σ.

10.2. Magnetohydrodynamics

In this section, we are interested in the motion of fluids that conduct electricity. Such situations occur, for instance, in the study of ionized gases, such as found in lasers; in geophysics (the earth's magnetic core or atmosphere at high altitude); in astronautics (for the reentry of space vehicles into the atmosphere); or in the study of seawater subjected to electric currents or magnetic fields, which is an approach currently considered to reduce turbulence and resistance to motion for certain ships and submarines.

The equations of magnetohydrodynamics thus consist of the Maxwell equations coupled with the equations of fluid mechanics to which we must add the electromagnetic forces.

As in the previous section, we will be content with writing the equations of magnetohydrodynamics without justifying their derivation via mathematical modeling of the physical phenomena. The reader interested in more details is referred, for instance, to Cabannes (1970) cited in the reference list.

As already mentioned, these equations consist of the equations of motion of the fluid (we keep here the notation introduced in the first part of this book) coupled with the Maxwell equations as follows:

$$\rho \left[\frac{\partial U}{\partial t} + \text{curl } U \wedge U + \text{grad} \left(p + \frac{U^2}{2} \right) \right] = \rho f + \Phi + \text{div } \Sigma, \tag{10.20}$$

$$\frac{\partial \rho}{\partial t} + \text{div} (\rho U) = 0, \tag{10.21}$$

$$\frac{\partial}{\partial t} \left[\rho \left(e + \frac{U^2}{2} \right) \right] + \text{div} \left[\rho U \left(e + \frac{U^2}{2} \right) \right]$$
$$= \rho f \cdot U + e_m + \text{div} (TU) - \text{div } \theta, \tag{10.22}$$

$$\frac{\partial B}{\partial t} + \text{curl } E = 0, \tag{10.23}$$

$$\frac{\partial D}{\partial t} - J = H, \tag{10.24}$$

$$\text{div } D = q, \tag{10.25}$$

where f denotes the external volume forces (other than electromagnetic forces).

These fundamental equations (conservation equations) are supplemented by the constitutive laws as follows:

$$J = \sigma_m(E + U \wedge B), \tag{10.26}$$

$$D = \varepsilon_m(\rho, T)E, \tag{10.27}$$

$$B = \mu_m(\rho, T)E. \tag{10.28}$$

We note that Ohm's law is extended from Eqs. (10.14′) to (10.26) in the case of moving media and that we allow ε_m and μ_m to depend on the density ρ and on the temperature T for more generality. Furthermore, the following notations have been used:

$$\Phi = qE + J \wedge B,$$

$$e_m = J \cdot E,$$

$$T = \frac{\partial e}{\partial s},$$

$$p = -\frac{\partial e}{\partial \tau},$$

$$\Sigma = [-p + \lambda_f(T) \text{ div } U]I + \mu_f(T)(\nabla U + {}^t\nabla U),$$

$$\theta = -\lambda(T)\nabla T.$$

In the preceding equations, T denotes the temperature, s the entropy, and $\tau = 1/\rho$.

We thus obtain a system of 21 scalar equations (Eqs. (10.20) to (10.28)) with 21 independent scalar unknowns, ρ, q, T or e, U, D, B, J, E, and H, to which we must add, as usual, the initial conditions and the boundary conditions. Of course, there is also the unknown p, but this supplementary unknown is compensated by the equation of state $p = \varphi(\rho, \tau)$ or $\varphi(\rho, e)$ for compressible flows or by the equation $\rho = \rho_0$ for incompressible flows, exactly as for nonconducting fluids.

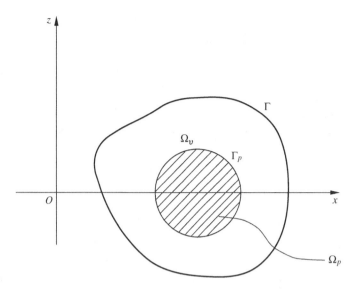

Figure 10.1 The Tokamak machine.

10.3. The Tokamak machine

As we said in the introduction to this chapter, the Tokamak is one of the devices considered to realize thermonuclear fusion, which in the long term could be used as a source of nonpolluting energy. Schematically, it consists of a torus partially filled with plasma (conductive material submitted to high pressures and temperatures and to intense magnetic fields); the magnetic fields are produced by electromagnetic coils surrounding the torus.

In what follows, the plasma is assumed to be at rest, and we thus want to write the equations that characterize the medium at equilibrium in an axisymmetric geometry.[3]

We represent the Tokamak machine as an axisymmetric torus of axis Oz (see Figure 10.1). We further assume that the cross section of the Tokamak by the plane Oxz is an open set Ω of boundary Γ and that the plasma fills a subdomain Ω_p of Ω with boundary Γ_p. Finally, the domain $\Omega_v = \Omega \backslash (\Omega_p \cup \Gamma_p)$ is assumed to be empty.

---

[3] Such an equilibrium is rather unstable in reality. At this time, actually maintaining the plasma in such a state of equilibrium is a major problem of physics (see the remarks on the stability of stationary solutions in Chapter 9, Section 9.5).

The electromagnetic state of the medium is described, in the empty region Ω_v, by the Maxwell equations that reduce here to

$$\operatorname{div} B = 0, \quad \text{in } \Omega_v, \tag{10.29}$$

$$\operatorname{curl} B = 0, \quad \text{in } \Omega_v. \tag{10.30}$$

In the plasma, the electromagnetic state is governed by the equations of magnetohydrodynamics. Because $U = 0$ and the system is time independent, they reduce to

$$\operatorname{div} B = 0, \quad \text{in } \Omega_p, \tag{10.29'}$$

$$\operatorname{curl} B = \mu_0 J, \quad \text{in } \Omega_p, \tag{10.30'}$$

$$\operatorname{grad} p = J \wedge B, \quad \text{in } \Omega_p. \tag{10.31}$$

It is natural, in view of the geometry of the problem, to introduce the cylindrical coordinates r, θ, z, with e_r, e_θ, and e_z denoting the corresponding local orthonormal system. Equation (10.29) (and also Eq. (10.29')) can be rewritten, setting $B = B_r e_r + B_\theta e_\theta + B_z e_z$ and $J = J_r e_r + J_\theta e_\theta + J_z e_z$ as

$$\frac{1}{r}\frac{\partial}{\partial r}(r B_r) + \frac{\partial B_z}{\partial z} = 0,$$

because, owing to the axisymmetry, B_r, B_θ, B_z, J_r, J_θ, J_z, and p do not depend on θ. This yields the existence, locally, of a function Φ, called the flux function, satisfying

$$B_r = -\frac{1}{r}\frac{\partial \Phi}{\partial z}, \quad B_z = \frac{1}{r}\frac{\partial \Phi}{\partial r}.$$

One can show that this function Φ is defined and single valued in the whole domain Ω thanks to the boundary conditions described below. By setting $h = h(r, z) = r B_\theta$, it then follows that

$$B = \nabla_c \Phi \wedge \frac{e_\theta}{r} + \frac{1}{r} h e_\theta, \tag{10.32}$$

where $\nabla_c = (\partial/\partial r)e_r + (\partial/\partial z)e_z$. We can then rewrite Eqs. (10.30) and (10.30') in the form

$$\mathcal{L}\Phi = 0, \quad \text{in } \Omega_v, \tag{10.33}$$

$$-\mathcal{L}\Phi e_\theta + \nabla_c h \wedge \frac{e_\theta}{r} = \mu_0 J, \quad \text{in } \Omega_p, \tag{10.33'}$$

where we have set

$$\mathcal{L} = \frac{\partial}{\partial r}\left(\frac{1}{r}\frac{\partial}{\partial r}\right) + \frac{1}{r}\frac{\partial^2}{\partial z^2}.$$

Finally, Eq. (10.31) becomes

$$\mu_0 \frac{\partial p}{\partial r} = -\frac{1}{r}\mathcal{L}\Phi \cdot \frac{\partial \Phi}{\partial r} + \frac{1}{2r^2}\frac{\partial H^2}{\partial r},$$

$$0 = -\frac{1}{r^2}\left(\frac{\partial \Phi}{\partial z}\frac{\partial h}{\partial r} - \frac{\partial h}{\partial z}\frac{\partial \Phi}{\partial r}\right), \qquad (10.34)$$

$$\mu_0 \frac{\partial p}{\partial z} = -\frac{1}{r}\mathcal{L}\Phi \cdot \frac{\partial \Phi}{\partial z} - \frac{1}{2r^2}\frac{\partial h^2}{\partial z}.$$

A first consequence of Eq. (10.34) is that, in Ω_p, $\nabla_c h$ is parallel to $\nabla_c \Phi$, and thus h depends only on Φ. More precisely, we set

$$h^2 = g_0(\Phi),$$

which yields

$$\nabla_c h^2 = g_0' \cdot \nabla \Phi,$$

where we have written $g_0' = dg_0/d\Phi$. It then follows that

$$\mu_0 \nabla_c p = \left(-\frac{1}{r}\mathcal{L}\Phi - \frac{1}{2r^2}g_0'\right)\nabla_c \Phi.$$

Similarly, because $\nabla_c p$ is parallel to $\nabla_c \Phi$, p only depends on Φ, and we set

$$p = g_1(\Phi), \quad \nabla_c p = g_1' \cdot \nabla_c \Phi,$$

and thus the three equations (Eq. (10.34)) finally reduce to the single equation

$$\mathcal{L}\Phi = -\mu_0 r \, g_1'(\Phi) - \frac{1}{2r}g_0'(\Phi), \quad \text{in } \Omega_p. \qquad (10.35)$$

Thus, the unknowns of the problem are all functions of Φ. However, the Maxwell equations are not sufficient to determine the functions g_0 and g_1, and additional modeling work is necessary to determine these functions in a satisfactorily approximate way (see the following paragraphs).

Denoting by n and τ the unit vectors, respectively, normal and tangent to Γ and Γ_p, we supplement the previous equations by the following boundary conditions (we assume for instance that the boundary Γ is conductive):

$$B \cdot n = 0, \quad \text{on } \Gamma \text{ and } \Gamma_p, \qquad (10.36)$$

$$B \cdot \tau \text{ is continuous at the crossing of } \Gamma_p. \qquad (10.37)$$

Because $B \cdot n = -(1/r)(\partial \Phi/\partial \tau)$ and $B \cdot \tau = (1/r)(\partial \Phi/\partial \tau)$, we obtain that (locally) $\partial \Phi/\partial \tau = 0$, on Γ and Γ_p, which yields that Φ is constant on Γ

and Γ_p and is therefore defined and single valued in all of Ω. Furthermore, Φ being defined up to an additive constant, we will take

$$\Phi = 0, \quad \text{on } \Gamma_p, \tag{10.38}$$

$$\Phi = \gamma \quad \text{(unknown constant), on } \Gamma, \tag{10.39}$$

and Eq. (10.37) then reduces to

$$\frac{\partial \Phi}{\partial n} = 0, \quad \text{on } \Gamma_p. \tag{10.40}$$

We also make the simplifying hypotheses

$$p = J_\theta = 0, \quad \text{on } \Gamma_p, \tag{10.41}$$

$$J_\theta \neq 0, \quad \text{in } \Omega_p, \tag{10.42}$$

which is considered very realistic, and we then deduce from Eq. (10.38) that $g_1(0) = 0$. Moreover, Eq. (10.33') yields

$$\mu_0 J_\theta = -\mathcal{L}\Phi = \mu_0 r g_1'(\Phi) + \frac{1}{2r} g_0'(\Phi),$$

and it follows that

$$g_0'(0) = g_1'(0) = 0.$$

Finally, thanks to Eq. (10.42), we deduce that Φ does not vanish in Ω_p.

One of the simplest models is obtained by taking for g_0 and g_1 (and thus for p and h^2) quadratic functions of Φ as follows:

$$g_0(\Phi) = b_0 + b_2 \Phi^2,$$

$$g_1(\Phi) = a_2 \Phi^2,$$

where a_2, b_0, and b_2 are positive constants, the pressure and h^2 being positive. In this particular case, Eq. (10.35) is rewritten as

$$\mathcal{L}\Phi = -\left(2\mu_0 r a_2 + \frac{b_2}{r}\right)\Phi, \quad \text{in } \Omega_p. \tag{10.43}$$

In general, we also assume that the total current is fixed in the plasma, which is expressed by

$$\mu_0 \int_{\Omega_p} J_\theta \, dr dz = -\int_{\Omega_p} \mathcal{L}\Phi \, dr dz = \int_{\Gamma_p} \frac{1}{r} \frac{\partial \Phi}{\partial n} d\Gamma = I \mu_0. \tag{10.44}$$

The set of equations for the problem reduces to the determination of a single scalar function Φ defined in Ω and continuous at the crossing of Γ_p such that Eqs. (10.33), (10.43), (10.38), (10.39), and (10.44) hold with I given and γ

unknown. This problem is nonlinear even though Eqs. (10.33) and (10.43) are linear equations; indeed, the curve Γ_p defining the shape of the plasma at equilibrium is also unknown, which makes the problem nonlinear. For fixed Ω, g_0, g_1, the number I is the only data of the problem, and the function $\{I \mapsto \Phi\}$ is not a linear function, as one may easily notice.

Remark 10.3: We conclude this chapter with several remarks at various levels of generality:

1. Concerning the Tokamak machine, it is interesting to consider functions g_0 and g_1 other than the quadratic functions introduced above.

    One may also study nonaxisymmetric equilibria, or even evolution problems, but the equations remain very complicated; there are no simplifications similar to those in the axisymmetric stationary case.
2. Geometries other than the torus may be considered for the machines intended to achieve thermonuclear fusion. Let us mention in particular the Stellerator machine of cylindrical shape.
3. As indicated above, other approaches are also considered for thermonuclear fusion using, in particular, power lasers.
4. Many media (of natural origin or industrially produced) are in the plasma state. Plasma has even been called the fourth state of matter (with solids, liquids, and gases), that would actually be the most common. For all these plasmas, the basic equations are those of magnetohydrodynamics (Section 10.2) with one or several fluids; these equations are very complicated, but they can be simplified, as in Section 10.3, by taking into account the specificity of the problem under study.

Exercises

1. Derive from (10.23)–(10.28) an equation for the magnetic field H.
2. Consider the equation obtained in Exercise 1, and assume that there is no fluid motion and that the electric induction D vanishes. Consider a volume Ω.

    a) Show that

$$\frac{1}{2}\frac{d}{dt}\int_{\Omega}|H|^2 dx = -\int_{\Omega} H \cdot \mathrm{curl}\,(\eta\,\mathrm{curl}\,H)dx, \quad \text{where } \eta = \frac{1}{\mu_m \sigma_m}.$$

b) Show that

$$\frac{1}{2}\frac{d}{dt}\int_\Omega |H|^2 dx = -\int_\Omega \eta|\text{curl } H|^2 dx + \int_{\partial\Omega} \eta(H \wedge \text{curl } H) \cdot nd\sigma.$$

c) Show that

$$\frac{d\mathcal{M}}{dt} = -\int_\Omega \frac{|J|^2}{\sigma_m} dx + \int_{\partial\Omega} \frac{1}{\sigma_m}(H \wedge J) \cdot nd\sigma,$$

$$\frac{d\mathcal{M}}{dt} = -\int_\Omega \frac{|J|^2}{\sigma_m} dx + \int_{\partial\Omega} (H \wedge E) \cdot nd\sigma,$$

where $\mathcal{M} = (\mu/2)\int_\Omega |H|^2 dx$ is the magnetic energy (the above equalities mean that the change in the magnetic energy inside Ω consists of a loss of energy resulting from the Joule effect by the currents flowing into the conductors and of the Poynting flux of energy flowing from the external field into the conductor).

3. Consider the equation for the magnetic field H for an infinite resistance σ_m and constant magnetic permeability μ_m

$$\frac{\partial H}{\partial t} - \text{curl } (u \wedge H) = 0.$$

We further assume that the electric induction D vanishes.

a) Show that $\int_{\Sigma_t} H \cdot nd\sigma$ =const., for every surface Σ_t that moves with the flow.

b) Show that, for every volume Ω,

$$\frac{d\mathcal{M}}{dt} = \mu_m \int_\Omega H \cdot \text{curl } (u \wedge H) dx.$$

c) Assume that the fluid is confined into the volume Ω (i.e., $u \cdot n = 0$ on $\partial\Omega$). Show that

$$\frac{d\mathcal{M}}{dt} = \mu_m \int_\Omega u \cdot (H \wedge J) dx + \mu_m \int_{\partial\Omega} [(H \cdot u)H] \cdot nd\sigma.$$

4. Consider the following *MHD* equations for an incompressible, inviscid fluid in a potential field:

$$\frac{\partial u_i}{\partial t} + u_j\frac{\partial u_i}{\partial x_j} - \frac{\mu_m}{\rho}H_j\frac{\partial H_i}{\partial x_j} = -\frac{\partial}{\partial x_i}(\frac{p}{\rho} + \mu_m\frac{|H|^2}{2\rho} + V),$$

$$\frac{\partial H_i}{\partial t} + u_j\frac{\partial H_i}{\partial x_j} = H_j\frac{\partial u_i}{\partial x_j},$$

$$\text{div} u = 0,$$

$$\text{div} H = 0.$$

Show that the relations

$$u_i = \pm \left(\frac{\mu_m}{\rho} \right)^{\frac{1}{2}} H_i,$$

$$\frac{\partial}{\partial x_i} \left(\frac{p}{\rho} + \mu_m \frac{|H|^2}{2\rho} + V \right) = 0,$$

define a solution of these equations.

CHAPTER ELEVEN

Combustion

The objective of this chapter is the study of mixtures of reactive gaseous or liquid fluids, that is to say fluids undergoing chemical reactions (fuel, oil and oxygen, various chemical species, etc.). The equations and models that we present are commonly used nowadays for the study of phenomena as different as the propagation of chemical pollution in the atmosphere or in water (seas, rivers, estuaries) or the combustion of mixtures of fuel, oil, and oxygen in cars, planes, or rocket engines. The number N of chemical species taken into account can vary from a few units to several hundreds.

We start, in Section 11.1, by describing the framework for mixtures of non-necessarily reactive fluids. Section 11.2 is devoted to the equations of chemical kinetics. We then formulate in Section 11.3 the general equations of combustion and describe a typical problem of combustion. In Section 11.4, we introduce the Stefan–Maxwell equations corresponding to an even more complex problem of chemical kinetics and combustion. Finally, we finish, in Section 11.5, with an overview of the case in which the mixture only contains two chemical species; although much simpler, this case already leads to very complex mathematical problems that we will be only able to describe briefly.

11.1. Equations for mixtures of fluids

We consider in this section a mixture composed of N different miscible species. Each fluid may be considered separately; it then has its own characteristics, density, velocity, and pressure denoted respectively by ρ_i, u_i, p_i in the Eulerian description ($\rho_i = \rho_i(x, t)$, etc.), $i = 1, \ldots, N$. In a complete description of the flow, we would write the Navier–Stokes equations for each fluid with viscosity coefficients μ_i (incompressible case) or μ_i and λ_i (compressible case), $i = 1, \ldots, N$, specific to each fluid.[1]

Most often, we will be content with a description of the motion of the mixture treated as a single fluid; the density of the mixture will then be the sum of the partial densities ρ_i,

$$\rho = \sum_{i=1}^{N} \rho_i; \tag{11.1}$$

the pressure will be the sum of the partial pressures p_i,

$$p = \sum_{i=1}^{N} p_i; \tag{11.2}$$

and the velocity will be the weighted mean velocity

$$u = \sum_{i=1}^{N} \frac{\rho_i}{\rho} u_i. \tag{11.3}$$

In this case, we will also need to define the viscosity coefficient μ (or the coefficients μ and λ if the fluid is compressible) for the mixture corresponding to an appropriate weighting of the μ_i (or the μ_i and λ_i): the definition (the choice) of these coefficients necessitates undertaking some modeling work. Once the coefficients μ, λ (or μ alone) are defined, we write the Navier–Stokes or the Euler equations of the mixture treated as a single fluid.

Similarly, we will be able to write the energy or the temperature equation for each fluid. If we consider the mixture as a single fluid, we will write the energy equation in a form similar to Eq. (6.5) of Chapter 6 or the temperature equation in a form similar to Eq. (6.6) of Chapter 6. In this last case, we will need to define the coefficients C_V and κ of the mixture from the coefficients $C_{V,i}$, and κ_i of each species; it is customary to set

$$C_V = \sum_{i=1}^{N} C_{V,i},$$

whereas additional modeling work will be necessary for κ.

The quantity

$$Y_i = \frac{\rho_i}{\rho}, \quad i = 1, \ldots, N, \tag{11.4}$$

which always remains between 0 and 1, is called the mass fraction of the fluid

---

[1] The interested reader can find a detailed study of the equations for mixtures, taking thermodynamics into account, in Appendices C and D of Williams (1985).

or species i, and the sum of the Y_i is equal to one:

$$
\begin{cases}
0 \leq Y_i \leq 1, \quad i = 1, \ldots, N, \\
\displaystyle\sum_{i=1}^{N} Y_i = 1.
\end{cases}
\tag{11.5}
$$

The most common model of a mixture of N chemical species, as we undertake it here, consists of writing, on the one hand, the equations describing the evolution of each mass fraction Y_i and, on the other hand, writing the fluid equations of the mixture considered as a single fluid with density ρ, pressure p, temperature T, velocity u, and viscosity μ. We also consider, in a second step, the mixture as an homogeneous incompressible fluid, and thus div $u = 0$ and $\rho = \rho_0$ is constant.

11.2. Equations of chemical kinetics

We study the mixture of N chemical species considered, as indicated above, as a single fluid. Our first objective is to write the equations governing the evolution of the mass fractions $Y_i, i = 1, \ldots, N$.

We call $\omega_i = \omega_i(x, t)$ the production rate per unit volume, at point x and at time t, of species i (its expression is given below) and assume that the total mass is constant, mass being only transfered from one species to another; this is expressed by the relation[2]

$$
\sum_{i=1}^{N} \omega_i = 0.
\tag{11.6}
$$

Let us consider a volume Ω_t of species i that we follow in its motion; the corresponding mass conservation for species i reads

$$
\frac{d}{dt} \int_{\Omega_t} \rho_i \, dx = \int_{\Omega_t} \omega_i \, dx.
$$

Thanks to Eq. (11.4) and to the formula for differentiation of a volume integral given in Chapter 1, this yields

$$
\int_{\Omega_t} \left[\frac{\partial}{\partial t}(\rho Y_i) + \mathrm{div}(\rho Y_i u_i) - \omega_i \right] dx = 0.
$$

Because Ω_t is an arbitrary open set, we deduce, by a reasoning already used

---

[2] Of course, $\omega_i > 0$ in the case of production and $\omega_i < 0$ in the case of removal.

several times, the equation

$$\frac{\partial}{\partial t}(\rho Y_i) + \text{div}(\rho Y_i u_i) = \omega_i, \quad 1 \leq i \leq N. \tag{11.7}$$

We note that Eqs. (11.7) are not all independent: indeed, by summation in i and taking Eqs. (11.3), (11.4), and (11.6) into account, we find

$$\frac{\partial \rho}{\partial t} + \text{div}(\rho u) = 0;$$

this is nothing else but the continuity equation for the mixture that will reappear below.

For the time being, we rewrite Eqs. (11.7) in the following way: We introduce the velocities

$$V_i = u_i - u, \quad i = 1, \ldots, N, \tag{11.8}$$

called the diffusion velocities of species i, and the fluxes F_i defined by

$$F_i = Y_i V_i, \quad i = 1, \ldots, N. \tag{11.9}$$

Taking into account Eqs. (11.3) and (11.8), we have

$$\sum_{i=1}^{N} F_i = 0. \tag{11.10}$$

Using the F_i, we can rewrite Eqs. (11.7) in the form

$$\frac{\partial}{\partial t}(\rho Y_i) + \text{div}(\rho Y_i u) + \text{div}(\rho F_i) = \omega_i, \quad 1 \leq i \leq N. \tag{11.11}$$

To complete the modeling of chemical kinetics, we need to specify the expression (functional form) of the ω_i and the F_i. We assume that the net production ω_i of species i is of the form

$$\begin{aligned} \omega_i &= \omega_i(T, Y_1, \ldots, Y_N) \\ &= \alpha_i(T, Y_1, \ldots, Y_N) - Y_i \beta_i(T, Y_1, \ldots, Y_N). \end{aligned} \tag{11.12}$$

Here, the functions α_i and β_i are defined for the physically acceptable values of T and of the Y_i, namely, for $T \geq 0$ and $0 \leq Y_i \leq 1, 1 \leq i \leq N$, and they are continuous positive and bounded on this domain. The general form of the functions α_i and β_i, stemming from chemical kinetics, is made precise in the next section.

Finally, the diffusive mass fluxes are given by the Fick diffusion law

$$F_i = V_i Y_i = -D_i \nabla Y_i, \tag{11.13}$$

where $D_i > 0$ is the diffusion coefficient of species i.

Owing to Eq. (11.13) and the ω_i being given by Eq. (11.12), Eqs. (11.11) become

$$\frac{\partial}{\partial t}(\rho Y_i) + \mathrm{div}(\rho Y_i u) - D_i \Delta Y_i = \omega_i, \quad 1 \le i \le N. \qquad (11.14)$$

11.3. The equations of combustion

We now combine the equations of chemical kinetics introduced in the previous section with the equations of fluid mechanics to obtain the equations of combustion. Next, by adding appropriate boundary conditions to these equations, we obtain a problem of combustion constituting a model of developed premixed laminar flame.

As indicated at the end of Section 11.1, in the model that we consider, we treat the mixture as a single fluid, and the equations of chemical kinetics describe on the other hand the evolution of the mass fractions of the various species.

To simplify, we now assume that the mixture is incompressible and homogeneous; the case of a compressible fluid is described in Section 11.4. In the Eulerian description, the mixture fills the domain Ω_t at time t, and the unknown functions, defined for $x \in \Omega_t$ and $t > 0$, are the velocity u of the mixture, its pressure p, its temperature T, and the mass densities $Y_1, \dots, Y_N$ of the N chemical species; the density $\rho = \rho_0$ is a positive constant.

We thus write the Navier–Stokes equations of incompressible fluids, and the temperature equation for the mixture, in a form similar to Eq. (6.6) of Chapter 6: the heat source r on the right-hand side is due here to the exothermic chemical reactions, and it is taken equal to

$$r = -\sum_{i=1}^{N} h_i \omega_i, \qquad (11.15)$$

where h_i is the enthalpy of species i with the following property being satisfied:

$$\sum_{i=1}^{N} h_i \omega_i \le 0$$

for every $T \ge 0$ and every $Y_1, \dots, Y_N, 0 \le Y_i \le 1$. We finally write Eq. (11.14) in which $\rho = \rho_0$ and $\mathrm{div}\, u = 0$. Hence, we find

$$\rho_0 \left[\frac{\partial u}{\partial t} + (u.\nabla)u \right] - \mu \Delta u + \mathrm{grad}\, p = f, \qquad (11.16)$$

$$\mathrm{div}\, u = 0, \qquad (11.17)$$

$$\rho_0 c_V \left[\frac{\partial T}{\partial t} + (u.\nabla)T \right] - \kappa \Delta T = -\sum_{i=1}^{N} h_i \omega_i, \qquad (11.18)$$

$$\rho_0 \left[\frac{\partial Y_i}{\partial t} + (u.\nabla)Y_i \right] - D_i \Delta Y_i = \omega_i, \quad 1 \le i \le N. \qquad (11.19)$$

The ω_i are given by the expressions of Eq. (11.12) that we do not reproduce.

These equations are the basic equations for a combustion problem. As was stated in the introduction, these equations are used for the study of a variety of phenomena, including the propagation of chemical pollutants in the air or in water,[3] or combustion in cars, planes, or rocket engines. Depending on the level of refinement desirable for the model, the number N of chemical species may vary from a few units to several hundreds, or even a thousand. In Section 11.4, some even more complicated models will be introduced, whereas Section 11.5 will be devoted to simplified versions of these equations.

As usual, Eqs. (11.16) to (11.19) must be supplemented by appropriate boundary conditions: we now derive these boundary conditions in the case of a model of premixed laminar flame.

Model of premixed laminar flame

In this problem, the domain $\Omega_t = \Omega$ filled by the mixture is a cylinder of axis parallel to Ox_3, of height h, and section $\mathcal{O}$, and thus $\Omega = \mathcal{O} \times (0, h)$. The chemical mixture enters this channel by the bottom of the cylinder and exits the channel by the top, the section $x_3 = h$ of the cylinder. We also assume that the lateral boundary of the channel is thermally insulated.

A set of physically reasonable boundary conditions for this problem is given below; note that these boundary conditions are not all absolutely un-questionable. A certain flexibility is actually left to the modeler, who will decide according to the specific aspects of the problem (or experiment) and the phenomena most important in the investigation. Here is the proposed set of boundary conditions:

• *Base of the channel, $x_3 = 0$.*

$$u_1 = u_2 = 0, \quad u_3(x_1, x_2) = U(x_1, x_2),$$
$$T = T_u \ge 0,$$
$$Y_i = Y_{i,u} > 0, \quad i = 1, \dots, N, \quad \sum_{i=1}^{N} Y_{i,u} = 1,$$

the function U and the constants $Y_{i,u}$, T_u being given.

---

[3] For this last point, see also Chapter 12.

178   *Mathematical Modeling in Continuum Mechanics*

- *Lateral boundary of the channel, $x \in \partial\mathcal{O}, 0 < x_3 < h$.*

$$u = 0, \quad \frac{\partial T}{\partial n} = 0,$$

$$\frac{\partial Y_i}{\partial n} = 0, \quad 1 \leq i \leq N.$$

- *Exit of the channel, $x_3 = h$.*

$$u_i = 0, \quad i = 1, 2, \quad \frac{\partial u_3}{\partial x_3} = 0,$$

$$\frac{\partial T}{\partial x_3} = 0, \quad \frac{\partial Y_i}{\partial x_3} = 0, \quad 1 \leq i \leq N.$$

- *Initial conditions.*

Finally, we are given initial conditions for u, T, and the Y_i; typically

$$u(x, 0) = u_0(x),$$
$$T(x, 0) = T_0(x) \geq 0,$$
$$Y_i(x, 0) = Y_{i,0}(x), \quad Y_{i,0}(x) \geq 0, \quad \sum_{i=1}^{N} Y_{i,0}(x) = 1.$$

Problems like this one are the object of many current theoretical and numerical studies in combustion theory.

11.4. Stefan–Maxwell equations

We describe in this section a more general and more complicated model of combustion by using the so-called Stefan–Maxwell equations of chemical kinetics. We also consider a compressible fluid, which is necessary for explosion and detonation phenomena.

In this model, we first need to introduce other chemical quantities. We denote by $c_i = c_i(x, t), i = 1, \ldots, N$, the molar concentration of chemical species i at x at time t; that is to say, the number of moles of chemical species i per unit volume at x at time t; the total molar concentration c is given by

$$c = \sum_{i=1}^{N} c_i. \tag{11.20}$$

If M_i denotes the molar mass of species i, then

$$c_i = \frac{\rho_i}{M_i}. \tag{11.21}$$

On the other hand, besides the mass fraction Y_i of species i, we introduce its molar fraction

$$X_i = \frac{c_i}{c}. \tag{11.22}$$

It is easy to see that, as in Eq. (11.5)

$$\begin{cases} 0 \le X_i \le 1, & i = 1, \dots, N, \\ \displaystyle\sum_{i=1}^{N} X_i = 1. \end{cases} \tag{11.23}$$

We have

$$\begin{aligned} X_i &= \frac{\rho_i}{c M_i} = \frac{\rho_i}{\rho} \frac{\rho}{c M_i} = \frac{Y_i}{M_i} \frac{\rho}{c} \\ &= \frac{Y_i/M_i}{c/\rho} = \frac{Y_i/M_i}{\sum_{j=1}^{N} (\rho_j/(M_j \rho))}. \end{aligned}$$

Thus the X_i and Y_i are related by the following equations, and we can easily express one set of quantities in terms of the other:

$$\begin{cases} X_i = \dfrac{1}{Y_M} \dfrac{Y_i}{M_i}, & Y_i = \dfrac{X_i M_i}{X_M}, \\ Y_M = \displaystyle\sum_{j=1}^{N} \dfrac{Y_j}{M_j}, & X_M = \displaystyle\sum_{j=1}^{N} X_j M_j, \quad X_M Y_M = 1. \end{cases} \tag{11.24}$$

We now return to Eqs. (11.12) and (11.13) for the ω_i and the F_i.

The net production ω_i of species i is given by phenomenological relations of chemical kinetics related to the Arrhenius law. A more precise form of Eq. (11.12) is the following:

$$\begin{aligned} \omega_i = M_i \sum_{k=1}^{N} (v_{i,k}'' - v_{i,k}') B_k T^{\alpha_k} \exp(-E_k/(RT)) \\ \times \prod_{j=1}^{N} \left(\frac{X_{j,p}}{RT} \right)^{v_{j,k}'}, \quad i = 1, \dots, N. \end{aligned} \tag{11.25}$$

We recall and complete the notations: M_i is the molar mass of species i; R the constant of perfect gases; T the temperature; p the pressure; B_k, α_k, E_k constants, $B_k \ge 0, -1 < \alpha_k \le 2$; E_k the activation energy; the constants $v_{i,k}'$ and $v_{i,k}''$ are various constants (stoichiometric coefficients) related to the interactions of species i and k; and finally we recall that X_j is the molar fraction of species (of fluid) j.

In the model considered here, the Stefan–Maxwell laws below replace Eq. (11.13) for the determination of the mass diffusive fluxes:

$$\nabla X_i = \sum_{j=1}^{N} \frac{X_i X_j}{D_{ij}} (V_j - V_i) + (Y_i - X_i) \frac{\nabla p}{p} + \frac{\rho}{p} \sum_{j=1}^{N} Y_i Y_j (f_i - f_j)$$

$$+ \sum_{j=1}^{N} \frac{X_i X_j}{\rho D_{ij}} \left(\frac{D_{T,j}}{Y_j} - \frac{D_{T,i}}{Y_i} \right) \left(\frac{\nabla T}{T} \right). \tag{11.26}$$

Here D_{ij} is the binary diffusion coefficient for species i and j, $D_{T,i}$ the thermal diffusion coefficient for the species, f_i is the volume density of external forces applied to species i, and all other quantities have already been defined.

Under proper assumptions, Eq. (11.26) is simplified by only keeping the first term on the right-hand side, which gives

$$\nabla X_i = \sum_{j=1}^{N} \frac{X_i X_j}{D_{ij}} (V_j - V_i), \quad 1 \le i \le N. \tag{11.27}$$

At a point (x, t), where the $Y_i(x, t)$ (or $X_i(x, t)$) are all nonzero (strictly positive), we determine the F_i as follows: Eqs. (11.27) (or (11.26)) are considered as a system of linear equations in the V_i; these equations are not independent but are compatible (note that $\sum_{i=1}^{N} \nabla X_i = 0$ because $\sum_{i=1}^{N} X_i = 1$). We can show that, by supplementing them with Eq. (11.10) rewritten in the form

$$\sum_{i=1}^{N} Y_i V_i = 0, \tag{11.28}$$

we obtain a linear system that uniquely defines the $V_i = V_i(x, t)$ and then the $F_i = V_i Y_i$. We can also show that, at a point (x, t) where some of the Y_j vanish, Eqs. (11.27) or (11.26) expressed in terms of the F_j allow, with Eqs. (11.10) and (11.24), to express the F_i, in a unique way, as linear functions of the ∇Y_j with coefficients that are rational functions of the Y_k and functions as well of the other quantities.

Substituting these expressions for F_i in Eq. (11.11), we obtain, for the Y_i, nonlinear partial differential equations of a great level of complexity. Nevertheless, such equations are taken into account in numerous studies.

Let us now summarize what the combustion model based on the Stefan–Maxwell equations consists of, without rewriting the equations:

1. The unknowns are u and p, ρ for a compressible fluid, T (temperature) or e (internal energy), and the $Y_i, i = 1, \ldots, N$.
2. The equations are

- The Navier–Stokes and mass conservation equations for an incompressible or a compressible fluid, depending on the type of phenomenon under study.
- The temperature equation (Eq. (11.18)) for an incompressible fluid or, for a compressible fluid, the equation of state of the fluid, and the energy equation (Eq. (6.5), Chapter 6). In this last case, the usual expression for q is the following:

$$q = -\kappa \nabla T + \rho \sum_{i=1}^{N} h_i Y_i V_i$$

$$+ RT \sum_{i,j=1}^{N} \frac{X_j D_{T,i}}{M_i D_{ij}} (V_i - V_j) + q_R, \qquad (11.29)$$

where q_R is a radiant heat flux (which is provided) and h_i the mean enthalpy per unit of mass for species i, which is assumed constant for simplicity; all other quantities have already been described.

- Equations (11.11) for the Y_i with the expression of the fluxes resulting from previous considerations.

Additional and substantial developments concerning combustion may be found in Williams (1985).

11.5. A simplified problem: the two-species model

We assume, for simplicity, that the fluid is incompressible, and we consider the case in which only two chemical species are present. The equations are thus (11.16) to (11.19) with $N = 2$. Equation (11.19) is not independent because $Y_1 + Y_2 \equiv 1$. Actually, an interesting case is that in which the ratio

$$Le = D_2/D_1,$$

called the Lewis number, is equal to 1 ($D_1 = D_2$). It is then sufficient to consider one single equation (11.19) such as that of Y_1 denoted by Y: if the velocity u of the fluid is given, it only remains to consider two equations for the temperature T and the mass fraction Y. Hence, assuming, for simplicity, that $D_1 = 1$ and $\rho_0 = 1$, we have

$$\frac{\partial T}{\partial t} + (u.\nabla)T - \kappa \Delta T = -h\omega, \qquad (11.30)$$

$$\frac{\partial Y}{\partial t} + (u.\nabla)Y - D\Delta Y = \omega. \qquad (11.31)$$

Here $\omega = \omega_1$ takes the form (related to Eq. (11.25)) of

$$\omega = BY(1 - Y)T \exp(-E/RT),$$

and $h = h_1 - h_2$ because $\omega_2 = -\omega_1 = -\omega$ and

$$h_1\omega_1 + h_2\omega_2 = (h_1 - h_2)\omega.$$

One of the numerous problems raised by Eqs. (11.30) and (11.31) is the search of traveling waves and the study of their stability in connection with the study of the propagation of the flame; one then looks for solutions in the whole space $\mathbb{R}^3_x \times \mathbb{R}_t$ of the form

$$T(x, t) = T(x - Ut), \quad Y(x, t) = \mathcal{Y}(x - Ut),$$

where the constant vector $U = (U_1, U_2, U_3)$, which is the propagation velocity of the front of the flame, is unknown as well as T and Y. We then infer, for the functions $T = T(\xi)$ and $\mathcal{Y} = \mathcal{Y}(\xi)$ ($\xi \in \mathbb{R}^3$), the equations

$$[(u - U) \cdot \nabla_\xi]T - \kappa \Delta_\xi T = -h\omega, \tag{11.32}$$

$$[(u - U) \cdot \nabla_\xi]\mathcal{Y} - D\Delta_\xi \mathcal{Y} = \omega. \tag{11.33}$$

These equations are already very complex and very rich, even in space dimension 1; that is to say, when T and Y only depend on x_1 (or, similarly, when T and $\mathcal{Y}$ only depend on ξ_1). A large part of the book of A. I. Volpert, V. A. Volpert, and V. A. Volpert (1994) is devoted to their study.

Exercises[1]

1. **Semenov's theory of heat explosion.**

   If we assume that the gas in the reactor is well mixed and that the temperature is homogeneous in space, then at the initial stage of the reaction where the consumption of the reactants can be neglected, the temperature evolution can be described by the equation

   $$\frac{dT}{dt} = qke^{-E/RT} - \sigma(T - T_0), T(0) = T_0.$$

   This is the so-called Semenov's model of heat explosion. The Frank-Kamenteskii transformation

   $$\frac{E}{RT} \approx \frac{E}{RT_0} - \frac{E}{RT_0^2}(T - T_0)$$

[1] These exercises are intended for more advanced readers. We thank V. A. Volpert for providing us with them.

allows its reduction to the model

$$\frac{d\theta}{dt} = \beta e^{Z\theta} - \sigma\theta, \theta(0) = 0, \tag{11.34}$$

where $\theta = (T - T_0)/q$ is the dimensionless temperature, $Z = qE/(RT_0^2)$ is the Zeldovich number, β and σ are positive parameters, β characterizes the heat production due to the reaction, and σ the heat loss through the wall of the reactor.

a) Let $\alpha = \sigma/\beta$. Show that there exists a critical value α_* of the parameter α such that Eq. (11.34) has two stationary solutions for $\alpha > \alpha_*$ and it does not have any solution for $\alpha < \alpha_*$.

b) In the case where there are two solutions, θ_1 and $\theta_2, \theta_1 < \theta_2$, verify that the solution θ_1 is stable with respect to the nonstationary problem (11.34), and the solution θ_2 is unstable.

c) What happens with the solution of the nonstationary problem (11.34) if $\alpha < \alpha_*$? What is the physical interpretation of this solution?

2. **Frank-Kamenteskii's theory of heat explosion.**
   If we do not suppose that the temperature is homogeneous in space, then instead of Eq. (11.34) from Exercise 1, we have the Frank-Kamenteskii model of heat explosion. In the one-dimensional case, it has the form

$$\frac{\partial\theta}{\partial t} = \frac{\partial^2\theta}{\partial x^2} + \beta e^{Z\theta}, 0 < x < 2L, \theta(0) = \theta(2L) = 0. \tag{11.35}$$

Assuming that the stationary solution is symmetric with respect to the middle of the interval, we obtain the following problem

$$\theta'' + \beta e^{Z\theta} = 0, 0 < x < L, \theta(0) = \theta'(L) = 0. \tag{11.36}$$

a) Reduce (11.36) to a problem on the interval $0 \le x \le 1$ and find its explicit solution.

b) Show that there exists a critical value α_* of the parameter $\alpha = L^2\beta$ such that there exist two solutions for $\alpha < \alpha_*$, and there are no solutions for $\alpha > \alpha_*$.

c) Consider the value $\theta_m = \theta(L)$ as a parameter and show all solutions on the parameter plane (α, θ_m).

3. **Existence of gaseous flames.**
   Consider the system of two reaction-diffusion equations

$$\kappa T'' + cT' + qK(T)(1 - \alpha) = 0,$$
$$d\alpha'' + c\alpha' + K(T)(1 - \alpha) = 0,$$

describing stationary one-dimensional flames in the moving coordinate frame. Here T is the temperature, α the depth of conversion, the kinetic function $(1 - \alpha)$ corresponds to the reaction of the first order.

a) Show that if the Lewis number $Le = \kappa/d$ equals 1, then this system of two equations can be reduced to the single equation

$$\kappa T'' + cT' + F(T) = 0. \tag{11.37}$$

We will look for solutions of this problem having limits $T(+\infty) = T_0$, $T(-\infty) = T_a$, where T_a is a zero of the function $F(T)$. To avoid the so-called "cold boundary difficulty", we assume that $F(T) = 0$ for $T \leq T_*$ for some $T_* < T_a$. The function $F(T)$ is positive on the interval $T_* < T < T_a$. To find a solution of this problem means to find a value of the parameter c (the wave speed) such that there exists a solution of Eq. (11.37) with the given limits at infinity.

b) Reduce Eq. (11.37) to the system of two first order equations

$$T' = p, \ p' = \frac{1}{\kappa}(-cp - F(T)), \tag{11.38}$$

and find its stationary points.

c) Show that the stationary point $T = T_a$, $p = 0$ is a saddle. Find the corresponding eigenvectors and prove that there exists a trajectory leaving the stationary point into the quarter plane $T < T_a$, $p < 0$.

d) Show that this trajectory "descends" when c is decreased.

e) Show that there exists a value of c such that system (11.38) has a trajectory connecting two stationary points, $T = T_a$, $p = 0$, and $T = T_0$, $p = 0$.

CHAPTER TWELVE

Equations of the atmosphere and of the ocean

Weather forecasting in meteorology is based on two complementary approaches:

1. Accumulating a very large amount of data and interpreting these statistically (wind velocity, humidity, and temperature measured over very large intervals of time and over large regions of the earth). Here, the mathematical techniques that are necessary for the assimilation and the exploitation of these data are those of statistics and of stochastic processes.

2. Modeling of atmospheric phenomena by ordinary and partial differential equations and the numerical simulation of these equations. From the computational point of view, one obtains, by discretization of such partial differential equations, systems of equations with millions, even billions, of unknowns whose numerical resolution could saturate the most powerful computers currently available. The memory size and computational speed capacities needed for such calculations are very high even in the context of the teraflop (10^{12} operations per second), which is the next step in the foreseeable future.

In this chapter, we are interested in the second approach, and we intend to give a very modest description of the most fundamental equations widely accepted in the field. The atmosphere is a fluid whose state is described by the velocity vector, the temperature, the density, and the pressure at every point in the Eulerian description. The equations are essentially variants of the Navier–Stokes and of the temperature equations that take into account the particular aspects of the problem.

Section 12.1 is devoted to various preliminaries. Section 12.2 gives the fundamental equations of the atmosphere, which are called primitive equations of the atmosphere (PEs). In Section 12.3 we proceed to a similar study of the ocean; apart from the intrinsic interest of studying the motion of oceans, we know at present (considering, for example, the El Niño phenomenon) that the ocean plays a fundamental role in the study of meteorology through the thermal and dynamic exchanges between the ocean and the atmosphere.

Finally, Section 12.4 is a brief introduction to even more complex (and important) phenomena: the study of the equations governing the concentration in the atmosphere of water and gases and rare gases such as those related to pollution, namely ozone, carbon dioxide, chlorinated fluorocarbons (CFCs), nitrous oxides, sulfides, and so forth. Their concentrations are governed by the equations of chemical kinetics, which are very close to the equations considered for combustion in the previous chapter; a brief description of a relevant model is given in Section 12.4.

Of course, these problems give rise to very complex studies that rally important teams in large laboratories all over the world owing to the human and economic interests at stake. One should thus not lose sight of the fact that the study presented here is extremely superficial. It would be misleading to believe, after reading this chapter, that geophysical fluid dynamics consists just of the juxtaposition of well-known equations. This field has its own rationale and raises its own problems, which are not present at all in this chapter. We thought, nevertheless, that it would be useful to show how these studies are related to the preceding chapters.

12.1. Preliminaries

The fundamental law of dynamics on a frame attached to the earth

In the study of meteorological phenomena, one assumes that the earth is spherical (of radius a); that its center is fixed, as well as the line joining the poles; and that the earth rotates uniformly around the line of the poles with respect to a fixed (Galilean) frame. Except in the study of specific problems such as the tides, which are very much influenced by the moon and the sun, the earth is generally assumed isolated and thus is not subjected to any external force.

Hence, in a frame related to the earth, the only forces are those resulting from the Coriolis acceleration and the transport acceleration (see Chapter 2, Section 2.5); we now want to make these forces explicit. We denote by u the velocity of the wind with respect to the earth; hence, according to Eq. (2.16) of Chapter 2, the mass density of the Coriolis force is

$$-2\omega \wedge u.$$

The same formula gives the transport acceleration, which reads, because $\dot{\omega} = 0$,

$$\gamma_e = \omega \wedge (\omega \wedge OM).$$

We easily check that (with $OM = x$)

$$\gamma_e = \frac{1}{2} \mathrm{grad} |\omega \wedge x|^2.$$

It is then possible to add this force to the pressure, and we are led to consider the total (or augmented) pressure:

$$P = p + \frac{1}{2} |\omega \wedge x|^2, \tag{12.1}$$

($0x_3$ upward vertical). In the sequel, the pressure P will be denoted by p.

Static equations of the atmosphere

We have derived in Chapter 7 the simplified equations of the atmosphere considered as a static barotropic fluid: p is function of the altitude x_3 only (and of the time t) and

$$\frac{\partial p}{\partial x_3} = -\rho g. \tag{12.2}$$

With an equation of state of barotropic fluid

$$p = \varphi(\rho), \tag{12.3}$$

$\varphi' > 0$, $\varphi'' \geq 0$, we can express both p and ρ as functions of x_3. For instance, if it is assumed that the air is a perfect fluid with constant specific heats, $p = k\rho$ and then

$$p = p_0 \exp\left(-\frac{1}{k} g x_3\right); \tag{12.4}$$

$x_3 = 0$ corresponding to the surface of the earth, and p_0 and ρ_0 are, respectively, the pressure and density at the surface of the earth.

The differential operators on the sphere

Our aim in the following paragraphs is to write the equations of the atmosphere and ocean around the whole globe. It is thus natural to introduce the spherical coordinates (r, θ, φ), where $0 \leq \theta \leq \pi$ is here the colatitude of the earth, $0 \leq \varphi \leq 2\pi$ is the latitude, and r is the distance to the center of the earth. We further introduce the variable $z = r - a$, where a denotes the radius of the earth, that measures the altitude with respect to the sea level. We also introduce the corresponding local orthonormal system e_r, e_θ, e_φ, and, for a vector v, we write $v = v_r e_r + v_\theta e_\theta + v_\varphi e_\varphi$.

The height of the atmosphere (20 to 100 km, depending on its definition) is small when compared with its horizontal dimension and the radius of the earth ($\simeq$6000 km). Consequently, and this is confirmed by meteorological observations, the vertical scales of the terrestrial atmosphere are small when compared with the horizontal scales. A first consequence of this is that we will neglect the variations of the vertical variable z and will replace, wherever legitimate, the variable r by the earth's radius a. In particular, we replace the operator of total differentiation

$$\frac{D}{Dt} = \frac{\partial}{\partial t} + \frac{v_\theta}{r}\frac{\partial}{\partial \theta} + \frac{v_\varphi}{r \sin\theta}\frac{\partial}{\partial \varphi} + \frac{\partial}{\partial z}$$

by the operator

$$\frac{d}{dt} = \frac{\partial}{\partial t} + \frac{v_\theta}{a}\frac{\partial}{\partial \theta} + \frac{v_\varphi}{a \sin\theta}\frac{\partial}{\partial \varphi} + \frac{\partial}{\partial z}.$$

This simplification consists essentially of replacing the domain actually filled by the atmosphere by the product of the sphere S^2 (representing the surface of the earth) and an interval for the variable z.

We will give, in the appendix at the end of this chapter, the expression in spherical coordinates of the differential operators that we use in this chapter.

12.2. Primitive equations of the atmosphere

The general equations of the atmosphere are those of a compressible fluid subjected to the Coriolis and gravitational forces, to which we add the temperature equation (a consequence of the first law of thermodynamics, see Chapter 6). The resulting equations are particularly rich and complex, and thus, they seem to defy, for the time being and in the foreseeable future, every possibility of analysis. It is thus necessary to make simplifying hypotheses.

Because the vertical scales of the atmosphere are small with respect to the horizontal scales, a dimensional analysis that corroborates meteorological observations and historical data shows that the large scales satisfy the hydrostatic equation (see Eq. (12.2) and below). This equation, as a result of its good precision, is acknowledged as fundamental and is considered as a starting point for the study of large classes of atmospheric phenomena. Furthermore, it furnishes an equation relating the pressure to the density. The general equations of the atmosphere then reduce to a set of equations called primitive equations of the atmosphere (PEs).

In a moving frame rotating with the earth, the fundamental law of dynamics for a mobile frame (see Section 12.1) furnishes the equation of conservation of momentum (in vector form):

$$\frac{DV_3}{Dt} = \text{(total) pressure gradient} + \text{gravity forces}$$
$$+ \text{Coriolis forces} + \text{dissipative forces};$$

that is,

$$\frac{DV_3}{Dt} = -\frac{1}{\rho}\text{grad}_3\, p + G - 2\omega \wedge V_3 + D, \tag{12.5}$$

where V_3 denotes the velocity. The general equations of the atmosphere are thus made of the momentum equation (Eq. (12.5)); the continuity equation

$$\frac{D\rho}{Dt} + \rho\, \text{div}_3 V_3 = 0; \tag{12.6}$$

the first law of thermodynamics,

$$c_p \frac{DT}{Dt} - \frac{RT}{\rho}\frac{Dp}{Dt} = \frac{DQ}{Dt}; \tag{12.7}$$

and the equation of state

$$p = R\rho T. \tag{12.8}$$

In these equations, $G = -ge_3$ represents the gravity forces, D represents viscosity terms that we will make precise in Section 12.4 (see also, for the interested reader, Lions, Temam, and Wang (1993)), DQ/Dt is the heat flux by unit volume and in a unit time interval (mainly through solar heating), and ω is the angular velocity of the earth. Moreover, grad_3 and div_3 denote, respectively, the gradient and the divergence in three dimensions.

As indicated above, we are led, in order to study these equations, to make simplifying assumptions. In particular, we make the (fundamental) hydrostatic assumption, which furnishes Eq. (12.2) (called the hydrostatic equation) rewritten in the form

$$\frac{\partial p}{\partial r} = -\rho g, \tag{12.9}$$

which relates the variables p and ρ. This equation replaces the equation of conservation of momentum (Eq. (12.5)) projected on the vertical, and it represents an approximate form of that equation (owing to the small height of the atmosphere). Also, as noted, we neglect the variations of the vertical variable z, which amounts to replacing r by the earth's radius a when appropriate (and in particular in the differential operators). Finally, we decompose V_3 into the form

$$V_3 = v + w, \tag{12.10}$$

where v denotes the horizontal velocity and w the vertical velocity. In view of the preceding simplifications, we obtain the following system of equations, which are called the primitive equations of the atmosphere (we refer the reader to the chapter's appendix for the definition of the differential operators appearing in the equations):

$$\frac{\partial v}{\partial t} + \nabla_v v + w \frac{\partial v}{\partial z} + \frac{1}{\rho} \text{grad } p + 2\omega \cos\theta \, \vec{k} \wedge v = D, \tag{12.11}$$

$$\frac{\partial p}{\partial z} = -\rho g, \tag{12.12}$$

$$\frac{d\rho}{dt} + \rho \left(\text{div } v + \frac{\partial w}{\partial z} \right) = 0, \tag{12.13}$$

$$c_p \frac{dT}{dt} - \frac{RT}{p} \frac{dp}{dt} = \frac{dQ}{dt}, \tag{12.14}$$

$$p = R\rho T, \tag{12.15}$$

where $\vec{k}$ denotes the unit vertical vector pointing upward.[1] Here, we have written the primitive equations in the system of coordinates (t, θ, φ, z).

Remark 12.1: In the primitive equations (12.11) to (12.15), we have not taken into account the humidity of the atmosphere. This, and other atmospheric components, will be considered in Section 12.4.

A straightforward consequence of the hydrostatic equation (Eq. (12.12)) is that the (total) pressure p is a strictly decreasing function of the vertical variable z. It is thus possible, by an appropriate change of variables, to rewrite the equations in the system of coordinates (t, θ, φ, p), called the pressure coordinates. We will further assume that the pressure p is restricted to an interval of the form $[p_0, P]$, where P is the pressure at the surface[2] of the earth and $p_0 > 0$ is the pressure at some high isobar in the atmosphere. Strictly speaking, the region between the isobar $p = P$ and the earth is not taken into

---

[1] The definition of ∇_v is given at the end of the chapter's appendix. It is the proper mathematical (geometrical) definition of the term $(v.\nabla)$ used in Cartesian coordinates.
[2] An average pressure at some small altitude above the earth.

account, as well as the region $p < p_0$; the latter is, however, made of rarefied and ionized gases (plasmas), and the physics is different.

In the pressure system of variables (t, θ, φ, p), the altitude z is an unknown function of t, θ, φ, and p. It is customary to introduce the function

$$\Phi = gz, \tag{12.16}$$

which is called the geopotential; on the other hand, we set

$$\varpi = \frac{dp}{dt}. \tag{12.17}$$

We then obtain, thanks to this change of coordinates, the new system

$$\frac{\partial v}{\partial t} + \nabla_v v + \varpi \frac{\partial v}{\partial p} + 2\omega \cos \theta \, k \wedge v + \text{grad } \Phi = D, \tag{12.18}$$

$$\frac{\partial \Phi}{\partial p} + \frac{RT}{p} = 0, \tag{12.19}$$

$$\text{div } v + \frac{\partial \varpi}{\partial p} = 0, \tag{12.20}$$

$$\frac{R^2}{c^2} \left(\frac{\partial T}{\partial t} + \nabla_v T + \omega \frac{\partial T}{\partial p} \right) - \frac{R\omega}{p} - \frac{R^2}{c^2} \omega \frac{\partial \bar{T}(p_a)}{\partial p_a}$$

$$= \mu_T \Delta T + \nu_T \frac{\partial}{\partial p} \left[\left(\frac{gp}{R\bar{T}} \right)^2 \frac{\partial T}{\partial p} \right] + Q_T. \tag{12.21}$$

In Eq. (12.21), the function $\bar{T} \in C^\infty([p_0, P])$ is known (given). It satisfies a relation of the form

$$c^2 = R \left(\frac{R\bar{T}}{cp} - p \frac{\partial \bar{T}}{\partial p} \right) = \text{const.} \tag{12.22}$$

The temperature $\bar{T}(p)$ represents the mean climatic temperature on the isobar p; it is determined by meteorological measurements. Moreover, we have

$$\mu_T = \frac{R^2}{c^2} \frac{\tilde{\mu}_T}{c_p}, \quad \nu_T = \frac{R^2}{c^2} \frac{\tilde{\nu}_T}{c_p}, \tag{12.23}$$

$$Q_T = \frac{R^2}{c^2} \frac{\mathcal{E}}{c_p}, \tag{12.24}$$

where $\tilde{\mu}_T$ and $\tilde{\nu}_T$ are diffusion coefficients and $\mathcal{E}$ is the diabatic warming of the air. We refrain from expressing here the right-hand side D of Eq. (12.18), but it is similar to the right-hand side of Eq. (12.21).

Equations (12.18) to (12.24) constitute the primitive equations of the atmosphere in pressure variables. They must be supplemented by boundary conditions on the isobars $p = p_0$ and $p = P$, for which we refer the interested reader to the books and articles cited in the references.

12.3. Primitive equations of the ocean

It is generally assumed that the ocean is a slightly compressible fluid subjected to the Coriolis and gravity forces for which we make the Boussinesq approximation described in Chapter 9. The general equations of the ocean consist thus of the momentum and continuity equations:

$$\rho_0 \frac{DV_3}{Dt} + 2\rho_0\, \omega \vec{k} \wedge V_3 + \text{grad}_3\, p + \rho g = D, \qquad (12.25)$$

$$\text{div}_3\, V_3 = 0; \qquad (12.26)$$

of the equation for the temperature

$$\frac{DT}{Dt} = Q_T + \text{diffusion}; \qquad (12.27)$$

of the equation for the salinity (concentration of salt):

$$\frac{DS}{DT} = \text{diffusion}; \qquad (12.28)$$

and of the following equation of state expressing that the density is a linear function of temperature and salinity:

$$\rho = \rho_0 + \alpha(T - T_0) - \beta(S - S_0), \qquad (12.29)$$

where α and β are constants and ρ_0, T_0, S_0 are, respectively, reference density, temperature, and salinity.

As for the atmosphere, we note that the vertical scales of the ocean are small in comparison with the horizontal scales. We are thus led to make the hydrostatic approximation[3]

$$\frac{\partial p}{\partial r} = -\rho g. \qquad (12.30)$$

As in the atmosphere, we neglect the variations of the variable $z = r - a$ in the ocean and, wherever this is legitimate, we replace the variable r by the earth's radius a.

In view of these simplifications, and writing again $V_3 = v + w$, where v is the horizontal velocity of the water and w the vertical velocity, we deduce

---

[3] Note that, in the Boussinesq approximation (see Chapter 9), ρ is everywhere constant in the equations, $\rho = \rho_0$, except in the gravity term, which thus produces the so-called buoyancy force.

from the general equations above the following system of equations:

$$\frac{\partial v}{\partial t} + \nabla_v v + w \frac{\partial v}{\partial z} + \frac{1}{\rho_0} \nabla p + 2\omega \cos \theta \, \vec{k} \wedge \vec{v} = D, \qquad (12.31)$$

$$\frac{\partial p}{\partial z} = -\rho g, \qquad (12.32)$$

$$\operatorname{div} v + \frac{\partial w}{\partial z} = 0, \qquad (12.33)$$

$$\frac{\partial T}{\partial t} + \nabla_v T + w \frac{\partial T}{\partial z} - \mu_T \Delta T - \nu_T \frac{\partial^2 T}{\partial z^2} = Q_T, \qquad (12.34)$$

$$\frac{\partial S}{\partial t} + \nabla_v S + w \frac{\partial S}{\partial z} - \mu_S \Delta S - \nu_S \frac{\partial^2 S}{\partial z^2} = 0, \qquad (12.35)$$

$$\rho = \rho_0 + \alpha(T - T_0) - \beta(S - S_0), \qquad (12.36)$$

where μ_T, ν_T, μ_S, and ν_S are diffusion coefficients. Equations (12.31) to (12.36) are called the primitive equations of the ocean. They must be supplemented by boundary conditions (at the surface, the bottom, and on the lateral boundaries of the ocean) for which we refer the interested reader to the books and articles cited in the references. The primitive equations of the ocean constitute the fundamental equations of the ocean.

12.4. Chemistry of the atmosphere and the ocean

The study of the propagation of pollutants in the atmosphere or in the oceans is based on the equations that we will now describe (or on some simplified version of these equations). These equations result essentially from the coupling of the primitive equations of the atmosphere and of the ocean with the equations of chemical kinetics quite similar to those considered in the previous chapter for the study of combustion.

Our aim here is only to give an overview of the methodology because the subject is very complex, and one can indeed consider a large variety of problems, depending on the objectives. Roughly speaking, the equations of fluid mechanics (motion of the air or of water) must be coupled with an energy or temperature equation and with equations describing the spatiotemporal evolution of the chemical pollutant concentrations such as ozone, carbon dioxide, sulfurous compounds, and so forth. One may also be led to make local and global studies. Local or small-scale studies are those that would concern for instance a city, a region, or an estuary. Global or large-scale studies are those that would concern, for instance, a whole ocean, or the entire earth's atmosphere (for studying for example the ozone holes above the poles). Because of the very limited pretensions of this section and of this chapter (see Remark 12.2), we will be content with writing the chemistry equations for the atmosphere only.

The vertical variable being the pressure variable p, as in Section 12.2, all the unknown quantities are functions of t, φ, θ, p (time t, spherical coordinates longitude, and colatitude φ, θ).

The unknown functions are

- The velocity of the wind V_3, where $V_3 = v + \varpi$, v is the horizontal velocity, and ϖ is the vertical velocity (in p coordinates);
- The density ρ, the geopotential $\Phi = gz$ (the height z is a function of t, φ, θ, p);
- The temperature T;
- The mass fractions $Y_1, \ldots, Y_N$, of the N chemical species contained in the atmosphere, one of them, say Y_1, being water.

The equations governing the spatiotemporal evolution of these quantities are

- Equations (12.18) to (12.21) of this chapter (with Eqs. (12.22) to (12.24));
- Equations (11.19) of Chapter 11 with the expressions of the ω_i given by Eqs. (11.12) and (11.25) of chapter 11.

With an explicit expression of D in Eq. (12.18) (similar to that on the right-hand side of Eq. (12.21)), the following system is obtained in which the underlined term is usually neglected:

$$
\frac{\partial v}{\partial t} + \nabla_v v + \omega \frac{\partial v}{\partial p} + 2\Omega \cos\theta \, k \wedge v + \nabla\Phi
$$

$$
- \mu_v \Delta v - \nu_v \frac{\partial}{\partial p}\left[\left(\frac{gp}{RT}\right)^2 \frac{\partial v}{\partial p}\right] = 0, \tag{12.37}
$$

$$
\frac{R^2}{C^2}\left\{\frac{\partial T}{\partial t} + \nabla_v T + \omega\frac{\partial T}{\partial p}\right\} - \frac{R\omega}{p} - \mu_T \Delta T - \nu_T \frac{\partial}{\partial p}\left[\left(\frac{gp}{RT}\right)^2 \frac{\partial T}{\partial p}\right]
$$

$$
= \frac{R^2}{C^2}\left[\frac{\mathcal{E}}{c_p} - \sum_{i=1}^{N} h^i w^i(T, Y^1, \ldots, Y^N)\right], \tag{12.38}
$$

$$
\frac{\partial Y^i}{\partial t} + \nabla_v Y^i + \omega\frac{\partial Y^i}{\partial p} - \mu_{Y_i}\Delta Y^i - \nu_{Y_i}\frac{\partial}{\partial p}\left[\left(\frac{gp}{RT}\right)^2 \frac{\partial Y^i}{\partial p}\right]
$$

$$
= \omega^i(T, Y^1, \ldots, Y^N), \quad i = 1, \ldots, N, \tag{12.39}
$$

$$\frac{\partial \Phi}{\partial p} + \frac{RT}{p} = 0, \tag{12.40}$$

$$\operatorname{div} v + \frac{\partial \varpi}{\partial p} = 0. \tag{12.41}$$

All the quantities have been defined; h^i is the enthalpy of species i, as in Eq. (11.15) of Chapter 11.

All these equations must be supplemented with initial and boundary conditions that we do not describe here.

The unknown functions divide into two groups: the prognostic variables, for which one defines evolution equations and prescribes initial conditions, and the diagnostic variables that are determined at each time in terms of the prognostic variables by using the equations and the boundary conditions (not given here): v, T, and the Y_i are the prognostic unknowns; Φ, ϖ, and ρ $(= p/RT$ thanks to Eq. (12.8)) are the diagnostic unknowns.

Remark 12.2: As mentioned earlier, we have just touched upon the problems and difficulties related to the modeling of the atmosphere and oceans. Numerous problems, such as the boundary conditions, the coupling of the atmosphere and oceans, or taking into account other phenomena (for instance radiation, the earth's topography, and the vegetation) have not been developed, and it is of course impossible to treat them, even superficially, within the scope of this book.

Appendix. The differential operators in spherical coordinates

Let $(e_\theta, e_\varphi, e_r)$ be the local canonical basis in spherical coordinates, let $V_3 = v_\theta e_\theta + v_\varphi e_\varphi + v_r e_r$ be a vector and F a scalar. Our aim in this appendix is to write a list of formulas giving the expressions of the gradient, divergence, curl, and Laplacian operators in spherical coordinates. For the Laplacian operators, we have to distinguish between the scalar and vector cases because the Laplacian of a vector function on a manifold is not defined in a unique way. We obtain, after some calculations

$$\operatorname{grad}_3 F = \frac{1}{r}\frac{\partial F}{\partial \theta} e_\theta + \frac{1}{r \sin \theta}\frac{\partial F}{\partial \varphi} e_\varphi + \frac{\partial F}{\partial r} e_r, \tag{12.42}$$

$$\operatorname{div}_3 V_3 = \frac{1}{r \sin \theta}\frac{\partial(\sin \theta \, v_\theta)}{\partial \theta} + \frac{1}{r \sin \theta}\frac{\partial v_\varphi}{\partial \varphi} + \frac{1}{r^2}\frac{\partial(r^2 v_r)}{\partial r}, \tag{12.43}$$

$$\operatorname{curl}_3 V_3 = \frac{1}{r}\left(\frac{1}{\sin\theta}\frac{\partial v_r}{\partial\varphi} - \frac{\partial(rv_\varphi)}{\partial r}\right)e_\theta + \frac{1}{r}\left(\frac{\partial(rv_\theta)}{\partial r} - \frac{\partial v_r}{\partial\theta}\right)e_\varphi$$

$$+ \frac{1}{r\sin\theta}\left(\frac{\partial(v_\varphi\sin\theta)}{\partial\theta} - \frac{\partial v_\theta}{\partial\varphi}\right)e_r,$$

$$\Delta_3 F = \frac{1}{r^2\sin\theta}\frac{\partial}{\partial\theta}\left(\sin\theta\frac{\partial F}{\partial\theta}\right) + \frac{1}{r^2\sin^2\theta}\frac{\partial^2 F}{\partial\varphi^2}$$

$$+ \frac{1}{r^2}\frac{\partial}{\partial r}\left(r^2\frac{\partial F}{\partial r}\right),$$

$$\Delta V_3 = \left(\Delta v_\theta + \frac{2}{r^2}\frac{\partial v_r}{\partial\theta} - \frac{v_\theta}{r^2\sin\theta} - \frac{2}{r^2\sin\theta}\frac{\partial v_\varphi}{\partial\varphi}\right)e_\theta$$

$$+ \left(\Delta v_\varphi + \frac{2}{r^2\sin\theta}\frac{\partial v_r}{\partial\varphi} + \frac{2\cos\theta}{r^2\sin^2\theta}\frac{\partial v_\theta}{\partial\varphi} - \frac{v_\varphi}{r^2\sin^2\theta}\right)e_\varphi$$

$$+ \left(\Delta v_r - \frac{2v_r}{r^2} - \frac{2}{r^2\sin\theta}\frac{\partial(v_\theta\sin\theta)}{\partial\theta} - \frac{2}{r^2\sin\theta}\frac{\partial v_\varphi}{\partial\varphi}\right)e_r.$$

In view of the approximations made in the preceding paragraphs, which consist of replacing the domain filled by the atmosphere (or by the ocean) by the product of the sphere S_a^2 with an interval (i.e., of making the approximation $r \simeq a$), we will also need to define these operators on the sphere S_a^2. Thus, if $v = v_\theta e_\theta + v_\varphi e_\varphi$ is a vector tangent to the sphere and if F is a scalar, we obtain the following expressions for the two-dimensional operators below:

$$\operatorname{grad} F = \frac{1}{a}\frac{\partial F}{\partial\theta}e_\theta + \frac{1}{a\sin\theta}\frac{\partial F}{\partial\varphi}e_\varphi,$$

$$\operatorname{div} v = \frac{1}{a\sin\theta}\frac{\partial(\sin\theta v_\theta)}{\partial\theta} + \frac{1}{a\sin\theta}\frac{\partial v_\varphi}{\partial\varphi},$$

$$\Delta F = \frac{1}{a^2\sin\theta}\frac{\partial}{\partial\theta}\left(\sin\theta\frac{\partial F}{\partial\theta}\right) + \frac{1}{a^2\sin^2\theta}\frac{\partial^2 F}{\partial\varphi^2},$$

$$\Delta v = \left(\Delta v_\theta - \frac{v_\theta}{a^2\sin^2\theta} - \frac{2\cos\theta}{a^2\sin^2\theta}\frac{\partial v_\varphi}{\partial\varphi}\right)e_\theta$$

$$+ \left(\Delta v_\varphi + \frac{2\cos\theta}{a^2\sin^2\theta}\frac{\partial v_\theta}{\partial\varphi} - \frac{v_\varphi}{a^2\sin^2\theta}\right)e_\varphi.$$

Furthermore, we set, for $v = v_\theta e_\theta + v_\varphi e_\varphi$, F denoting a scalar function and

$w = w_\theta e_\theta + w_\varphi e_\varphi$ a vector function tangent to the sphere:

$$\nabla_v F = \frac{v_\theta}{a}\frac{\partial F}{\partial \theta} + \frac{v_\varphi}{a\sin\theta}\frac{\partial F}{\partial \varphi},$$

$$\nabla_v w = \left[v \cdot \operatorname{grad} w_\theta - \frac{v_\varphi w_\varphi}{a}\cot\theta\right]e_\theta + \left[(v \cdot \operatorname{grad})w_\varphi + \frac{v_\varphi w_\theta}{a}\cot\theta\right]e_\varphi.$$

These functions are called, in differential geometry, the covariant derivatives of F and w in the v direction.

To go further:

Much more important developments in fluid mechanics can be found in the books cited in the bibliography: the recent book by Candel in French and, in English, the classical books by Batchelor, Chorin and Marsden, Lamb, and Landau and Lifschitz; see also the recent book by Majda and Bertozzi, with a particular emphasis on Euler equations and vortices.

The books by Hinze and Schlichting are classical references, in turbulence and boundary layers (possibly turbulent) respectively. The books by Chandrasekhar and by Drazin and Reed are devoted to the study of the stability in fluid mechanics (and in magnetohydrodynamics for Chandrasekhar's).

The mathematical aspects of the Navier-Stokes equations are developed in the books, cited, by Doering and Gibbon (more physically oriented), Ladyzhenskaya, Temam, in the large article by Serrin, and in many other references quoted therein.

The book by Cabannes is a very complete book on magnetohydrodynamics (see also that of Chandrasekhar). The book by Williams is a reference in combustion; the mathematical developments can be found in the book by the Volperts.

Among the numerous references concerning the equations of the atmosphere and the ocean, one can find in the bibliography the classical books by Pedloski and by Washington and Parkinson; mathematical developments appear in the cited articles by Lions, Temam and Wang, and in a book in preparation.

Of course, for all these huge subjects, the reader can also refer to the bibliography of the books and articles cited.

PART III

SOLID MECHANICS

CHAPTER THIRTEEN

The general equations of linear elasticity

Our aim in this chapter is to study in more detail the equations of linear elasticity as well as the boundary conditions that are associated with them.

Throughout Part 3 of this book, we change our notations and call x the Lagrangian variable and x' the Eulerian variable. Hence, $x' = \Phi(x, t) = x + u(x, t)$ represents the position at time $t > 0$ of the particle occupying the position x at time 0, $u(x, t)$ denoting the displacement of this particle.

13.1. Back to the stress–strain law of linear elasticity: the elasticity coefficients of a material

We recall that for an elastic material, and under the small deformations assumption, the stress–strain law, which is linear, is

$$\sigma_{ij} = 2\mu\varepsilon_{ij} + \lambda\varepsilon_{kk}\delta_{ij},$$

where $\varepsilon_{ij} = \frac{1}{2}(u_{i,j} + u_{j,i})$, u being the displacement and $u_{i,j} = \partial u_i/\partial x_j$ denoting the derivative of u_i with respect to the (Lagrangian) variable x_j.

The quantities λ and μ are the Lamé coefficients of the material, and the second principle of thermodynamics implies that $\mu \geq 0$ and $3\lambda + 2\mu \geq 0$ (as for fluids; see Chapter 7). These coefficients may be functions of the position, the time, the temperature, or even of other quantities. However, in general and at first approximation, they are considered to be constant, which we will assume hereafter.

We consider the "deviatoric" parts ε^D and σ^D of the tensors ε and σ, whose components are

$$\varepsilon_{ij}^D = e_{ij} = \varepsilon_{ij} - \frac{1}{3}\varepsilon_{kk}\delta_{ij} = \varepsilon_{ij} - e\,\delta_{ij},$$

$$\sigma_{ij}^D = s_{ij} = \sigma_{ij} - \frac{1}{3}\sigma_{kk}\delta_{ij} = \sigma_{ij} - s\,\delta_{ij},$$

where

$$e = \frac{1}{3}\varepsilon_{kk}, \quad s = \frac{1}{3}\sigma_{kk}.$$

Thus,

$$s_{ij} + s\,\delta_{ij} = 2\mu(e_{ij} + e\,\delta_{ij}) + 3\lambda\,e\,\delta_{ij},$$

and hence, by summation ($s_{jj} = e_{jj} = 0$):

$$s = (3\lambda + 2\mu)e,$$

so that

$$s_{ij} = 2\mu\,e_{ij}.$$

Other elasticity coefficients: the Young and Poisson moduli

By inverting the previous relations, other coefficients (related to λ and μ) that have a physical significance appear. We only write here the algebraic formulas, and we will come back later to the physical meaning of the coefficients introduced.

We first write

$$3\kappa = 3\lambda + 2\mu \geq 0;$$

κ is called the rigidity to compression modulus (see the experiment of uniform compression of a rod in Chapter 14, Section 14.2). We then set

$$\frac{1}{E} = \frac{\lambda + \mu}{\mu(3\lambda + 2\mu)}, \quad \nu = \frac{\lambda}{2(\lambda + \mu)}.$$

Because $3\lambda + 2\mu \geq 0$ and $\mu \geq 0$, it follows that $\lambda \geq -2\mu/3$, $\lambda + \mu \geq \mu/3 \geq 0$, and then $E \geq 0$; E is the Young modulus and ν the Poisson coefficient of the material.

Furthermore, $3\kappa = 3\lambda + 2\mu = E/(1 - 2\nu)$, which yields $1 - 2\nu \geq 0$, that is, $\nu \leq 1/2$.

We easily check that

$$\mu = \frac{E}{2(1 + \nu)}, \quad \lambda = \frac{\nu E}{(1 - 2\nu)(1 + \nu)},$$

and because $\mu \geq 0$ and $E \geq 0$, it follows that $1 + \nu \geq 0$. Thus,

$$-1 \leq \nu \leq \frac{1}{2}, \quad E \geq 0.$$

By inverting the stress–strain laws, we then find

$$s = 3\kappa e = \frac{E}{1 - 2v} e,$$

and because $e_{ij} = (1/2\mu)s_{ij}$,

$$\varepsilon_{ij} - e\,\delta_{ij} = \frac{1}{2\mu}(\sigma_{ij} - s\,\delta_{ij}).$$

Consequently,

$$\varepsilon_{ij} = \frac{\sigma_{ij}}{2\mu} + \left(\frac{1 - 2v}{E}s - \frac{s}{2\mu}\right)\delta_{ij}$$

$$= \frac{1 + v}{E}\sigma_{ij} + \left(\frac{1 - 2v}{E} - \frac{1 + v}{E}\right)s\,\delta_{ij}$$

$$= \frac{1 + v}{E}\sigma_{ij} - \frac{3v}{E}s\,\delta_{ij},$$

and finally

$$\boxed{\varepsilon_{ij} = \frac{1 + v}{E}\sigma_{ij} - \frac{v}{E}\sigma_{kk}\delta_{ij}.}$$

13.2. Boundary value problems in linear elasticity: the linearization principle

We start this section by briefly recalling the general concepts of linear elasticity.

For the study of a problem of linear elasticity, we have the fundamental law of dynamics

$$\rho\gamma_i = f_i + \tilde{\sigma}_{ij,j}, \qquad (13.1)$$

and the linear stress–strain law

$$\sigma_{ij} = 2\mu\varepsilon_{ij} + \lambda\varepsilon_{kk}\delta_{ij},$$
$$\varepsilon_{ij} = \frac{1}{2}(u_{i,j} + u_{j,i}), \qquad (13.2)$$

where u is the displacement, at our disposal. The notations in this paragraph are not the usual notations. In the formulas above, $\tilde{\sigma} = \tilde{\sigma}(x', t)$ is the stress tensor expressed in Eulerian variable x', and the derivative $\tilde{\sigma}_{ij,j}$ is relative to the x'_j variable. However, in the stress–strain law, the displacement

$u_i = u_i(x, t)$ is expressed in the Lagrangian variable x, and the derivative $u_{i,j}$ is relative to the Lagrangian variable x_j. For consistency, we also express the stress tensor in (2) in the Lagrangian variable, and denote it by σ. This complication will disappear in linear elasticity thanks to the small displacements hypothesis that will allow us to approximate the equations (and the boundary conditions) by linearizing them with respect to the displacements. Eventually, *all the equations and boundary conditions will be written in the nondeformed (Lagrangian) variable x*.

Remark 13.1: It is of course necessary to check the validity a posteriori of the approximations made. We will come back to this problem in Section 13.4 where we will study the limit of elasticity criteria, giving empirical rules (which we will be content with but which cannot replace rigorous convergence theorems that are difficult to prove and not always available).

Linearization of the equations

We consider an elastic body filling the domain Ω_0 at time 0 and the domain Ω_t at time t. As already mentioned, we will denote by $x' = \Phi(x, t)$ the position at time $t > 0$ of the particle located at x at time 0 ($x \in \Omega_0$, $x' \in \Omega_t$). We call $\tilde{u}(x', t) = \tilde{u}[\Phi(x, t), t] = u(x, t)$ the displacement. Thus, according to the definitions and results of Chapter 1:

$$\gamma_i = \frac{\partial^2 u_i}{\partial t^2}(x, t),$$

$$\tilde{\sigma}_{ij,k} = \frac{\partial}{\partial x'_k}\tilde{\sigma}_{ij}(x', t) = \frac{\partial}{\partial x'_k}\sigma_{ij}(\psi(x', t), t)$$

$$= \frac{\partial \sigma_{ij}}{\partial x_\ell} \cdot \frac{\partial \psi_\ell}{\partial x'_k},$$

where we have set $x = \psi(x', t)$ and $\sigma(x, t) = \tilde{\sigma}[\Phi(x, t), t]$. We have

$$\frac{\partial \Phi_j}{\partial x_k} = \delta_{jk} + \frac{\partial u_j}{\partial x_k}.$$

Because the displacement and its derivatives are small (infinitesimally small at first order) and ψ is the inverse function of the displacement Φ,

$$\frac{\partial \psi_j}{\partial x'_k} = \delta_{jk} - \frac{\partial u_j}{\partial x_k} + o(\eta) \simeq \delta_{jk} - \frac{\partial u_j}{\partial x_k},$$

$\partial \psi_j / \partial x'_k$ being equal to the element (j, k) of the inverse matrix of $\nabla \Phi$. Consequently,

$$\tilde{\sigma}_{ij,k} = \frac{\partial \sigma_{ij}}{\partial x_\ell} \cdot \left(\delta_{\ell k} - \frac{\partial u_\ell}{\partial x_k} \right) + o(\eta);$$

hence, to first order (i.e., to order $o(1)$):

$$\sigma_{ij,j} = \tilde{\sigma}_{ij,j}.$$

This means, in an informal way, that one can replace $\tilde{\sigma}_{ij}$ by σ_{ij} in Eq. (13.1) or else that we can differentiate with respect to x_j rather than to x'_j; this solves one of the difficulties mentioned above.

The linearized momentum equation then becomes

$$\rho \frac{\partial^2 u_i(x, t)}{\partial t^2} = f_i(x, t) + \frac{\partial \sigma_{ij}}{\partial x_j}(x, t).$$

We again transform this equation by using the relation

$$\varepsilon_{ij} = \frac{1}{2} \left(\frac{\partial u_i}{\partial x_j} + \frac{\partial u_j}{\partial x_i} \right),$$

and the stress–strain law:

$$\sigma_{ij} = \mu \left(\frac{\partial u_i}{\partial x_j} + \frac{\partial u_j}{\partial x_i} \right) + \lambda \frac{\partial u_k}{\partial x_k} \delta_{ij}.$$

We deduce the following equation for the displacements, which is also called the Navier equation:

$$\boxed{\rho \frac{\partial^2 u}{\partial t^2} = f + \mu \, \Delta u + (\lambda + \mu) \, \text{grad div } u.}$$

We recall that u and f depend on x and t, $t \geq 0$, $x \in \Omega$, which is the name that we give, from now on, to the undeformed state Ω_0.

The next step is the linearization of the boundary conditions.

Linearization of the boundary conditions

We are given two disjoint parts of $\partial \Omega$, Γ_u and Γ_T such that $\Gamma_u \cup \Gamma_T = \partial \Omega$ ($= \partial \Omega_0$), and we assume that u is prescribed on Γ_u and that $T = \sigma \cdot n$ is given on the image of Γ_T by Φ. This corresponds to the natural boundary conditions for a problem of solid mechanics. Hence, part of the boundary conditions relates to the undeformed state and part relates to the deformed state. As for the equations, this complication, inevitable in general, will be removed here by the linearization of the boundary conditions.

a) Prescribed displacement

We set

$$u = \varphi \ (\text{given}), \quad \text{on } \Gamma_u.$$

As already noted, this boundary condition is automatically prescribed in the undeformed geometry.

b) Prescribed stress

We provisionally return to the notation with $\tilde{\ }$ and use the variable x' to write the boundary condition on $\Phi(\Gamma_T)$:

$$\tilde{\sigma}(x', t) \cdot \tilde{n}(x', t) = F(x', t) \ (\text{given}), \quad x' \in \Phi(\Gamma_T, t).$$

Here, $\tilde{n}$ denotes the unit outward normal to $\Phi(\Gamma_T)$, but we easily check that, to first approximation, $\tilde{n}(x', t) = \tilde{n}[\Phi(x, t), t] = n(x, t)$, where n is the unit outward normal to $\partial\Omega_0$. Indeed, if the equation of $\partial\Omega_t$ is $\tilde{g}(x', t) = 0$, then $\tilde{g}[\Phi(x, t), t] \equiv g(x, t) = 0$ is the equation of $\partial\Omega_0$, and

$$\tilde{n} = \frac{\nabla_{x'}\tilde{g}}{|\nabla_{x'}\tilde{g}|},$$

$$\frac{\partial\tilde{g}}{\partial x_i'} = \frac{\partial g}{\partial x_\ell}\frac{\partial\psi_\ell}{\partial x_i'}$$

$$\simeq \frac{\partial g}{\partial x_i} + o(1).$$

Therefore, to first order

$$\tilde{n} = \frac{\nabla_{x'}\tilde{g}}{|\nabla_{x'}\tilde{g}|} \simeq \frac{\nabla_x g}{|\nabla_x g|} = n.$$

Using $\sigma(x, t) = \tilde{\sigma}[\Phi(x, t), t]$, the boundary condition on $\Phi(\Gamma_T)$ becomes, after linearization

$$(\sigma_{ij}n_j)(x, t) = F_i[\Phi(x, t), t] = F_i(x, t) + o(1), \quad \text{on } \Gamma_T.$$

Returning to the notation without tildes, we will then retain the condition

$$\sigma_{ij} \cdot n_j = F_i, \quad \text{on } \Gamma_T,$$

σ_{ij} being expressed in terms of the derivatives of u, by using the stress–strain law of the medium and the definition of the $\varepsilon_{k\ell}$.

In summary, the fundamental equations of linear elasticity for a time-dependent problem are

$$\rho \frac{\partial^2 u}{\partial t^2} = f + \mu \, \Delta u + (\lambda + \mu) \, \mathrm{grad} \, \mathrm{div} \, u, \quad x \in \Omega, \; t \geq 0, \quad (13.3)$$

$$u = \varphi, \quad \text{on } \Gamma_u, \quad (13.4)$$

$$\sigma \cdot n = F, \quad \text{on } \Gamma_T. \quad (13.5)$$

Particular cases

a) Statics

In this case, u being independent of t, $\partial^2 u / \partial t^2 = 0$, and the displacement equation (Eq. (13.3)) becomes

$$f + \mu \, \Delta u + (\lambda + \mu) \, \mathrm{grad} \, \mathrm{div} \, u = 0. \quad (13.6)$$

b) Evolutionary case

In this case, we have to prescribe, in addition to f, φ, and F, the initial displacements and velocities θ_0, θ_1

$$u(x, 0) = \theta_0(x), \quad x \in \Omega, \quad (13.7)$$

$$\frac{\partial u}{\partial t}(x, 0) = \theta_1(x), \quad x \in \Omega. \quad (13.8)$$

Remark 13.2: We can also consider the limit cases where $\Gamma_u = \partial \Omega$ and $\Gamma_T = \emptyset$ or where $\Gamma_T = \partial \Omega$ and $\Gamma_u = \emptyset$. In the static case, and if $\Gamma_u = \emptyset$, the solution to the displacement problem is not unique; it is defined up to a rigid body displacement.

Boundary value problems

One can prove that, with proper hypotheses on the data Ω, f, φ, F, and possibly θ_0, θ_1 (regularity assumptions, in particular), the boundary value problems of elastostatics and of elastodynamics possess a unique solution in a proper class of functions. These boundary value problems are as follows:

a) Elastostatics

An elastostatic problem consists of the study of Eq. (13.6) and of the boundary conditions defined by Eqs. (13.4) and (13.5). This problem is related to the projection theorem and to the variational formulation of elliptic boundary value problems. We will come back to this in more detail in the following chapters. The problem possesses a unique solution except for the case mentioned in Remark 13.2. The uniqueness of solution in elastostatics is proved in Chapter 15, Section 15.4.

b) Elastodynamics

We study in this case the boundary value problem consisting of Eq. (13.3), of the boundary conditions defined by Eqs. (13.4) and (13.5), and of the initial conditions set by Eqs. (13.7) and (13.8). One can prove the existence and uniqueness of solutions for this problem in a suitable mathematical setting; further remarks concerning this case are given hereafter.

c) Linearity

We easily observe that the solution u (and thus σ and ε) is a linear function of the data, that is, the function $(f, \varphi, F) \mapsto (u, \sigma, \varepsilon)$ is linear. Thus, if $(u^1, \varepsilon^1, \sigma^1)$ (respectively, $(u^2, \varepsilon^2, \sigma^2)$) is the solution that corresponds to (f^1, φ^1, F^1) (respectively, (f^2, φ^2, F^2)), and if α and β are real numbers, then $(\alpha u^1 + \beta u^2, \alpha \varepsilon^1 + \beta \varepsilon^2, \alpha \sigma^1 + \beta \sigma^2)$ is the solution that corresponds to $(\alpha f^1 + \beta f^2, \alpha \varphi^1 + \beta \varphi^2, \alpha F^1 + \beta F^2)$; this is the additivity principle of linear elasticity.

13.3. Other equations

We will supplement the set of equations for u obtained in the previous section with other equations on the deformations and on the stresses that are sometimes useful. Of course, these equations are not independent of the previous ones.

Compatibility equations (for the deformations)

We have

$$\varepsilon_{ij} = \frac{1}{2}(u_{i,j} + u_{j,i}), \tag{13.9}$$

$$\omega_{ij} = \frac{1}{2}(u_{i,j} - u_{j,i}), \tag{13.10}$$

where ω is the rotation tensor. Consequently,

$$\omega_{ij,\ell} = \frac{1}{2}(u_{i,j\ell} - u_{j,i\ell})$$

$$= \frac{1}{2}(u_{i,j\ell} + u_{\ell,ij}) - \frac{1}{2}(u_{\ell,ij} + u_{j,i\ell}),$$

and thus

$$\omega_{ij,\ell} = (\varepsilon_{i\ell,j} - \varepsilon_{j\ell,i}). \tag{13.11}$$

Then, because $\omega_{ij,\ell k} = \omega_{ij,k\ell}$, it follows that

$$\varepsilon_{i\ell,jk} - \varepsilon_{j\ell,ik} = \varepsilon_{ik,j\ell} - \varepsilon_{jk,i\ell}. \tag{13.12}$$

Remark 13.3: Equation (13.12) is necessary and sufficient for a given tensor field $\varepsilon = (\varepsilon_{ij})$ to be locally the deformation tensor of a vector field u. Indeed, if Eq. (13.12) holds, we can conversely recover the u_i from the ε_{ij}. We first recover the ω_{ij} using Eq. (13.11) and then the u_i from

$$u_{i,j} = \varepsilon_{ij} + \omega_{ij} \tag{13.13}$$

if the necessary and sufficient Schwarz conditions are satisfied for Eqs. (13.11) and (13.13). They are satisfied for Eq. (13.11): the Schwarz conditions are precisely the relations of Eq. (13.12). It remains to check that $u_{i,jk} = u_{i,kj}$, that is,

$$\varepsilon_{ij,k} + \omega_{ij,k} = \varepsilon_{ik,j} + \omega_{ik,j},$$

which is straightforward with Eq. (13.11). We notice that if it possesses a solution, the problem of the determination of the u_i by the ε_{ij} also possesses other solutions that are deduced from one another by the addition of a rigid displacement.

We now rewrite Eq. (13.12) in the form

$$0 = \varepsilon_{i\ell,jk} + \varepsilon_{jk,i\ell} - (\varepsilon_{j\ell,ik} + \varepsilon_{ik,j\ell})$$

and notice that the right-hand side is antisymmetric with respect to the pair (i, j) and to the pair (ℓ, k). Thus, we can limit ourselves to the pairs (i, j) and (k, ℓ) chosen among the pairs (1,2), (1,3), and (2,3). The expression is also invariant by exchange of the pair (i, j) and of the pair (k, ℓ). Consequently, we would obtain all the relations by taking

$$(i, j) = (1, 2), \quad (k, \ell) = (1, 2), (1, 3), (2, 3);$$

$$(i, j) = (1, 3), \quad (k, \ell) = (1, 3), (2, 3);$$

$$(i, j) = (2, 3), \quad (k, \ell) = (2, 3);$$

this gives six independent relations.

Now, if for every i, ℓ we take $j = k = 1, 2, 3$ in Eq. (13.12), we obtain by summation

$$\Delta\varepsilon_{i\ell} + (3e)_{,i\ell} - (\varepsilon_{ik,\ell k} + \varepsilon_{\ell k,ik}) = 0. \tag{13.14}$$

These six equations are independent and form a system equivalent to the initial equations.

The equations obtained above are the compatibility equations for the ε_{ij}. As mentioned before, they are necessary and sufficient for the ε_{ij} to be locally a deformation tensor (just as the Schwarz relations are necessary and sufficient conditions for a vector field to be locally a gradient).

The Beltrami equations

The preceding equations were purely kinematic. We now obtain other equations by also taking into account the fundamental law of dynamics.

We saw that

$$e = \frac{1 - 2v}{E}s,$$

$$\varepsilon_{ij} = \frac{1 + v}{E}\sigma_{ij} - \frac{3v}{E}s\,\delta_{ij}.$$

We immediately deduce from the compatibility equations (13.14) that

$$\frac{1 + v}{E}\Delta\sigma_{i\ell} - \frac{3v}{E}(\Delta s)\,\delta_{i\ell} + \frac{3(1 - 2v)}{E}s_{,i\ell}$$

$$-\frac{1 + v}{E}(\sigma_{ik,\ell k} + \sigma_{\ell k,ik}) + \frac{3v}{E}[(s\delta_{ik})_{,\ell k} + (s\delta_{\ell k})_{,ik}] = 0,$$

and, because in statics,

$$\sigma_{ik,k} = -f_i,$$

it follows that

$$\Delta\sigma_{i\ell} - \frac{3v}{v + 1}(\Delta s)\,\delta_{i\ell} + \frac{3(1 - 2v)}{1 + v}s_{,i\ell} + (f_{i,\ell} + f_{\ell,i}) + \frac{6v}{v + 1}s_{,i\ell} = 0.$$

Because $-3\Delta s = [(1 + v)/(1 - v)]$ div f (this relation is proven independently hereafter), we deduce the Beltrami equation, which is valid in elastostatics:

$$\Delta\sigma_{i\ell} + \frac{3}{1 + v}s_{,i\ell} + \frac{v}{1 - v}f_{k,k}\delta_{i\ell} + f_{i,\ell} + f_{\ell,i} = 0.$$

If the external forces are constant, this reduces to

$$\Delta\sigma_{i\ell} + \frac{3}{1 + v}s_{,i\ell} = 0.$$

Other equations

We deduce from the Navier equation that

$$\rho \frac{\partial^2}{\partial t^2} \text{div } u = \text{div } f + \mu \, \Delta \text{ div } u + (\lambda + \mu) \Delta \text{ div } u,$$

that is

$$\rho \frac{\partial^2}{\partial t^2} \text{div } u = \text{div } f + (\lambda + 2\mu) \Delta \text{ div } u.$$

Because div $u = \varepsilon_{kk} = 3e$, it follows that

$$\rho \frac{\partial^2 e}{\partial t^2} - (\lambda + 2\mu) \Delta e = \frac{1}{3} \text{div } f,$$

or else, by introducing the function $s = [E/(1 - 2v)]e$,

$$\rho \frac{\partial^2 s}{\partial t^2} - (\lambda + 2\mu) \Delta s = \frac{E}{1 - 2v} \frac{\text{div } f}{3}.$$

In statics, we have

$$-3 \Delta s = \frac{E}{1 - 2v} \frac{1}{\lambda + 2\mu} \text{div } f.$$

Now,

$$\lambda + 2\mu = \frac{vE}{(1 - 2v)(1 + v)} + \frac{E}{1 + v}$$

$$= \frac{(1 - v)E}{(1 - 2v)(1 + v)}.$$

Therefore, we recover the equation mentioned above as follows:

$$-3 \Delta s = \frac{1 + v}{1 - v} \text{div } f.$$

13.4. The limit of elasticity criteria

The fundamental hypothesis in linear elasticity is that the displacements remain small. In general, we are unable to prove convergence or approximation theorems rigorously, and it is necessary, at least, to check a posteriori the validity of this assumption. To do so, we have at our disposal empirical criteria that are commonly used in mechanical engineering; the Tresca and von Mises criteria are the most well known. We recall that elastic deformations are reversible, contrary to nonelastic deformations, and that, in general, engineers want structures working in the elastic domain so that they can predict their behavior at all times.[1]

The Tresca criterion

We write that the modulus of the tangential stress remains, at every point x and at every time t, lower than a constant denoted by $g/2$, $g > 0$. One can show, using the so-called Mohr circles theory (not presented in this book), that

$$\text{Max }|T_t| = \text{Max}\frac{1}{2}|\sigma_i - \sigma_j|,$$

where the σ_i are the principal normal stresses. The first maximum is taken for all directions n ($T = \sigma \cdot n$); the second one is for $i, j = 1, 2, 3$. We then have to write

$$\text{Max}|\sigma_i - \sigma_j| \leq g,$$

at all points $x \in \Omega$ and all time t, that is

$$\text{Max}|s_i - s_j| \leq g,$$

where the s_i are the eigenvalues of the deviator σ^D of σ, so that $s_i - s_j = \sigma_i - \sigma_j, \forall_{i,j}$. The last maximum is taken with respect to i and j and with respect to $x, t, x \in \Omega, t \geq 0$.

The von Mises criterion

We write in that case

$$s_{ij}s_{ij} \leq \text{const.} = g'^2;$$

that is to say,

$$s_1^2 + s_2^2 + s_3^2 \leq g'^2.$$

Again, this relation must be satisfied at every point $x \in \Omega$ and at every time $t \geq 0$ if we are dealing with a time-dependent problem. We will see later how one can experimentally determine the values of g and g'. The Tresca and von Mises criteria are in good agreement with experimental results.

---

[1] If a media becomes unelastic after undergoing large deformations or after being subjected to important forces, or both, then its mechanical state is not fully known because it depends on the history of the loading; its response to future forcing becomes unpredictable.

Exercises

1. a) Show that the components $\sigma_{11}, \sigma_{22}, \sigma_{12}$ of the stress tensor corresponding to a planar deformation field can be expressed, when the volume forces vanish, in terms of a function $\chi(x_1, x_2)$ (solution of the bi-Laplacian equation $\Delta(\Delta\chi) = 0$) as follows:

$$\sigma_{11} = \frac{\partial^2 \chi}{\partial x_2^2}, \sigma_{22} = \frac{\partial^2 \chi}{\partial x_1^2}, \sigma_{12} = -\frac{\partial^2 \chi}{\partial x_1 \partial x_2}.$$

   b) We consider the functions χ of the form $\chi = f(x_2)\cos(\omega x_1)$, $\omega > 0$. Give the most general form of the function f.

2. An elastic body is at equilibrium with respect to an orthonormal frame $(0x_1x_2x_3)$. We set $\vec{r} = \overrightarrow{OM}$.

   a) We look for solutions of the Navier equation of the form

   $$\vec{X} = \vec{X_0} + \vec{B} + \nabla\chi,$$

   where $\vec{X_0}$ is a particular solution of the Navier equation and $\vec{B}$ is a harmonic vector (i.e., such that $\Delta\vec{B} = 0$). Show that

   $$\nabla(\Delta\chi + \frac{\lambda + \mu}{\lambda + 2\mu} \operatorname{div}\vec{B}) = 0,$$

   and that χ is of the form

   $$\chi = -\frac{\lambda + \mu}{2(\lambda + 2\mu)}(\vec{r} \cdot \vec{B} + B_0) + b\vec{r}^2,$$

   where B_0 is harmonic.

   b) Show that these solutions of the Navier equation are of the form

   $$\vec{X} = \vec{X_1} + \vec{B} - \frac{\lambda + \mu}{2(\lambda + 2\mu)}\nabla(\vec{r} \cdot \vec{B} + B_0),$$

   X_1 being a particular solution.

3. The stress field of an elastic medium is defined by

   $$\sigma_{11} = x_2^2 + k(x_1^2 - x_2^2), \sigma_{12} = -2kx_1x_2, \sigma_{22} = x_1^2 + k(x_2^2 - x_1^2),$$
   $$\sigma_{13} = \sigma_{23} = 0, \sigma_{33} = k(x_1^2 + x_2^2).$$

   Compute the associated force field. Is this force field compatible with such a stress field in an elastic medium?

4. We consider an elastic medium at equilibrium and we assume that the volume forces vanish and that the stress tensor is defined by

$$\sigma_{11} = \sigma_{22} = \sigma_{12} = 0, \sigma_{13} = \frac{\partial F}{\partial x_2} - \frac{mx_1^2}{2},$$

$$\sigma_{23} = -\frac{\partial F}{\partial x_1} - \frac{nx_2^2}{2}, \sigma_{33} = mx_1x_3 + rx_2x_3.$$

Show that we may assume that, without loss of generality, F is independent of x_3 and show that ΔF is affine with respect to x_1 and x_2.

5. Reconsider Exercise 3, Chapter 4, assuming that the body is an elastic material (i.e., that σ satisfies the stress-strain law of linear elasticity).

CHAPTER FOURTEEN

Classical problems of elastostatics

Our aim in this chapter is to treat several classical problems of elastostatics. Strictly speaking, elasticity problems such as those described below cannot be solved exactly in general: they can be solved exactly in very particular cases (e.g., special geometry); otherwise, approximate numerical solutions are obtained using computers. However, in the examples treated below, we are going to find approximate solutions giving an idea of the exact solution under some reasonable conditions that will be made precise in each case (by using, in particular, the Saint-Venant principle described in Section 14.7).

For each of the mechanical problems that we will consider, we will find (by guessing) an exact solution for a modified problem related to the one under consideration. By the uniqueness theorem for elastostatics, there is no other solution to the modified problem. Then, the relation between the solutions of the initial and modified problems is made precise using the Saint-Venant principle. We will also interpret the mathematical results from the mechanical point of view, which leads in general, but not always, to conclusions that are consistent with practical intuition.

14.1. Longitudinal traction–compression of a cylindrical bar

We consider an elongated cylindrical bar in traction (or in compression). We assume that the axis of the cylinder is parallel to Ox_1 (see Figure 14.1).

For the study of the problem, we formulate the following simplifying assumptions that are realistic when the bar is long enough and when we remain far enough from its ends:

- The volume forces are negligible;
- The external forces on the lateral surface vanish;

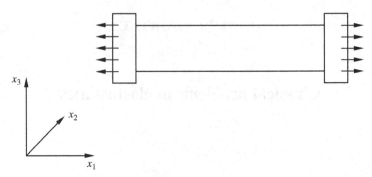

Figure 14.1 A cylindrical bar in traction – compression.

- The stresses on the faces Σ_0 and Σ_1 with respective equations $x_1 = 0$ and $x_1 = L$ are of the form

$$\sigma_{11} = F, \quad \sigma_{21} = \sigma_{31} = 0, \text{ on } x_1 = L,$$
$$\sigma_{11} = -F, \sigma_{21} = \sigma_{31} = 0, \text{ on } x_1 = 0.$$

We then propose the following solution for this elastostatic problem concerning the stress tensor:

$$\sigma_{11} = F \quad \text{and} \quad \sigma_{ij} = 0, \quad \text{for all the other } i \text{ and } j,$$

which yields

$$\sigma_{kk} = F, \quad \varepsilon_{ij} = \frac{1+\nu}{E}\sigma_{ij} - \frac{\nu}{E}F\,\delta_{ij}, \quad \forall i, j.$$

The quantity

$$\varepsilon_{11} = \frac{1+\nu}{E}F - \frac{\nu F}{E} = \frac{F}{E}$$

is called the elongation rate by unit length in the Ox_1 direction. We have furthermore

$$\varepsilon_{22} = \varepsilon_{33} = -\frac{\nu}{E}F, \quad \varepsilon_{ij} = 0, \quad \text{if } i \neq j.$$

The equilibrium equations are satisfied, as well as the compatibility equations for the ε_{ij}. Thus, there exists a displacement field u such that $u_{i,j} + u_{j,i} = 2\varepsilon_{ij}$; we find for instance the field

$$u_1 = \frac{F}{E}x_1, \quad u_2 = -\frac{\nu}{E}Fx_2, \quad u_3 = -\frac{\nu}{E}Fx_3.$$

Every other solution (displacement field) can be obtained by adding a rigid displacement to this vector field.

For the chosen displacement field, the section Σ_0 is fixed, whereas the section Σ_1 has moved in the Ox_1 direction of a distance FL/E; this last quantity, proportional to L and F, is the elongation of the bar. Moreover, every section undergoes a translation parallel to Ox_1, and a homothety in the directions Ox_2 and Ox_3 of ratio $(1 - \nu F/E)$. Its area is thus multiplied by $(1 - \nu F/E)^2$. We will speak of elongation and contraction when $F > 0$, and of shortening and dilation when $F < 0$ (it is assumed here that $\nu \geq 0$).

Remark 14.1: According to the Saint-Venant approximation principle that we describe in Section 14.7, the solution just obtained is a good approximation to the exact solution in the case of an elongated bar having one extremity, say Σ_0, fixed, and if we are sufficiently far from Σ_0 and from Σ_1.

Comparison with experiment

This experiment is a basic one in continuum mechanics. It allows us to check Hooke's law experimentally: $E\varepsilon_{11} = F$, where ε_{11} is the elongation rate; alternatively, for new materials whose behavior is unknown, the experiment allows us to determine the Young modulus E. We will see subsequently that it also allows us to relate the Tresca and von Mises limit-of-elasticity criteria by comparing the constants g and g' appearing in these criteria.

For a given rod, the relation between the traction F and the elongation by unit of length ε_{11} is that described in Figure 14.2. As long as the absolute value of F remains smaller than a limit value k, this relation is linear, as assumed by linear elasticity. It stops being linear for $|F| > k$.

The experimental determination of k allows us to make the values of the coefficients g and g' appearing in the Tresca and von Mises criteria precise; these criteria can then be applied, by extrapolation, to other situations.

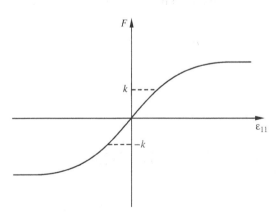

Figure 14.2 Hooke's law.

The Tresca criterion gives, with $\sigma_1 = F$, $\sigma_2 = \sigma_3 = 0$

$$\text{Max}|\sigma_i - \sigma_j| = |F| \leq g,$$

and we thus choose $g = k$. Similarly, for the von Mises criterion, we calculate $(\sigma_i^D = s_i)$:

$$s_1 = \frac{2}{3}F, \quad s_2 = s_3 = -\frac{F}{3};$$

hence,

$$s_1^2 + s_2^2 + s_3^2 = \frac{2}{3}F^2 \leq g'^2.$$

This is equivalent to $|F| \leq \sqrt{3/2}\,g'$ and, for consistency, we choose for the von Mises criterion $g' = \sqrt{2/3}k$ (and thus, $g' = \sqrt{2/3}g$).

14.2. Uniform compression of an arbitrary body

We assume that a body S is subjected, on its boundary, to a uniform compression $F = -pn$, where p denotes the pressure, n is the unit outward normal, and that it is not subjected to volume forces ($f = 0$).

We propose, for the stress field, the solution

$$\sigma_{11} = \sigma_{22} = \sigma_{33} = -p, \quad \text{and} \quad \sigma_{ij} = 0 \quad \text{if } i \neq j.$$

The equilibrium equations, as well as the boundary conditions, are satisfied. This solution is thus appropriate (there are no boundary conditions on u in this problem). The corresponding deformation tensor field is given below; it defines a displacement u unique up to the addition of a rigid displacement. Indeed,

$$\sigma^D = 0, \quad \varepsilon^D = 0,$$

which yields

$$\sigma_I = (3\lambda + 2\mu)\varepsilon_I,$$

and then

$$\varepsilon_{ii} = -\frac{3p}{(3\lambda + 2\mu)}$$

and

$$\varepsilon_{11} = \varepsilon_{22} = \varepsilon_{33} = -\frac{p}{3\lambda + 2\mu} = -\frac{p}{3\kappa}.$$

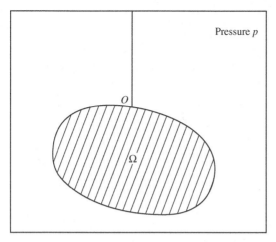

Figure 14.3 Uniform compression of an arbitrary body: practical realization.

We finally obtain a displacement of the form

$$u(x) = -\frac{p}{3\kappa}x.$$

In practice, the body S will generally be fixed at one of its points (vanishing displacement and rotation at this point), which specifies the rigid displacement to be added to u (in what precedes, S is fixed at O; see Figure 14.3).

The limit-of-elasticity criteria imply that the behavior remains elastic for arbitrarily large values of p (because $s_{ij} = \sigma_{ij}^D = 0$). This somehow unexpected fact is well confirmed by experimentation. This experiment is also used in practice to determine κ for new materials (S being, for instance, a sphere fixed at one point of its surface).

14.3. Equilibrium of a spherical container subjected to external and internal pressures

We consider a body limited by two concentric spheres of radius R_1 and R_2, $R_1 < R_2$. We assume that the volume forces vanish and that the body is subjected to uniform normal pressures, p_1 at the inner surface and p_2 at the outer surface (Figure 14.4).

We look (which is reasonable) for a radial displacement of the form

$$u(x) = g(r)x, \quad r = |x|.$$

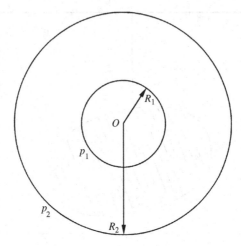

Figure 14.4 Spherical container.

Because $\partial r / \partial x_i = x_i / r$, it follows that

$$\operatorname{curl} u = \operatorname{grad} g \wedge x + g \operatorname{curl} x$$
$$= g'(r) \frac{x}{r} \wedge x = 0.$$

One easily checks that

$$\Delta u = \operatorname{grad} \operatorname{div} u - \operatorname{curl} \operatorname{curl} u,$$

and thus the Navier equation may be rewritten in the form

$$\mu \operatorname{curl} \operatorname{curl} u - (\lambda + 2\mu) \operatorname{grad} \operatorname{div} u = f.$$

The Navier equation thus reduces to $\operatorname{grad} \operatorname{div} u = 0$, which gives

$$\operatorname{div} u = r g'(r) + 3 g(r) = \text{const.} \quad \text{(denoted by } 3\alpha\text{)}.$$

By integration, we infer that

$$g(r) = \alpha + \beta \frac{1}{r^3},$$

where α and β are constants.

Consequently, the expression for the ε_{ij} is

$$\varepsilon_{ij} = u_{i,j} = g(r) \delta_{ij} + \frac{1}{r} g'(r) x_i x_j,$$

and, for the stress tensor, we have

$$\sigma_I = (3\lambda + 2\mu)\varepsilon_I = 3\alpha(3\lambda + 2\mu),$$

$$\sigma_{ij} = 2\mu\varepsilon_{ij} + \lambda\varepsilon_I\delta_{ij}$$

$$= 2\mu\frac{g'(r)}{r}x_i x_j + [2\mu g(r) + 3\lambda\alpha]\,\delta_{ij}.$$

The principal directions of stresses at x are Ox and all the orthogonal directions with eigenvalues

$$\sigma_1 \text{ (corresponding to } Ox) = 3\alpha\kappa - \frac{4\mu\beta}{r^3},$$

$$\sigma_2 = \sigma_3 = 3\alpha\kappa + \frac{2\mu\beta}{r^3}.$$

Boundary conditions

On the surface $r = R_2$, the unit outward normal is $n = x/r$, and thus

$$\sigma n = \sigma_1 n = -p_2 n;$$

hence, the boundary condition

$$\sigma_1|_{r=R_2} = -p_2,$$

that is to say,

$$3\alpha\kappa - \frac{4\mu\beta}{R_2^3} = -p_2.$$

Similarly, for $r = R_1$, we have

$$\sigma_1|_{r=R_1} = -p_1,$$

$$3\alpha\kappa - \frac{4\mu\beta}{R_1^3} = -p_1.$$

It then follows that

$$\beta = \frac{p_2 - p_1}{4\mu}\frac{R_1^3 R_2^3}{R_1^3 - R_2^3},$$

$$3\kappa\alpha = \frac{p_1 R_1^3 - p_2 R_2^3}{R_2^3 - R_1^3}.$$

As a result, σ, ε, and u are entirely determined, u being known up to a rigid displacement.

Limit of elasticity

It is particularly interesting, for this example, to make explicit the limit of
elasticity criteria.

For that purpose, we first compute the principal normal stresses σ_i. At an
arbitrary point x, we have

$$\sigma_1 = -p_1 \frac{R_1^3}{r^3} \frac{R_2^3 - r^3}{R_2^3 - R_1^3} - p_2 \frac{R_2^3}{r^3} \frac{r^3 - R_1^3}{R_2^3 - R_1^3},$$

$$\sigma_2 = p_1 \frac{R_1^3}{2r^3} \frac{R_2^3 + 2r^3}{R_2^3 - R_1^3} - p_2 \frac{R_2^3}{2r^3} \frac{R_1^3 + 2r^3}{R_2^3 - R_1^3}.$$

Therefore,

$$|\sigma_1 - \sigma_2| = \left|\frac{6\mu\beta}{r^3}\right| = \frac{3}{2}|p_1 - p_2| \frac{R_2^3 R_1^3}{r^3 \left(R_2^3 - R_1^3\right)},$$

$$\sigma_{II}^D = \frac{1}{2}\sigma_{ij}^D \sigma_{ij}^D = \frac{12\mu^3\beta^3}{r^6}$$

$$= \frac{3}{4}\frac{(p_1 - p_2)^2}{r^6} \frac{R_2^6 R_1^6}{\left(R_2^3 - R_1^3\right)^2}.$$

The maxima of $|\sigma_1 - \sigma_2|$ and of σ_{II}^D are both reached on the inner surface
$r = R_1$, and they are equal to

$$\frac{3}{2}|p_1 - p_2| \frac{R_2^3}{R_2^3 - R_1^3} \quad \text{and} \quad \frac{3}{4}|p_1 - p_2|^2 \frac{R_2^6}{\left(R_2^3 - R_1^3\right)^2}.$$

By taking, as in Section 14.1, $g' = \sqrt{2/3}g = \sqrt{2/3}k$, we see that the Tresca
and von Mises criteria coincide, and they give

$$|p_1 - p_2| \le \frac{2}{3}k \left(1 - \frac{R_1^3}{R_2^3}\right).$$

The right-hand side of the last inequality tends to $2k/3$, as $R_2 \to \infty$. This
result means that it is not necessarily useful to increase the thickness of
the container too much; for a container of given capacity (R_1 fixed) submitted to inner and outer pressures p_1 and p_2, there is no point in increasing the width ($R_2 \to +\infty$) indefinitely because the difference of pressures
that the container can bear in the elastic regime remains limited to $2k/3$, as
$R_2 \to \infty$.

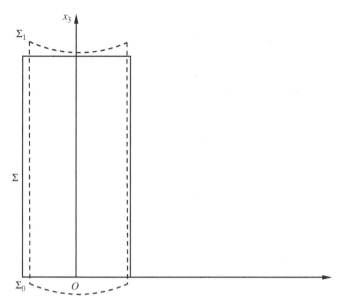

Figure 14.5 Vertical body subjected to gravitation.

14.4. Deformation of a vertical cylindrical body under the action of its weight

We now consider a vertical cylindrical body deforming solely under the action of gravity (Figure 14.5).

We assume that Ox_3 is the upward vertical and that this axis coincides with the axis of the cylinder and with the line joining the centers of mass of the sections (of the cylinder). Let Σ be the lateral surface, Σ_1 the upper section, and Σ_0 the lower section. We will specify later what happens on Σ_1 (we will assume that a "small" region around A is fixed).

The volume forces (gravity forces) are given by

$$f = -\rho g e_3.$$

We propose, for the stress tensor, a solution of the form

$$\sigma_{33} = \rho g x_3,$$
$$\sigma_{ij} = 0 \quad \text{for the other pairs } (i, j).$$

Then, all the equilibrium relations are satisfied. Furthermore, the boundary condition $\sigma \cdot n = 0$ is satisfied on Σ and Σ_0.

We thus obtain

$$\varepsilon_{11} = \varepsilon_{22} = -\frac{\nu\rho g}{E}x_3, \qquad \varepsilon_{33} = \rho g \frac{x_3}{E},$$

$$\varepsilon_{ij} = 0 \quad \text{if } i \neq j.$$

We then need to determine u (the existence of u is guaranteed because the compatibility conditions for the ε_{ij} are satisfied). Because $u_{i,i} = \varepsilon_{ii}$, we obtain

$$u_1 = -\frac{\nu\rho g}{E}x_1 x_3 + \varphi_1(x_2, x_3),$$

$$u_2 = -\frac{\nu\rho g}{E}x_2 x_3 + \varphi_2(x_3, x_1),$$

$$u_3 = \frac{\rho g}{2E}x_3^2 + \varphi_3(x_1, x_2);$$

the functions $\varphi_1, \varphi_2, \varphi_3$ remain to be determined. The relations $u_{i,j} + u_{j,i} = 0$, $i \neq j$, give

$$\varphi_{1,2} + \varphi_{2,1} = 0,$$

$$\varphi_{2,3} + \varphi_{3,2} = \frac{\nu\rho g}{E}x_2,$$

$$\varphi_{1,3} + \varphi_{3,1} = \frac{\nu\rho g}{E}x_1.$$

We set $\psi_1 = \varphi_1$, $\psi_2 = \varphi_2$, $\psi_3 = -\frac{1}{2}(\nu\rho g/E)(x_1^2 + x_2^2) + \varphi_3$, and we find

$$\psi_{1,2} = -\psi_{2,1}, \qquad \psi_{2,3} = -\psi_{3,2}, \qquad \psi_{3,1} = -\psi_{1,3}.$$

Because, moreover, $\varphi_{1,1} = \varphi_{2,2} = \varphi_{3,3} = 0$, we have, similarly, $\psi_{1,1} = \psi_{2,2} = \psi_{3,3} = 0$. Consequently, $\varepsilon_{ij}(\psi) = 0$, with $\psi = (\psi_1, \psi_2, \psi_3)$; that is to say, ψ is a rigid displacement. This yields $\psi = \alpha \wedge x + \beta$, where α and β are constant vectors, that is,

$$\psi_1 = \alpha_2 x_3 - \alpha_3 x_2 + \beta_1,$$

$$\psi_2 = \alpha_3 x_1 - \alpha_1 x_3 + \beta_2,$$

$$\psi_3 = \alpha_1 x_2 - \alpha_2 x_1 + \beta_3.$$

We now assume that the displacement and the rotation at the point $A = (0, 0, \ell)$ vanish, which yields finally

$$u_1 = -\frac{\nu\rho g}{E}x_1 x_3,$$

$$u_2 = -\frac{\nu\rho g}{E}x_2 x_3,$$

$$u_3 = \frac{\rho g}{2E}\left[x_3^2 + \nu\left(x_1^2 + x_2^2\right) - \ell^2\right].$$

On the section Σ_1, we have already imposed the condition that the displacement and the rotation vanish at A. Furthermore, observing that

$$\sigma \cdot n = \rho g \ell e_3, \quad \text{on } \Sigma_1,$$

we see that the resultant of the external forces on Σ_1, equal to $\rho g \ell e_3 \times \text{area}(\Sigma_1)$, is opposed to the gravity force and the resulting momentum at A vanishes. This situation does not correspond to reality but, by the Saint-Venant principle (see Section 14.7), and if the bar is long enough ($\ell \gg$ diameter of Σ_1), the solution found is a rather good approximation to the exact solution if we are far enough from Σ_1.

For a point on the Ox_3 axis, the displacement is downward:

$$u_1 = u_2 = 0,$$

$$u_3 = -\frac{\rho g}{2E} \left(\ell^2 - x_3^2 \right).$$

The section $x_3 = c$ becomes, after deformation,

$$x_1' = x_1 \left(1 - \frac{\nu \rho g c}{E} \right),$$

$$x_2' = x_2 \left(1 - \frac{\nu \rho g c}{E} \right),$$

$$x_3' = c + \frac{\rho g}{2E}(c^2 - \ell^2) + \frac{\nu \rho g}{2E} \left(x_1^2 + x_2^2 \right)$$

$$\simeq \text{(to first order)}$$

$$\simeq c + \frac{\rho g}{2E}(c^2 - \ell^2) + \frac{\nu \rho g}{2E} \left(x_1'^2 + x_2'^2 \right).$$

The deformed surface of the section $x_3 = c$ is a revolution paraboloid with axis Ox_3. Finally, the contraction of the lateral dimensions is proportional to x_3 (the distance to the basis).

14.5. Simple bending of a cylindrical beam

We consider in this section a cylindrical bar of axis Ox_3 whose lateral surface Σ is not submitted to any force, the face Σ_1 being submitted to a torque $M_2 e_2$ orthogonal to Ox_3, and the face Σ_0 to the opposite torque (see Figure 14.6).

We make the following simplifying hypotheses that lead to a realistic solution if the bar is long enough and if we are far enough from the ends of

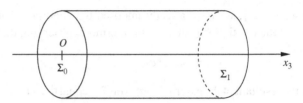

Figure 14.6 Bending of a cylindrical beam.

the bar (see also the Saint-Venant principle):

- The volume forces, including gravity, are negligible;
- The external forces on the surface Σ are neglected;
- The stresses on the surfaces Σ_1 and Σ_0 are of the form

$$F = \begin{cases} -cx_1e_3, & \text{on} \quad \Sigma_1, \\ cx_1e_3, & \text{on} \quad \Sigma_0, \end{cases}$$

and the axes $Ox_1x_2x_3$ are chosen as follows:

- $\int_{\Sigma_i} x_j \, dx_1 \, dx_2 = 0, i = 0, 1, \ j = 1, 2$, that is, $(0, 0)$ is the center of mass of Σ_0 (and Σ_1);
- e_1 and e_2 are the principal directions of the inertia tensor at O of Σ_0 (and Σ_1), that is,

$$\int_{\Sigma_i} x_1 x_2 \, dx_1 \, dx_2 = 0, \quad i = 0, 1,$$

and we denote by I_j the moment of inertia with respect to $Ox_j, \ j = 1, 2$.

The reduction elements of the helicoidal vector field associated with the external forces are then

$$R = \int_{\Sigma_0} F \, dx_1 \, dx_2 = ce_3 \int_{\Sigma_0} x_1 \, dx_1 \, dx_2 = 0,$$

and

$$\mathcal{M} = \int_{\Sigma_0} x \wedge (cx_1 e_3) \, dx_1 \, dx_2$$

$$= -c \int_{\Sigma_0} x_1^2 e_2 \, dx_1 \, dx_2 + c \int_{\Sigma_0} x_1 x_2 e_1 \, dx_1 \, dx_2$$

$$= -cI_2 e_2 = -M_2 e_2.$$

Similarly, on Σ_1, the resultant vanishes, and the resulting moment is equal to

$$M_2 e_2 = cI_2 e_2.$$

This problem can be solved for stresses only because there are no boundary conditions on the displacements. We will check that there exists a solution of the form

$$\sigma_{33} = -\frac{M_2}{I_2} x_1,$$

$$\sigma_{ij} = 0 \quad \text{for the other pairs } (i, j).$$

We have, of course,

$$\sigma_{ij,j} = 0 = -f_i,$$

and

$$\sigma_{ij} n_j = 0, \quad \text{on } \Sigma.$$

Furthermore, on Σ_1,

$$F_i = \sigma_{ij} n_j = \sigma_{13};$$

hence,

$$F = \sigma_{33} e_3 = -\frac{M_2}{I_2} x_1, \quad \text{on } \Sigma_1,$$

and similarly

$$F = -\sigma_{33} e_3 = \frac{M_2}{I_2} x_1, \quad \text{on } \Sigma_0.$$

It remains to determine the displacements. Thanks to the preceding formulas and to the stress–strain laws,

$$\varepsilon_{ij} = \frac{1 + \nu}{E} \sigma_{ij} - \frac{\nu}{E} \sigma_I \delta_{ij};$$

thus,

$$\varepsilon_{ij} = 0 \quad \text{for } i \neq j, \text{ and, for } i = j,$$

$$\varepsilon_{11} = \varepsilon_{22} = \frac{\nu}{E} \frac{M_2}{I_2} x_1 = u_{1,1} = u_{2,2},$$

$$\varepsilon_{33} = -\frac{M_2}{E I_2} x_1 = u_{3,3},$$

$$u_{1,2} + u_{2,1} = u_{1,3} + u_{3,1} = u_{2,3} + u_{3,2} = 0.$$

We can then take for u

$$u_1 = \frac{M_2}{2EI_2}\left[x_3^2 + v\left(x_1^2 - x_2^2\right)\right],$$

$$u_2 = \frac{vM_2}{EI_2}x_1x_2,$$

$$u_3 = -\frac{M_2}{EI_2}x_1x_3.$$

As we have already said, this result is a good approximation to the actual solution for a sufficiently long beam fixed at O or on Σ_0 far enough from Σ_0.

Remark 14.2: If we apply a more general torque $M_1e_1 + M_2e_2$ whose components on the principal inertia axes of the section are M_1 and M_2, one would obtain the same type of result. We will treat separately the case of a torque of the form M_3e_3 (torsion problem) in the next section.

Discussion of the result

We assume in what follows that the section Σ_0 is fixed.

1. Every part (or every segment) of the beam corresponding to a fixed x_1 and x_2 is called a fiber. A fiber is elongated or stretched if $x_1 < 0$, that is $u_3 > 0$; it is contracted or shortened if $x_1 > 0$, that is $u_3 < 0$.

2. The neutral fiber, corresponding to $x_1 = x_2 = 0$, becomes, after deformation

$$x_2' = x_2 = 0, \quad x_3' = x_3,$$

$$x_1' = \frac{M_2}{2EI_2}x_3^2 = \text{(since } x_3' = x_3)\,\frac{M_2}{2EI_2}x_3'^2.$$

It is a parabola whose curvature is equal to $1/R = M_2/EI_2$ to first order.

Because we are dealing with small deformations, the arc of the parabola can be assimilated to the tangent circle at O of radius R (the osculatory circle at O).

This result is known as the Euler–Bernoulli law: the bending torque is proportional to the curvature of the beam. The coefficient of proportionality EI_2 is called the rigidity-to-bending modulus for the axis Ox_2. To increase EI_2, one has to increase $I_2 = \int_{\Sigma_0} x_1^2\,dx_1\,dx_2$.

Let us interpret this last point: For a given section area (prescribed amount of material volume), one increases the resistance to bending by reducing I_2; this leads to the classical I-shaped beams used in high-rise buildings and more generally in civil engineering.

3. The section $\mathcal{D}_a = \{x_3 = a\}$ of the initial state is found, after deformation, in the plane

$$x_3' = x_3 + u_3(x) = a\left(1 - \frac{M_2}{EI}x_1\right),$$

$$x_3 = a\left(1 - \frac{x_1}{R}\right).$$

This plane is perpendicular to the parabola corresponding to the deformation of the neutral fiber.

These results are classical ones in the theory of the elasticity of curvilinear materials in civil engineering.

14.6. Torsion of cylindrical shafts

A cylindrical shaft with axis Ox_3 is subjected to torques $-Me_3$ and Me_3 (see Figure 14.6) on its faces Σ_0 ($x_3 = 0$) and Σ_1 ($x_3 = \ell$).

We assume that there are no volume forces and no forces exerted on the lateral surface Σ. The actions on Σ_0 and Σ_1 will be specified later on. We will further assume that Ox_3 coincides with the line joining the centers of mass of the sections of the cylinder.

For the stresses, we look a priori for a solution for which all the σ_{ij} vanish except for σ_{13} and σ_{23}, which we will assume independent of x_3. The equilibrium equations then give

$$\sigma_{13,3} = 0, \quad \sigma_{23,3} = 0,$$

$$\sigma_{31,1} + \sigma_{32,2} = 0 = \sigma_{13,1} + \sigma_{23,2}.$$

Thus, $\sigma_{13,1} = -\sigma_{23,2}$, $(\sigma_{23}, -\sigma_{13})$ is then a gradient, and there exists a function $\theta = \theta(x_1, x_2)$ such that

$$\sigma_{13} = \mu\alpha\theta_{,2}, \quad \sigma_{23} = -\mu\alpha\theta_{,1},$$

where α is the rotation angle, which is assumed to be small. Because $s = \sigma_I = 0$, the Beltrami equation becomes

$$\Delta\sigma_{i\ell} = 0,$$

which yields

$$\Delta\theta_{,2} = \Delta\theta_{,1} = 0,$$

and thus

$$\Delta\theta = \text{const.} = C.$$

We obtain the strain tensor in the form

$$\varepsilon_{13} = \frac{\alpha}{2}\theta_{,2}, \quad \varepsilon_{23} = -\frac{\alpha}{2}\theta_{,1},$$

$$\varepsilon_{ij} = 0 \quad \text{for all the other pairs } (i, j).$$

Finally, for the displacements, we are going to show that there exists a solution of the form

$$u_1 = -\alpha x_2 x_3, \quad u_2 = \alpha x_1 x_3, \quad u_3 = \alpha\varphi(x_1, x_2, x_3),$$

where φ, as well as θ and the constant C, remain to be determined.

Because $u_{3,3} = 0$, we deduce that φ only depends on x_1 and x_2. Furthermore,

$$\varepsilon_{13} = \frac{1}{2}(-\alpha x_2 + \alpha\varphi_{,1}) = \frac{\alpha}{2}\theta_{,2},$$

$$\varepsilon_{23} = \frac{1}{2}(\alpha x_1 + \alpha\varphi_{,2}) = -\frac{\alpha}{2}\theta_{,1};$$

hence,

$$2(\varepsilon_{13,2} - \varepsilon_{23,1}) = -2\alpha = \alpha\Delta\theta,$$

and thus $C = -2$, and

$$\Delta\theta + 2 = 0.$$

Let us now consider the function $\psi = \theta + \frac{1}{2}(x_1^2 + x_2^2)$. It satisfies

$$\Delta\psi = \Delta\theta + 2 = 0$$

and

$$\psi_{,2} = \varphi_{,1},$$

$$\psi_{,1} = -\varphi_{,2},$$

which yields

$$\Delta\varphi = 0.$$

The function $\varphi + i\psi$ is holomorphic, and we are thus able to use the theory of complex analysis. Furthermore, we will have the complete solution of the problem as soon as we know one of the functions θ, φ, or ψ.

Boundary conditions

On Σ, the condition $\sigma_{ij} n_j = 0$ reads

$$\sigma_{31} n_1 + \sigma_{32} n_2 = 0,$$

and thus

$$\theta_{,2} n_1 - \theta_{,1} n_2 = 0.$$

Consequently,

$$\theta_{,2} \tau_2 + \theta_{,1} \tau_1 = 0;$$

that is to say,

$$\frac{\partial \theta}{\partial \tau} = 0;$$

that is, θ is constant on Σ. Because θ is defined up to an additive constant, we can assume that the constant vanishes and that

$$\theta = 0, \quad \text{on } \Sigma.$$

Thus, θ is the solution of the following boundary value problem, which is called the Dirichlet problem:

$$\begin{cases} \Delta\theta + 2 = 0, & \text{on } \Sigma_0, \\ \theta = 0, & \text{on } \partial\Sigma_0. \end{cases}$$

One can prove that this problem possesses a unique solution (that thus only depends on the section Σ_0 of the cylinder). The function θ is sometimes called the stress function of the torsion problem. It allows us to determine the displacements and the stresses.

To complete the mechanical study, we now specify more precisely the forces exerted on Σ_1, where $\sigma \cdot n = (\sigma_{13}, \sigma_{23}, 0)$. The resultant of the external forces on Σ_1 vanishes:

$$R = \int_{\Sigma_1} (\sigma_{13}, \sigma_{23}, 0) \, dx_1 \, dx_2 = \int_{\Sigma_1} \mu\alpha(\theta_{,2}, -\theta_{,1}, 0) \, dx_1 \, dx_2$$

$$= \int_{\partial\Sigma_1} \mu\alpha(n_2, -n_1, 0) \theta \, d\ell = 0.$$

Their resulting momentum at O is equal to $M e_3$, with

$$M = \int_{\Sigma_1} (x_1 \sigma_{23} - x_2 \sigma_{13}) \, dx_1 \, dx_2$$

$$= \int_{\Sigma_1} -\mu\alpha(x_1 \theta_{,1} + x_2 \theta_{,2}) \, dx_1 \, dx_2$$

$$= -\mu\alpha \int_{\Sigma_1} \left[\frac{\partial}{\partial x_1}(x_1\theta) + \frac{\partial}{\partial x_2}(x_2\theta) - 2\theta \right] dx_1\,dx_2$$

$$= -\mu\alpha \int_{\partial\Sigma} x \cdot n\theta\,d\ell + 2\mu\alpha \int_{\Sigma_1} \theta\,dx_1\,dx_2.$$

Therefore, setting $D = 2\mu \int_{\Sigma_1} \theta\,dx_1\,dx_2$, we find

$$M = D\alpha e_3.$$

The constant D is called the rigidity-to-torsion modulus of the shaft. We see that the moment of the torque applied to Σ_1 is proportional to the rotation angle α.

In the particular case of a cylindrical shaft of radius a, it is easy to see that $2\theta = a^2 - x_1^2 - x_2^2$ is a solution of the Dirichlet problem defining θ; then, by a straightforward calculation, we find

$$D = \frac{\pi a^4 \mu}{2}.$$

The rigidity-to-torsion modulus of a circular shaft is thus proportional to the fourth power of its radius.

Elasticity limit (for a cylindrical shaft of radius a)

We have

$$\sigma_{13}^2 + \sigma_{23}^2 = \sigma_{II}^D = \mu^2\alpha^2|\nabla\theta|^2$$
$$= \mu^2\alpha^2 \left(x_1^2 + x_2^2\right).$$

The maximum of σ_{II}^D is reached on $\partial\Sigma_0$ (it can be proved that this fact is true for a general cylinder), and it is equal to $\mu^2\alpha^2 a^2$. The von Mises criterion then gives

$$\mu^2\alpha^2 a^2 \leq \frac{k^2}{3};$$

that is,

$$\alpha \leq \frac{k}{\mu a\sqrt{3}}.$$

The maximal intensity of the torque that can be applied is given by

$$M \leq \frac{\pi a^3}{2\sqrt{3}}k.$$

14.7. The Saint-Venant principle

In this short section, we restrict ourselves to stating the Saint-Venant principle inasmuch as the previous sections contain numerous applications and illustrations of this principle.

According to this principle, if on some part Σ_0 of $\partial\Omega$, instead of the condition $u = 0$ on Σ_0, we write that the displacement $u(A)$ and the rotation $\omega(A)$, at a point A of Σ_0 vanish, or if $\Sigma_0 \subset \partial\Omega$ and we replace, on Σ_0, the surface forces of density F ($\sigma \cdot n = F$) by another set of surface forces producing the same helicoidal vector field (on Σ_0), then, **far from** Σ_0, the displacements and the stresses are unchanged to first approximation.

This principle justifies the simplifying hypotheses we have made in the previous sections. Several forms of it that are more precise in their formulation have been rigorously proven.

Exercises

1. We consider the tensor σ with components, with respect to an orthonormal frame, given by

$$\sigma_{11} = \sigma_{12} = \sigma_{22} = 0, \sigma_{13} = Ax_1^2 + Bx_2^2 + C,$$
$$\sigma_{23} = Dx_1x_2, \sigma_{33} = Ex_1 + Fx_1x_3.$$

a) Find the conditions that the constants A, B, C, D, E, F must satisfy for this tensor to be able to be the stress tensor of an elastic material with Poisson's coefficient ν, the volume forces being negligible.

b) We assume that these conditions are satisfied and that σ is the stress tensor of a cylindrical body $(\mathcal{C})$ defined by $x_1^2 + x_2^2 \le a$, $0 \le x_3 \le l$, and such that

• (i) the external forces on the lateral surface vanish;

• (ii) on the base $x_3 = l$, the HVF associated with the external forces is equivalent to a slider (a single force) with intensity P with support containing the center of the section (i.e., $\mathcal{R} = (P, 0, 0)$ and $\mathcal{M} = 0$); we consider the resulting moment at the center of the base.

Compute the constants A, B, C, D, E, and F.

2. We consider an elastic medium at equilibrium with respect to an orthonormal frame $(0x_1x_2x_3)$ such that the stress tensor has components

$$\sigma_{11} = \sigma_{22} = \sigma_{12} = \sigma_{13} = \sigma_{23} = 0,$$
$$\sigma_{33} = \alpha(l - x_3).$$

a) Compute the external forces.
b) Compute the components ε_{ij} of the deformation tensor in terms of the Young's modulus E and the Poisson's coefficient v.
c) Compute the displacement field u, assuming that the origin O belongs to the medium.

3. An elastic body with density ρ and with Lame's coefficients λ and μ is represented by a cylindrical pipe limited by 2 circular, coaxial cylinders with axis $(0x_3)$ and radii R_1 and R_2. We assume that the body is at equilibrium and that the volume forces are negligible. Let M be a point of the pipe, H its orthogonal projection onto $(0x_3)$ and set $HM = r$, $e_r = \dfrac{1}{r}\overrightarrow{HM}$.

a) We assume that the displacement is of the form

$$u = U(r)e_r.$$

Compute div u and curl u. Write the ordinary differential equation satisfied by U.

b) Solve this equation, assuming that the pressures inside and outside the pipe are P_1 and P_2 respectively.

CHAPTER FIFTEEN

Energy theorems, duality, and variational formulations

In this chapter, we present some principles of energy minimization that characterize equilibria in elastostatics; we also introduce two related concepts, namely the concept of duality (duality for the variational principles and duality between displacements and stresses), and the concept of variational formulations. Apart from their relevance to mechanics, these concepts also play an important role in the numerical approximation of the problems we consider (but we do not address this here of course), and, in other related forms, in many other areas of science.

15.1. Elastic energy of a material

In an isotropic elastic material, we can define a quadratic functional $w = w(\varepsilon)$, called the energy function, such that:

$$\frac{\partial w(\varepsilon)}{\partial \varepsilon_{ij}} = \sigma_{ij}, \quad \forall\, i, j, \tag{15.1}$$

for every tensor field ε, and thus Eq. (15.1) is then identical to the stress–strain law of the material. This function is thus defined by

$$w(\varepsilon) = \frac{1}{2}\{\lambda \varepsilon_{kk}\varepsilon_{\ell\ell} + 2\mu \varepsilon_{ij}\varepsilon_{ij}\};$$

hence,

$$2w(\varepsilon) = 2\mu \left(\varepsilon_{ij}^D + \frac{\varepsilon_{\ell\ell}}{3}\delta_{ij}\right)\left(\varepsilon_{ij}^D + \frac{\varepsilon_{kk}}{3}\delta_{ij}\right) + \lambda(\varepsilon_{kk})^2,$$

or else

$$w(\varepsilon) = \frac{1}{2}\left(\frac{2\mu}{3} + \lambda\right)(\varepsilon_{kk})^2 + \mu\varepsilon_{ij}^D\varepsilon_{ij}^D. \tag{15.2}$$

Consequently, w is a positive definite quadratic form in ε (we will always assume that μ and $\kappa = \lambda + 2\mu/3$ are strictly positive).

If we now consider a tensor field ε defined on Ω, $x \in \Omega \mapsto \varepsilon(x)$, we can associate with it the functional

$$W(\varepsilon) = \int_{\Omega} w[\varepsilon(x)]\, dx.$$

We have the following result, which provides a physical interpretation of W and justifies its introduction:

Proposition 15.1. *We consider an elastic system S that is in its natural state ($\varepsilon = 0$) and fills the volume Ω at time $t = 0$ and that is at time $t = t_1$ in the deformation state corresponding to the deformation tensor ε. Then, $-W(\varepsilon)$ is, to first approximation, the work of the internal forces between times 0 and t_1. Thus, $W(\varepsilon)$ is the amount of energy stored in stresses between 0 and t_1.*

Proof: We know that the power of the internal forces is, at each time, given by

$$\mathcal{P}_{\text{int}} = -\int_{\Omega_t} \tilde{\sigma}_{ij}(x', t)\tilde{\varepsilon}_{ij}[U(x', t)]\, dx'$$

$$= -\int_{\Omega_0} \tilde{\sigma}_{ij}[\Phi(x, t)]\tilde{\varepsilon}_{ij}\left[\frac{\partial \Phi(x, t)}{\partial t}\right] \frac{D\Phi}{Dx}\, dx.$$

In linear elasticity, we saw in Chapter 13 that $D\Phi/Dx \simeq 1$ up to first order, and thus, to first approximation

$$\mathcal{P}_{\text{int}} = -\int_{\Omega} \sigma_{ij}(x, t)\frac{\partial \varepsilon_{ij}}{\partial t}(x, t)\, dx,$$

where $\Omega = \Omega_0$ is the undeformed state (at $t = 0$). Consequently,

$$\mathcal{P}_{\text{int}} = -\int_{\Omega} \frac{\partial w[\varepsilon(x, t)]}{\partial \varepsilon_{ij}} \frac{\partial \varepsilon_{ij}(x, t)}{\partial t}\, dx$$

$$= -\frac{d}{dt} \int_{\Omega} w[\varepsilon(x, t)]\, dx$$

$$= -\frac{d}{dt} W[\varepsilon(t)].$$

The work of the internal forces between times 0 and t_1 is, by definition, the integral of $\mathcal{P}_{\text{int}}$ between 0 and t_1; it is thus equal to

$$-\int_0^{t_1} \frac{d}{dt}[W(\varepsilon(t))]\, dt = -W[\varepsilon(t_1)] + W[\varepsilon(0)]$$

$$= -W(\varepsilon),$$

because $\varepsilon = \varepsilon(t_1)$ and $\varepsilon(0) = 0$.

Taking this result into account, we adopt the following definition:

Definition 15.1. *If ε is a tensor field defined on Ω, we say that $W(\varepsilon)$ is the elastic deformation energy of the field ε; $w[\varepsilon(x)]$ is the volume rate of deformation energy at x, $x \in \Omega$.*

15.2. Duality – generalization

We can also consider the conjugate quadratic form of w that is defined by

$$w^*(\sigma) = \underset{\varepsilon_{ij}}{\text{Sup}}\{\varepsilon_{ij}\sigma_{ij} - w(\varepsilon)\}. \tag{15.3}$$

A straightforward calculation gives

$$w^*(\sigma) = \frac{1}{2}\left[\frac{1+\nu}{E}\sigma_{ij}\sigma_{ij} - \frac{\nu}{E}(\sigma_{\ell\ell})^2\right];$$

we note that, in a symmetric way,

$$\frac{\partial w^*(\sigma)}{\partial \sigma_{ij}} = \varepsilon_{ij}, \quad \forall\, i, j. \tag{15.4}$$

We can also write

$$2w^*(\sigma) = \frac{1+\nu}{E}\left(\sigma_{ij}^D + \frac{\sigma_{kk}}{3}\delta_{ij}\right)\left(\sigma_{ij}^D + \frac{\sigma_{\ell\ell}}{3}\delta_{ij}\right) - \frac{\nu}{E}(\sigma_{\ell\ell})^2$$

$$= \frac{1+\nu}{E}\sigma_{ij}^D\sigma_{ij}^D + \frac{1+\nu}{3E}(\sigma_{kk})^2 - \frac{\nu}{E}(\sigma_{\ell\ell})^2$$

$$= \frac{1+\nu}{E}\sigma_{ij}^D\sigma_{ij}^D + \frac{1-2\nu}{3E}(\sigma_{kk})^2.$$

Thus,

$$w^*(\sigma) = \frac{1+\nu}{E}\left(\sigma_{II}^D\right)^2 + \frac{1-2\nu}{3E}\sigma_I^2, \tag{15.5}$$

and, therefore, $w^*(\sigma)$ is also a positive definite quadratic form in σ (upon the assumption that $-1 < \nu < \frac{1}{2}$ and $E > 0$).

A direct consequence of Eq. (15.3) is that

$$w(\varepsilon) + w^*(\sigma) - \varepsilon_{ij}\sigma_{ij} \geq 0 \tag{15.6}$$

for every pair of tensors ε, σ. Furthermore, σ and ε are related by the stress–strain law if and only if

$$w(\varepsilon) = w^*(\sigma) = \frac{1}{2}\varepsilon_{ij}\sigma_{ij}. \tag{15.7}$$

Indeed, when Eq. (15.7) holds, $\varepsilon_{ij}\sigma_{ij} - w(\varepsilon) = w^*(\sigma)$, and thus the supremum in Eq. (15.3) is indeed reached at ε. By writing that the derivatives with respect to the ε_{ij} of $\varepsilon_{k\ell}\sigma_{k\ell} - w(\varepsilon)$ vanish, we obtain the following relations:

$$\sigma_{ij} = \frac{\partial w(\varepsilon)}{\partial \varepsilon_{ij}}, \quad \forall\, i, j; \tag{15.8}$$

this relation is identical to the stress–strain law. Conversely, because the function of ε, $\varepsilon \mapsto \varepsilon_{ij}\sigma_{ij} - w(\varepsilon)$ is convex, the conditions of Eq. (15.8) mean that this function reaches its maximum at ε.

Now, for every pair of tensor fields ε, σ defined on Ω, we set

$$W^*(\sigma) = \int_\Omega w^*[\sigma(x)]\,dx,$$

$$\langle \varepsilon, \sigma \rangle = \int_\Omega \varepsilon_{ij}(x)\sigma_{ij}(x)\,dx;$$

as a consequence of the relations of Eqs. (15.6), (15.7) and (15.8), we have the following.

Proposition 15.2. *For every pair of tensor fields σ, ε defined on Ω, we have*

$$W(\varepsilon) + W^*(\sigma) - \langle \varepsilon, \sigma \rangle \geq 0. \tag{15.9}$$

Furthermore, equality occurs in inequality (15.9) if and only if

$$W(\varepsilon) = W^*(\sigma) = \frac{1}{2}\langle \varepsilon, \sigma \rangle, \tag{15.10}$$

and, in this case,

$$\sigma_{ij}(x) = \frac{\partial w}{\partial \varepsilon_{ij}}[\varepsilon(x)], \tag{15.11}$$

for every $x \in \Omega$.

Proof: Inequality (15.9) is deduced from inequality (15.6) by integration. Furthermore, if equality occurs in (15.9), then

$$\int_\Omega [w(\varepsilon) + w^*(\sigma) - \varepsilon_{ij}\sigma_{ij}]\,dx = 0;$$

hence, according to inequality (15.6):

$$w[\varepsilon(x)] + w^*[\sigma(x)] - \varepsilon_{ij}(x)\sigma_{ij}(x) = 0, \quad \forall\, x \in \Omega.$$

Therefore, Eq. (15.7) is true for every x, which yields Eq. (15.10) by integration, and in this case Eq. (15.8) is true for every x, which is equivalent to saying that Eq. (15.11) holds. Conversely, if the condition of Eq. (15.10) takes place, we easily check that there is equality in inequality (15.9).

Definition 15.2. *Let σ be a tensor field defined on Ω. We say that $W^*(\sigma)$ is the elastic stress energy of the field σ; $w^*[\sigma(x)]$ is the volumetric rate of stress energy at x.*

Generalization

Until now, we have restricted ourselves, as far as linear elasticity is concerned, to homogeneous isotropic materials; that is to say, materials that possess the same behavior at every point (homogeneity) and in every direction (isotropy). When this is not so, the stress–strain law has the following more general form:

$$\sigma_{ij} = a_{ijhk}\varepsilon_{hk}, \tag{15.12}$$

where the elasticity moduli a_{ijhk} satisfy the symmetry conditions

$$a_{ijhk} = a_{jihk} = a_{ijkh},$$

and the positivity condition

$$a_{ijhk}\varepsilon_{ij}\varepsilon_{hk} \geq \alpha\varepsilon_{ij}\epsilon_{ij}, \quad \forall\, \varepsilon,$$

α being a strictly positive constant. The a_{ijhk} can depend on x, and the positivity condition would then be required at every point $x \in \Omega$.

We recover the isotropic case by taking

$$a_{ijkh} = \lambda\delta_{ij}\delta_{kh} + \mu(\delta_{ik}\delta_{jh} + \delta_{ih}\delta_{jk}).$$

By inverting the relations of Eq. (15.12), we can write

$$\varepsilon_{ij} = A_{ijkh}\sigma_{kh}, \tag{15.13}$$

the A_{ijhk} satisfying the same symmetry relations

$$A_{ijhk} = A_{jihk} = A_{ijkh},$$

and the positivity condition

$$A_{ijkh}\sigma_{ij}\sigma_{kh} \geq \alpha'\sigma_{ij}\sigma_{ij}, \quad \forall\, \sigma,$$

with $\alpha' > 0$.

All that precedes (and, in particular, Eqs. (15.1) and (15.4)) then extends to the nonisotropic case by taking

$$w(\varepsilon) = \frac{1}{2}a_{ijkh}\varepsilon_{ij}\varepsilon_{kh},$$

$$w^*(\sigma) = \frac{1}{2}A_{ijkh}\sigma_{ij}\sigma_{kh}.$$

Remark 15.1: We have restricted ourselves to linear elasticity. However, all that precedes can be extended to the case of *nonlinear* elasticity. The functions w, w^*, W, and W^* are then no longer quadratic forms of ε and σ but more general convex functions of ε and σ. A brief discussion of such situations will appear in Chapter 16.

15.3. The energy theorems

We consider, as in Chapter 13, the general problem of elastostatics: to find σ, ε, and u that are solutions of

$$(\mathcal{P}) \quad \begin{cases} \sigma_{ij,j} + f_i = 0, & \text{in } \Omega, \\ \sigma_{ij} = 2\mu\varepsilon_{ij} + \lambda\varepsilon_{kk}\delta_{ij}, \\ \varepsilon_{ij} = \dfrac{1}{2}(u_{i,j} + u_{j,i}), \\ \sigma_{ij}n_j = F_i, & \text{on } \Gamma_F, \\ u = U_d, & \text{on } \Gamma_u, \end{cases}$$

with $\Gamma_u \cup \Gamma_F = \partial\Omega$, $\Gamma_u \cap \Gamma_F = \emptyset$. We assume here that Γ_u and Γ_F are nonempty. Certain limit cases, corresponding to empty Γ_u or Γ_F will be studied separately.

Before stating the energy theorems, we give some definitions that will be useful hereafter.

Definition 15.3. *We say that a displacement field u defined on Ω is kinematically admissible for $(\mathcal{P})$ if it satisfies $u = U_d$ on Γ_u and if it satisfies certain regularity assumptions.*[1]

Definition 15.4. *We say that a stress tensor field σ defined on Ω is statically admissible for $(\mathcal{P})$ if it satisfies*

$$\sigma_{ij,j} + f_i = 0, \quad \text{in } \Omega,$$
$$\sigma_{ij}n_j = F_i, \quad \text{on } \Gamma_F,$$

as well as certain regularity assumptions.[2]

Definition 15.5. *We denote by $(\mathcal{P}_h)$ the homogeneous problem associated with $(\mathcal{P})$ corresponding to $f_i = 0$, $F_i = 0$, and $U_d = 0$. As in Definitions 15.3*

---

[1] In general, we will assume that u is piecewise C^1 or C^2. However, in modern functional analysis, we prefer to consider functions that belong to spaces called Sobolev spaces that are directly related here, from the physical point of view, to the set of *functions with finite energy*.

[2] As for u, we will consider, in general, continuous or piecewise C^1 tensor fields, but it may be desirable simply to consider tensor fields that are square integrable.

and 15.4, we define the admissible displacement and stress tensor fields for
$(\mathcal{P}_h)$.[3]

Definition 15.6.

1. *The potential energy of a kinematically admissible vector field u is defined by the expression*

$$V(u) = \frac{1}{2} W[\varepsilon(u)] - L(u),$$

where $\varepsilon(u)$ is the deformation tensor of elements

$$\varepsilon_{ij}(u) = \frac{1}{2}(u_{i,j} + u_{j,i}),$$

and

$$L(u) = \int_{\Omega} f \cdot u \, dx + \int_{\Gamma_F} F \cdot u \, d\Gamma.$$

2. *The potential energy of a statically admissible tensor field σ is defined by the expression*

$$V^*(\sigma) = -\frac{1}{2} W^*(\sigma) + K(\sigma),$$

where

$$K(\sigma) = \int_{\Gamma_u} \sigma_{ij} n_j (U_d)_i \, d\Gamma.$$

If u and σ are respectively kinematically and statically admissible for $(\mathcal{P})$, a straightforward calculation gives

$$L(u) + K(\sigma) = \int_{\Omega} f \cdot u \, dx + \int_{\partial\Omega} (\sigma \cdot n) u \, d\Gamma$$

$$= -\int_{\Omega} \sigma_{ij,j} u_i \, dx + \int_{\partial\Omega} (\sigma \cdot n) u \, d\Gamma$$

$$= \int_{\Omega} \sigma_{ij} u_{i,j} \, dx = \int_{\Omega} \varepsilon_{ij}(u) \sigma_{ij} \, dx.$$

We thus obtain the relation

$$L(u) + K(\sigma) = \langle \varepsilon(u), \sigma \rangle, \qquad (15.14)$$

a consequence of which is the following theorem.

Theorem 15.1. *Let u' be a kinematically admissible displacement field for $(\mathcal{P})$ and let σ' be a statically admissible stress field for $(\mathcal{P})$. Then,*

$$V^*(\sigma') \le V(u').$$

---

[3] The solution of problem $(\mathcal{P}_h)$ vanishes everywhere $(\sigma = \varepsilon = u = 0)$, but of course the admissible stress tensor or displacement fields for $(\mathcal{P}_h)$ do not all vanish.

If we further assume that problem ($\mathcal{P}$) *possesses solutions, and if* (u, σ) *is such a solution, then*

$$V^*(\sigma') \le V^*(\sigma) = V(u) \le V(u'). \tag{15.15}$$

It follows, in particular, that u achieves the minimum of $V(u')$ *among all the kinematically admissible displacements* u' *for* ($\mathcal{P}$), *and* σ *achieves the maximum of* $V^*(\sigma')$ *among all the statically admissible stress fields* σ' *for* ($\mathcal{P}$).

Proof: The relation $V^*(\sigma') \le V(u')$ immediately follows from Eq. (15.14) and the definition of V and V^*. For Eq. (15.15), we assume that the problem ($\mathcal{P}$) possesses solutions, and we denote such a solution by (u, σ). Of course, u is kinematically admissible, and σ is statically admissible. Furthermore, if u' is a kinematically admissible field,

$$W[\varepsilon(u')] \ge -W^*(\sigma) + \langle \varepsilon(u'), \sigma \rangle.$$

Because

$$\langle \varepsilon(u'), \sigma \rangle = L(u') + K(\sigma),$$

we see that

$$W[\varepsilon(u')] - L(u') \ge -W^*(\sigma) + K(\sigma);$$

hence,

$$V(u') \ge V^*(\sigma)$$

for every kinematically admissible field u'.

Similarly, if σ' is a statically admissible field,

$$\begin{aligned}
V^*(\sigma') &= -W^*(\sigma') + K(\sigma') \\
&= [\text{by Eq. (15.14)}] \\
&= -W^*(\sigma') + \langle \varepsilon(u), \sigma' \rangle - L(u) \\
&\le [\text{by inequality (15.9)}] \\
&\le W[\varepsilon(u)] - L(u) = V(u).
\end{aligned}$$

Let us further observe that

$$V(u) = V^*(\sigma)$$

because

$$W[\varepsilon(u)] + W^*(\sigma) = \langle \varepsilon(u), \sigma \rangle = L(u) + K(\sigma),$$

which yields

$$W[\varepsilon(u)] - L(u) = -W^*(\sigma) + K(\sigma).$$

Finally, we have proved that

$$V^*(\sigma') \le V^*(\sigma) = V(u) \le V(u')$$

for every kinematically admissible field u' and every statically admissible field σ'. This completes the proof of the theorem.

Remark 15.2: We infer from the preceding theorem that the solution of the problem of elastostatics $(\mathcal{P})$ is related to two variational problems concerning, respectively, the displacements and the stresses as follows:

To minimize $V(u')$ among the vector fields u' kinematically admissible for $(\mathcal{P})$;

To maximize $V^*(\sigma')$ among the tensor fields σ' statically admissible for $(\mathcal{P})$.

These two variational problems (problems of calculus of variations) are dual to each other in a sense that is defined in calculus of variations and in convex analysis. Theorem 15.1, and in particular Eq. (15.15), provide some of the relations existing between two general dual problems.

Let us finally notice that V and V^* are quadratic functions of u' and σ', respectively. From the physical point of view, Theorem 15.1 provides the energy principles (or theorems) that govern $(\mathcal{P})$.

Section 15.4 extends the study of the variational formulations related to $(\mathcal{P})$ in a sense that is useful for mechanics as well as for the mathematical theory and the numerical analysis of the problem.

15.4. Variational formulations

We introduce the symmetric bilinear forms associated with W and W^*, namely

$$\mathcal{W}(\varepsilon', \varepsilon'') = \frac{1}{2}[W(\varepsilon' + \varepsilon'') - W(\varepsilon') - W(\varepsilon'')] \quad \text{and}$$

$$\mathcal{W}^*(\sigma', \sigma'') = \frac{1}{2}[W^*(\sigma' + \sigma'') - W^*(\sigma') - W^*(\sigma'')].$$

(15.16)

Let (u, σ) be a solution of problem $(\mathcal{P})$. We saw in the previous section that

$$V(u) \le V(u')$$

for every kinematically admissible vector field u'; because V is convex, this relation is equivalent to

$$\frac{d}{d\lambda} V[u + \lambda(u' - u)]|_{\lambda=0} = 0, \quad \forall u',$$

or else

$$\mathcal{W}[\varepsilon(u), \varepsilon(u') - \varepsilon(u)] - L(u' - u) = 0 \qquad (15.17)$$

for every kinematically admissible vector field u'. Similarly, V^* being a concave function, the relation

$$V^*(\sigma') \leq V^*(\sigma), \qquad (15.18)$$

valid for every statically admissible vector field σ', is equivalent to

$$\mathcal{W}^*(\sigma, \sigma' - \sigma) - K(\sigma' - \sigma) = 0, \quad \forall \sigma'. \qquad (15.19)$$

On the other hand, we notice that Eq. (15.17) is equivalent to

$$\mathcal{W}[\varepsilon(u), \varepsilon(v)] - L(v) = 0 \qquad (15.17')$$

for every vector field v kinematically admissible for $\mathcal{P}_h$. Similarly, Eq. (15.19) is equivalent to

$$\mathcal{W}^*(\sigma, \tau) - K(\tau) = 0 \qquad (15.19')$$

for every tensor field τ statically admissible for $\mathcal{P}_h$.

We also notice that Eqs. (15.17) and (15.19) possess at most one solution. Indeed, if u' and u'' are two solutions of Eq. (15.17), we set $u = u' - u''$ and observe that

$$\mathcal{W}[\varepsilon(u), \varepsilon(u)] = W[\varepsilon(u)] = 0,$$

which yields

$$\varepsilon(u) = \varepsilon(u') - \varepsilon(u'') = 0$$

because $W(\varepsilon)$ is positive definite. Therefore, $u' = u''$ if $\Gamma_u \neq \emptyset$ and $u' = u'' +$ a rigid displacement if $\Gamma_u = \emptyset$. Similarly, if σ' and σ'' are two solutions of Eq. (15.19), and if $\sigma = \sigma' - \sigma''$, then

$$\mathcal{W}^*(\sigma, \sigma) = W^*(\sigma) = 0;$$

hence, $\sigma = 0$ and $\sigma' = \sigma''$.

Thus, Eq. (15.17) possesses at most one solution, namely u, and the same is true for Eq. (15.19). Furthermore, each of these equations is equivalent to problem $(\mathcal{P})$.

We could just as well derive this result by direct variational methods or, equivalently (for Eq. (15.17)), by writing the virtual power theorem for the kinematically admissible virtual velocity field v for $(\mathcal{P}_h)$ and by using the stress–strain laws. This will be the objective of Section 15.5.

We will now make Eqs. (15.17) and (15.19) explicit.

We saw that, under proper regularity assumptions, u is kinematically admissible if and only if $u = U_d$ on Γ_u. Therefore, Eq. (15.17) is equivalent to

$$2\mu \int_\Omega \varepsilon_{ij}^D(u)\varepsilon_{ij}^D(u - u')\,dx + \kappa \int_\Omega \varepsilon_{kk}(u)\varepsilon_{\ell\ell}(u - u')\,dx$$
$$= \int_\Omega f \cdot (u - u')\,dx + \int_{\Gamma_F} F \cdot (u - u')\,d\Gamma \tag{15.20}$$

for every u' such that $u' = U_d$ on Γ_u; hence, for $u = U_d$ on Γ_u and for every v such that $v = 0$ on Γ_u

$$2\mu \int_\Omega \varepsilon_{ij}^D(u)\varepsilon_{ij}^D(v)\,dx + \kappa \int_\Omega \varepsilon_{kk}(u)\varepsilon_{\ell\ell}(v)\,dx$$
$$= \int_\Omega f \cdot v\,dx + \int_{\Gamma_F} F \cdot v\,d\Gamma. \tag{15.20'}$$

Similarly, apart from the regularity properties, σ is statically admissible if and only if

$$\sigma_{ij,j} + f_i = 0, \quad \text{in } \Omega,$$
$$\sigma_{ij}n_j = F_i, \quad \text{on } \Gamma_F,$$

and Eq. (15.19) is thus equivalent to

$$\frac{1+\nu}{E} \int_\Omega \sigma_{ij}^D\left(\sigma_{ij}'^D - \sigma_{ij}^D\right)dx + \frac{1-2\nu}{3E} \int_\Omega \sigma_{jj}\left(\sigma_{kk}' - \sigma_{\ell\ell}\right)dx$$
$$= \int_{\Gamma_u} (\sigma_{ij}' - \sigma_{ij})\,n_j(U_d)_i\,d\Gamma \tag{15.21}$$

for every σ' such that $\sigma_{ij,j}' + f_i = 0$ in Ω, and $\sigma_{ij}'n_j = F_i$ on Γ_F; or else Eq. (15.19) is equivalent to

$$\frac{1+\nu}{E} \int_\Omega \sigma_{ij}^D\tau_{ij}^D\,dx + \frac{1-2\nu}{3E} \int_\Omega \sigma_{kk}\tau_{\ell\ell}\,dx$$
$$= \int_{\Gamma_u} \tau_{ij}n_j(U_d)_i\,d\Gamma \tag{15.21'}$$

for every τ such that $\tau_{ij,j} = 0$ in Ω and $\tau_{ij}n_j = 0$ on Γ_F.

In conclusion, we have established the following:

Theorem 15.2. *The solution* σ, ε, u *of problem* $(\mathcal{P})$ *is unique; if* $\Gamma_u = \emptyset$, *then u is unique up to a rigid displacement.*

This solution is characterized by Eqs. (15.20') (for u) and (15.21') (for σ).

We say that Eqs. (15.20') and (15.21') are the variational formulations associated with the variational problems introduced in Theorem 15.1 and Remark 15.2. In the calculus of variations, we also say that they are the Euler equations of the variational problem.

15.5. Virtual power theorem and variational formulations

We now show how the variational formulations may be directly deduced from the virtual power theorem. They can also be deduced from the equations of problem $(\mathcal{P})$ by direct mathematical calculations, but we will not do so.

Let $(u, \varepsilon = \varepsilon(u), \sigma)$ be the solution of problem $(\mathcal{P})$ and let u' be an arbitrary virtual velocity field on Ω, $\varepsilon' = \varepsilon(u')$ being the deformation rate field associated with u'. Then,

$$
\begin{aligned}
-\mathcal{W}(\varepsilon, \varepsilon') &= -2\mu \int_\Omega \varepsilon_{ij}^D(u)\varepsilon_{ij}^D(u')\,dx - \kappa \int_\Omega \varepsilon_{kk}(u)\varepsilon_{\ell\ell}(u')\,dx \\
&= -\int_\Omega \left[\varepsilon_{ij}^D(u') + \frac{\varepsilon_{\ell\ell}(u')}{3}\delta_{ij} \right]\left[2\mu\varepsilon_{ij}^D(u) + 3\kappa\frac{\varepsilon_{kk}}{3}(u) \right]dx \\
&= -\int_\Omega \varepsilon_{ij}(u') \cdot \sigma_{ij}\,dx.
\end{aligned}
$$

This last quantity is precisely the power produced by the stress field σ in the virtual velocity field u'. If u' is a displacement field (again arbitrary), we then find the work produced by these forces in the displacement field u'.

Thus, we deduce from the virtual power theorem that

$$
\mathcal{W}(\varepsilon, \varepsilon') = \int_\Omega f \cdot u'\,dx + \int_\Gamma (\sigma \cdot n) \cdot u'\,d\Gamma
$$

is the power of the external forces. We then recover Eqs. (15.17) and (15.19) if u' is kinematically admissible for $\mathcal{P}_{\mathrm{h}}$. Furthermore, for $u' = u$, we obtain

$$
\mathcal{W}(\varepsilon, \varepsilon) = 2W(\varepsilon) = \int_\Omega f \cdot u\,dx + \int_\Gamma (\sigma \cdot n) \cdot u\,d\Gamma,
$$

which means that, for an elastic system, the work of the external forces in the displacements $u(x)$ of the points of the system starting from the natural (rest) state is equal to twice the deformation energy.

We conclude with a mechanical interpretation of the variational formulations.

Reciprocity theorem

Let (u, ε, σ) and $(u', \varepsilon', \sigma')$ be two equilibrium states of an elastic system; then,

$$
\int_\Omega f \cdot u' \, dx + \int_\Gamma (\sigma \cdot n) \cdot u' \, d\Gamma = \mathcal{W}(\varepsilon, \varepsilon')
$$

$$
= \mathcal{W}(\varepsilon', \varepsilon)
$$

$$
= \int_\Omega f' \cdot u \, dx + \int_\Gamma (\sigma' \cdot n) \cdot u \, d\Gamma,
$$

and we deduce the following theorem.

Theorem 15.3 (Reciprocity Theorem). *Given two equilibrium states of the same elastic system, the work of the external forces of the first state in the displacement field of the second one is equal to the work of the external forces of the second one in the displacement field of the first one.*

CHAPTER SIXTEEN

Introduction to nonlinear constitutive
laws and to homogenization

Linear elasticity represents only a simplified and very particular behavior of solids. The purpose of this chapter is to present some simple examples of problems encountered when the constitutive laws are nonlinear, like some of the laws described in Chapter 5, when the stress tensor σ is a nonlinear function of the deformation tensor $\varepsilon(u)$ in the framework of nonlinear elasticity in small displacements (see Chapter 5). As a result, the corresponding equilibrium equations are nonlinear, contrary to the equations encountered in the previous chapters in Part 3 of this book. Nonlinear mechanical phenomena are at this time a very active domain of solid mechanics in connection with the research of new materials and with the study of their mechanical properties (polymers, composite materials, etc.).

This chapter will be rather short; we only consider stationary problems and sometimes limit ourselves to problems of mechanics that involve only one space variable.

In the first three sections of this chapter, the presentation is based on energy theorems similar to those of Chapter 15 and, as indicated in Remark 15.1, we consider energy functionals $w(\varepsilon)$ that are no longer quadratic functions of ε. In Section 16.1, in connection with nonlinear elasticity, we consider cases in which w is a strictly convex function of ε. In Section 16.2, in connection with plasticity, we consider energy functionals possessing some degeneracies: typically, $w(\varepsilon)$ is convex but not strictly convex, which may lead to discontinuities (cracks, sliding lines). Finally, in Section 16.3, we consider cases in which the law $\varepsilon \mapsto \sigma$ is not monotone, and then $w(\varepsilon)$ is a nonconvex function; this may lead to surprising and sometimes nonintuitive behaviors that have indeed been observed in materials. Section 16.4 is a brief introduction to nonhomogeneous materials (nonhomogeneities at the microscopic level); these materials have, to first approximation, a linear elastic behavior, but the study of their properties pertains to nonlinear analysis.

16.1. Nonlinear constitutive laws (nonlinear elasticity)

As indicated in the introduction, we consider in this section and in the two following ones materials whose equilibrium equations are given by energy principles similar to those presented in Chapter 15 (see also Section 5.3, Elastic media).

We start, as in Chapter 15, with a function $w = w(\varepsilon)$ such that

$$\frac{\partial w(\varepsilon)}{\partial \varepsilon_{ij}} = \sigma_{ij}, \quad \forall i, j, \tag{16.1}$$

which means that Eq. (16.1) is the constitutive law of the considered material (stress–strain law). The function $w(\varepsilon)$ will be nonquadratic in the case considered here, and thus Eq. (16.1) will be nonlinear. If, moreover, u is the displacement field of the material, and $\varepsilon(u)$ is the linearized displacement tensor, we also consider the quantity

$$W[\varepsilon(u)] = \int_\Omega w[\varepsilon(u)(x)]\,dx. \tag{16.2}$$

As in Chapter 15, the minimization of certain quantities related to w and W leads to the energy principles.

Before specifying the equilibrium equations and the associated energy principles, let us describe the type of functions $w(\varepsilon)$ that we consider in this section.

A typical example is given by a model of ice for glaciers for which

$$w(\varepsilon) = c \sum_{i,j} |\varepsilon_{ij}|^p, \quad p > 1, \tag{16.3}$$

where $c > 0$ is a physical constant to be specified, say $c = 1$. We then infer the constitutive law

$$\sigma_{ij} = \frac{\partial w(\varepsilon)}{\partial \varepsilon_{ij}} = p|\varepsilon_{ij}|^{p-2}\varepsilon_{ij}, \tag{16.4}$$

where $\varepsilon_{ij} = \frac{1}{2}(\partial u_i/\partial x_j + \partial u_j/\partial x_i)$.

Let us consider a nonlinear elastic body that would be in equilibrium under the action (as in Chapter 15) of volume forces f, of a prescribed displacement U_d on Γ_u, and of a prescribed traction $\sigma_{ij}n_j = F_i$ on Γ_F. Then, the equations of equilibrium are exactly the equations of the problem $(\mathcal{P})$ considered at the beginning of Section 15.3, Chapter 15, where the linear constitutive law is replaced by the law of Eq. (16.4). One can show, as in Chapter 15 (but the reasoning is now more difficult) that the displacement field at equilibrium minimizes the function

$$V(u) = \frac{1}{2}W[\varepsilon(u)] - L(u), \tag{16.5}$$

among all the displacement fields u that are kinematically admissible in the sense of Definition 15.3; here $W[\varepsilon(u)]$ has the expression given by Eqs. (16.2) and (16.3), and $L(u)$ is defined as in Chapter 15:

$$L(u) = \int_{\Omega} f \cdot u \, dx + \int_{\Gamma_F} F \cdot u \, d\Gamma. \tag{16.6}$$

We can also define an energy principle dual to the previous one. To do so, we introduce the conjugate function w^* of w (also called the Fenchel–Legendre transform of w):

$$w^*(\sigma) = \mathop{\mathrm{Sup}}_{\varepsilon_{ij}}\{\sigma_{ij}\varepsilon_{ij} - w(\varepsilon)\}, \tag{16.7}$$

where the supremum is taken among all the symmetric tensors ε_{ij}. For example, Eq. (16.7) yields, after a computation that we omit:

$$w^*(\sigma) = \frac{p^{p-1} - 1}{p^p} \sum_{i,j} |\sigma_{ij}|^{p/(p-1)}. \tag{16.8}$$

Also, given a stress–tensor field $\sigma = \sigma(x)$, we consider the integral

$$W^*(\sigma) = \int_{\Omega} w^*(\sigma(x)) \, dx. \tag{16.9}$$

With the same quantity $K(\sigma)$ as in Chapter 15,

$$K(\sigma) = \int_{\Gamma_u} \sigma_{ij} n_j (U_d)_i \, d\Gamma, \tag{16.10}$$

we can show that the stresses at equilibrium are solutions of the problem of minimizing

$$V^*(\sigma) = -\frac{1}{2}W^*(\sigma) + K(\sigma) \tag{16.11}$$

among all the stress fields σ that are statically admissible in the sense of Definition 15.4 of Chapter 15.

One can prove (and we assume this here) the existence and uniqueness of solutions for the equilibrium equations, or, equivalently, for each of the variational principles above, when the function $w(\varepsilon)$ is as given in Eq. (16.3) or for other classes of strictly convex functions $w(\varepsilon)$ that have a polynomial growth at infinity (in a suitable sense).

Remark 16.1: The analogue for fluids of the previous model corresponds, for instance, to the model of turbulence of Smagorinsky, where u is the velocity and σ is replaced by $\dot{\sigma} = \partial\sigma/\partial t$.

16.2. Nonlinear elasticity with a threshold (Henky's elastoplastic model)

What follows is a model of nonlinear elasticity with a threshold closely related to some problems of plasticity (see in Section 5.3 the parts related to Nonlinear elasticity and to Plasticity).[1]

We limit ourselves to the one-dimensional problem known as the antiplane shear problem. In this case, ε reduces to its first component ε_{11}, and thus we will write ε instead of ε_{11}.

We have [see Figure 16.1(a)]:

$$w(\varepsilon) = \begin{cases} \dfrac{\varepsilon^2}{2}, & \text{if } |\varepsilon| \le 1, \\[2mm] |\varepsilon| - \dfrac{1}{2}, & \text{if } |\varepsilon| \ge 1. \end{cases}$$

We note here that the function w is convex, but not strictly convex, contrary to the examples introduced in the previous section.

We deduce that $\sigma = \sigma_{11} = \partial w(\varepsilon)/\partial\varepsilon$ satisfies

$$\sigma(\varepsilon) = \begin{cases} \varepsilon & \text{if } |\varepsilon| \le 1, \\ \text{sign } \varepsilon & \text{if } |\varepsilon| \ge 1, \end{cases} \tag{16.12}$$

where sign $\varepsilon = 1$ if $\varepsilon > 0$, and -1 if $\varepsilon < 0$ [see Figure 16.1(b)]. We then compute the function

$$w^*(\sigma) = \mathop{\text{Sup}}_{\varepsilon \in \mathbb{R}} \{\varepsilon\sigma - w(\varepsilon)\};$$

it follows, by an elementary calculation, that

$$w^*(\sigma) = \begin{cases} \dfrac{\sigma^2}{2}, & \text{if } |\sigma| \le 1, \\[2mm] +\infty, & \text{if } |\sigma| > 1. \end{cases} \tag{16.13}$$

This type of constitutive law is connected with the elastoplastic constitutive laws introduced in Chapter 5: the constitutive law is linear below a certain

---

[1] The real plasticity phenomena are evolutive, nonrevertible phenomena that we do not consider here at all (see the Prandtl-Reuss law, section 5.3, Plasticity).

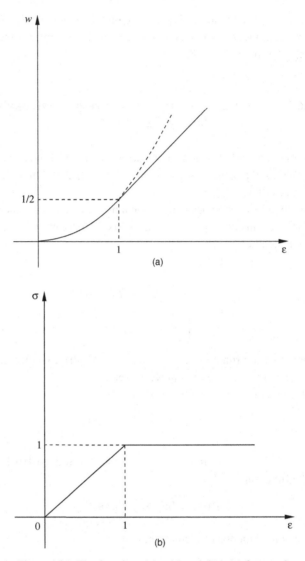

Figure 16.1 The functions (a) $w(\varepsilon)$ and (b) $\sigma(\varepsilon)$ for $\varepsilon \geq 0$.

threshold. We also recover the limit of elasticity criterion of von Mises: the material is elastic, linear for $|\sigma| < 1$. The condition $|\sigma| < 1$ thus corresponds to the limit of elasticity criterion of von Mises.

Let us consider the antiplane shear problem. We consider a section $0 < x = x_1 < 1$ of the material; the displacement occurs in the direction x_1 and is independent of x_2 and x_3. Concerning the boundary conditions, we assume, for instance, that the displacement $u = 0$ is given at $x = 0$ and that $\sigma = \sigma_{11}$ is given for $x = L$ ($\sigma = F$).

We then introduce the functions

$$W(\varepsilon) = \int_0^L w[\varepsilon(x)] \, dx,$$

$$W^*(\sigma) = \int_0^L w^*[\sigma(x)] \, dx,$$

for every tensor fields $\varepsilon = \varepsilon(u(x))$ and $\sigma(x)$ defined on $(0, L)$.

The analogues of the quantities L and K are

$$L(u) = Fu(1),$$

$$K(\sigma) = 0 \quad (U_d = 0 \text{ at } x = 0).$$

The energy principles associated with the problem are then exactly those from the previous section (or, as in Chapter 15):

To minimize $V(u)$ among all the u that are kinematically admissible and satisfy certain regularity assumptions, where

$$V(u) = W(\varepsilon(u)) - Fu(1);$$

and

To maximize $V^*(\sigma)$ among the σ that are statically admissible, where

$$V^*(\sigma) = W^*(\sigma).$$

The problem of the existence of solutions is difficult in this case; one of the new difficulties, if we make comparisons with the previous examples, is that the function w is no longer strictly convex but only convex. We refer the reader interested in existence problems for this type of functional to the specialized literature (see, e.g., Temam (1985) and the references therein).

16.3. Nonconvex energy functions

The stress-strain laws are generally deduced from the experiment of simple traction of a bar, such as described in linear elasticity in Chapter 14, Section 14.1. In the case of linear elasticity, as well as in the previous section in the case of nonlinear elasticity, the one-dimensional stress-strain law is monotonic,

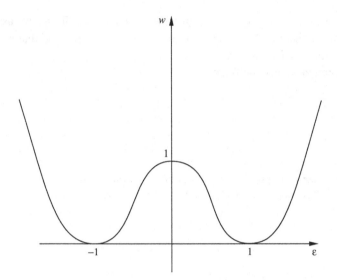

Figure 16.2 The function $w(\varepsilon)$.

yielding an energy functional $w(\varepsilon)$ that is convex with respect to ε. Even though this is not intuitive, the stress-strain law can be non-monotonic for some materials and for some regimes (the material then retracts when it is pulled); the fuction $w(\varepsilon)$ is nonconvex in that case. We give hereafter an overview of the complications that appear in the study of such materials.

We thus consider in this section nonlinear models of elasticity of the van der Waals type for which the energy functionals are no longer convex. For a one-dimensional problem for which $\varepsilon \simeq \varepsilon_{11}$ and $\sigma \simeq \sigma_{11}$, as in the previous section, we have, for instance, a function

$$w(\varepsilon) = (\varepsilon^2 - 1)^2, \tag{16.14}$$

(see Figure 16.2), and hence the constitutive law becomes

$$\sigma = 4\varepsilon(\varepsilon^2 - 1). \tag{16.15}$$

When the energy functional $w(\varepsilon)$ is no longer convex, the corresponding variational problems are more complicated, and the existence of solutions is not known (and is not true) in general.

Let us consider, for example, the problem of the minimization of

$$W[\varepsilon(u)] = \int_0^L w[u'(x)]\,dx = \int_0^L \left[\left| \frac{du}{dx} \right|^2 - 1 \right]^2 dx \tag{16.16}$$

among all the functions u that satisfy $u(0) = 0$ and $u(L) = 1$.

Introduction to nonlinear constitutive laws and to homogenization   255

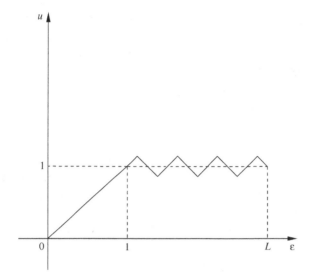

Figure 16.3 A minimizing sequence of Problem (16.16).

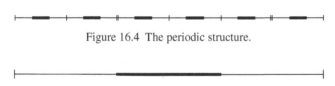

Figure 16.4 The periodic structure.

Figure 16.5 A period of the composite material.

One can easily check that when $L > 1$, the infimum is 0. However, this
infimum is not attained, but it can be approached arbitrarily closely $\{W[\varepsilon(u)]$
arbitrarily small$\}$ by, for example, the following functions u_n: $u_n(x) = x$ for
$0 < x < 1$, and for $1 < x < L$, a sawtooth behavior with amplitude $1/n$ and
$u'_n(x) = \pm 1$ alternately, as in Figure 16.3.

The reader interested in this problem (and in problems of buckling) may
consult the books of Antman (1995) and Ericksen (1998).

16.4. Composite materials: the problem of homogenization

We consider in this section a material with a periodic structure in one space
dimension, as in Figure 16.4. We further assume that each period is composed
of two or more different materials (see Figure 16.5), each one characterized by
a linear elastic constitutive law linking σ to ε. The principle of homogenization

consists typically of looking for a linear and homogeneous material equivalent, for small δ, to the periodic structure.

To illustrate the approach for the problem of homogenization, we consider the following "model" problem:

$$\begin{cases} \dfrac{d}{dx}\left[a\left(\dfrac{x}{\delta}\right)\dfrac{du^\delta}{dx}(x)\right] = f(x), \quad 0 < x < 1, \\[4mm] u^\delta(0) = u^\delta(1), \quad \dfrac{du^\delta}{dx}(0) = \dfrac{du^\delta}{dx}(1), \end{cases} \tag{16.17}$$

where $a = a(y)$ is periodic with period 1 and bounded from above and from below by $0 < \underline{a} \le a \le \bar{a} < +\infty$ and f is a function given and smooth on $(0,1)$. We assume, for the sake of simplicity, that $\delta = 1/N$, $N \in N^*$, and thus that $a(x/\delta)$ is periodic in x with period δ. Our aim is to find the limit of u^δ when $\delta \to 0$. We will actually obtain the limit u^0 of u^δ as the solution of an equation called the homogenized equation of Eq. (16.17). To do so, we set

$$b(x) = \frac{1}{a(x)}, \quad F(x) = \int_0^x f(\tau)\,d\tau,$$

and we have

$$p^\delta(x) = a\left(\frac{x}{\delta}\right)\frac{du^\delta}{dx} = F(x) - c_\delta; \tag{16.18}$$

hence,

$$\frac{du^\delta}{dx} = b\left(\frac{x}{\delta}\right)[F(x) - c_\delta], \tag{16.19}$$

where c_δ is given by the relation

$$0 = \int_0^1 \frac{du^\delta}{dx}\,dx = \int_0^1 b\left(\frac{x}{\delta}\right)[F(x) - c_\delta]\,dx. \tag{16.20}$$

It is clear that, as δ goes to 0, $b(x/\delta)$ converges in a weak sense (weak topology of $L^2(0, 1)$) to the average of b over the period denoted by $\langle b\rangle$, which does not vanish because it is between $\bar{a}^{-1}$ and $\underline{a}^{-1}$. It thus follows from Eq. (16.20) that

$$\lim_{\delta \to 0} c_\delta = \int_0^1 F(x)\,dx.$$

Then, we infer from Eqs. (16.18) and (16.19) that du^δ/dx and p^δ converge, in the same sense (weak topology of $L^2(0, 1)$), to the functions du^0/dx and

p^0 defined by

$$\frac{du^0}{dx}(x) = \langle b \rangle [F(x) - \langle F \rangle],$$

$$p^0(x) = F(x) - \langle F \rangle,$$

where

$$u^0(x) = \int_0^x \frac{du^0}{dx} dx,$$

and where $\langle \cdot \rangle$ denotes the average over the period. It follows that

$$p^0 = \langle b \rangle^{-1} \frac{du^0}{dx},$$

$$\frac{dp^0}{dx} = f,$$

and u^0 is the solution of the following linear problem:

$$\begin{cases} \dfrac{d}{dx}\left(\langle a^{-1} \rangle^{-1} \dfrac{du^0}{dx} \right) = f, \\[2mm] u^0(0) = u^0(1), \quad \dfrac{du^0}{dx}(0) = \dfrac{du^0}{dx}(1), \end{cases} \tag{16.21}$$

where $a^0 = \langle a^{-1} \rangle^{-1}$ is called the homogenized coefficient of the function a.

The reader who is interested in the subject may refer, among others, to the books on homogenization mentioned in the reference list, in particular the books by Bensoussan, Lions, and Papanicolaou (1978), and by Jikov, Kozlov, and Oleinik (1991).

Exercises

1. We consider the 1-periodic function $f(x) = \sin(2\pi x)$ defined on $\mathbb{R}$ and we set

$$f_\varepsilon(x) = f(\frac{x}{\varepsilon}), \quad x \in]a, b[, \quad a, b \in \mathbb{R}.$$

a) Show that f_ε cannot converge at every point.

b) Show that, for every interval $I =]\alpha, \beta[\subset]a, b[$,

$$\int_\alpha^\beta f_\varepsilon(x)\,dx \to 0 \text{ as } \varepsilon \to 0.$$

It follows from this result that f_ε converges to 0 in the weak topology of $L^2(a, b)$, since $\int_a^b f_\varepsilon(x)^2 dx$ is bounded independently of ε.

c) Show that $\int_a^b f_\varepsilon(x)^2 dx$ does not converge to 0 as $\varepsilon \to 0$ (i.e., f_ε does not converge to 0 in the strong topology of $L^2(a, b)$).

CHAPTER SEVENTEEN

Nonlinear elasticity and an application to biomechanics

In this chapter, we describe the equations of nonlinear elasticity in the context of large deformations. The particular case of hyperelasticity is then considered, and finally we present an application to biomechanics for the modeling of soft tissues. This chapter follows very closely the book by Ciarlet (1988), hereafter called [C88]; the models of soft tissues are borrowed from S. Jemiolo and J. J. Telega (2001).

17.1. The equations of nonlinear elasticity

We consider a system S corresponding to a nonlinearly elastic medium which occupies the domain $\Omega \subset \mathbb{R}^3$ in its reference state and the domain $\Omega_\Phi = \Phi(\Omega)$ in the deformed state. As usual Φ is the deformation, and we write $x = \Phi(a), a \in \Omega, x \in \Omega_\Phi; u(a) = \Phi(a) - a$ is the displacement of the point a.

Assuming that the body undertakes large deformations, we recall the general equations of motion in Lagrangian variables, introduced in Chapter 3:

$$f_a + \text{Div}_a \Pi = 0, \text{ in } \Omega, \tag{17.1}$$

where

$$f_a(a) = f(x) \det \mathbf{F}(a) = f(\Phi(a)) \det \mathbf{F}(a); \tag{17.2}$$

here, f is the density of the applied body forces per unit volume in the deformed configuration and f_a their density per unit volume in the reference configuration; $\mathbf{F}$ is the Jacobian matrix $\nabla_a \Phi = Dx/Da$, and $\text{Div}_a \Pi$ is the vector with ith component $\sum_{j=1}^3 \partial \Pi_{ij}/\partial a_j$. Furthermore, Π is the first

Piola-Kirchhoff tensor, defined by

$$\Pi = (\det \mathbf{F})\, \overline{\sigma} \cdot (\mathbf{F}^{-1})^T, \quad \Pi \cdot \mathbf{F}^T = \mathbf{F} \cdot \Pi^T, \qquad (17.3)$$

where $\overline{\sigma}(a) = \sigma(x) = \sigma(\Phi(a))$, σ being the Cauchy stress tensor in the deformed state. We also introduce the second Piola-Kirchhoff tensor

$$\mathbf{P} = \mathbf{F}^{-1} \cdot \Pi. \qquad (17.4)$$

We recall that P is symmetric, while Π is not in general.

Constitutive laws

It can be proven, using the principles of material indifference and isotropy,[1] that any constitutive law of nonlinear elasticity is of the form:

$$\mathbf{P} = \beta_0 I + \beta_1 C + \beta_2 C^2, \qquad (17.5)$$

where I is the identity tensor, $C = \mathbf{F}^T\mathbf{F}$ is the Cauchy–Green deformation tensor and β_0, β_1, and β_2 are functions of $a \in \Omega$ and C_I, C_{II}, C_{III} which are the three invariants of the tensor C. The reader interested in a proof of this assertion is referred to [C88]. As mentioned in Chapter 5, an essential complication here is that the equation of equilibrium (17.1) involves the first Piola-Kirchhoff tensor Π and not P, whereas the constitutive law (17.5) is for P; and there is no simple equivalent form of the constitutive law (17.5) for Π (see, however, Chapter 5, Section 3, for an equivalent equation involving the Cauchy stress tensor σ).

Reference configuration and natural state

In the reference state Ω, we have $\Phi = I, u = 0$ and $C = 0$. It thus follows from (17.5) that

$$\mathbf{P} = \beta_0 I, \qquad (17.6)$$

so that P is diagonal. This quantity is called the residual stress tensor at a point a of the reference state Ω and is denoted by $\mathbf{P}_R$. Now, since, in view of (17.5), P is not diagonal in general, it follows that isotropy only holds for particular reference configurations. Indeed, even though it is a priori possible to choose as a new reference configuration an arbitrary deformed configuration, the material will not be isotropic in general in such an arbitrary reference configuration.

---

[1] These are some of the general principles of rheology described in Chapter 3, Section 3.

Definition 17.1. *A reference configuration is called a natural state if the residual stress tensor* P_R *vanishes at each point.*

Remark 17.1: One can show (see [C88]) that, if a reference configuration is a natural state, then so is every reference configuration obtained from it by a rigid deformation.

Remark 17.2: Contrary to fluid mechanics, we assume in solid mechanics that that there exist natural reference states.

Constitutive laws near a reference configuration

It is natural to linearize the constitutive law (17.5) around X, where $X = \frac{1}{2}(C - I)$ is the deformation tensor. Indeed, in some sense, X measures the discrepancy between a given deformation Φ and a rigid deformation (for which $C = I$). We thus wish to linearize the stress tensor P corresponding to a deformed configuration Ω_Φ around the reference configuration corresponding to the particular rigid deformation I. One can show (see [C88]) that, thanks to isotropy, the linearization only involves two coefficients $\lambda = \lambda(a)$, $\mu = \mu(a)$, that is:

$$P = -pI + \lambda(TrX)I + 2\mu X + o(X), \tag{17.7}$$

where $\mu > 0$ and $3\lambda + 2\mu > 0$ from experimental tests (see [C88], Section 3.8).

   We then find that, near a natural state, $pI \equiv 0$, so that, neglecting the $o(X)$ term,

$$P = \lambda(TrX)I + 2\mu X. \tag{17.8}$$

This relation is precisely the one that defines linear elasticity in small deformations studied in Chapter 3.

Saint Venant-Kirchhoff material

A Saint Venant-Kirchhoff material is a material for which the stress tensor P is of the form

$$P = \lambda(TrX)I + 2\mu X, \tag{17.9}$$

where $I + 2X = C$.

   In what follows, we will consider the general constitutive law (17.5) and, sometimes, when indicated, Saint Venant-Kirchhoff materials. The Saint Venant-Kirchhoff materials are the simplest among the nonlinear models [C88], and they are quite popular in actual computations, where they are often used to model engineering structures in conjunction with finite elements methods.

17.2. Boundary conditions – boundary value problems

We consider, as in the previous chapters, a typical mechanical problem for which the exterior applied forces, defined in the deformed configuration, consist of:

   (i) given body forces with volume density f (in the deformed configuration);
  (ii) given surface traction forces with surface density g (in the deformed configuration)[2] on a part Γ_T of $\Gamma = \partial\Omega$;
       and,
 (iii) prescribed displacement on the complementary $\Gamma_u = \Gamma\backslash\Gamma_T$ of Γ_T.

An important difficulty appearing here is that (i) and (ii) will be easier to write in the deformed configuration, whereas (iii) will be simpler in the reference configuration. The same difficulty was mentioned in Chapter 13 for linear elasticity but we could, in that case, overcome the difficulty very easily through linearization.

Another difficulty is that some boundary value problems are global; this, typically, occurs when nonlocal surface or volumes forces are applied, that is the value of the force at one point depends on the values of the deformation at other points. We refer the interested reader to the book by Ciarlet ([C88], Section 2.7), for more details. Such situations can be encountered, e.g., with the balloon problem, for which the exterior boundary is subjected to a constant pressure load, whereas the interior boundary is subjected to a pressure which is a given function of the enclosed volume.

So, the typical mechanical problem will be described by the following boundary value problem:

$$\begin{cases} \mathrm{Div}_a \,\Pi + f_a = 0, \text{ in } \Omega, \\ \Pi\, n = g_a, \text{ on } \Gamma_T \subset \partial\Omega, \\ \Phi = \Phi_0, \text{ on } \Gamma_u = \Gamma\backslash\Gamma_T, \end{cases} \tag{17.10}$$

together with the constitutive law

$$P = \beta_0 I + \beta_1 C + \beta_2 C^2, \tag{17.11}$$

---

[2] In other parts of the book, the density of surface forces is denoted F instead of g, but we want here to avoid confusion with **F**.

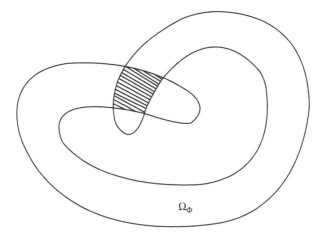

Figure 17.1 Superposition.

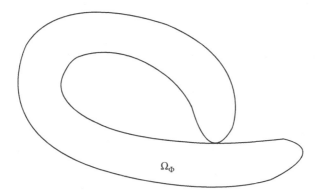

Figure 17.2 Contact.

where

$$\Pi = \mathbf{F} \cdot \mathbf{P}, \tag{17.12}$$

$$\mathbf{F} = \operatorname{grad}_a \Phi, \tag{17.13}$$

$$\Pi \cdot \mathbf{F}^T = \mathbf{F} \cdot \Pi^T. \tag{17.14}$$

We will also need some geometric conditions, in order to exclude inter-penetrations (see Figure 17.1), which, even though they are geometrically admissible, are unthinkable from the physical point of view. We will thus assume that Φ is one to one on $\overline{\Omega}$, and not only on Ω. Additional mechanical conditions will be needed in case of contacts (see Figure 17.2).

To conclude, we will call admissible a deformation field Φ such that

$$\det \operatorname{grad}_a \Phi > 0, \quad \text{in } \Omega, \tag{17.15}$$

$$\Phi = \Phi_0, \quad \text{on } \Gamma_u, \tag{17.16}$$

$$\Phi \text{ is one to one on } \overline{\Omega}. \tag{17.17}$$

17.3. Hyperelastic materials

A hyperelastic material is a material for which the first Piola-Kirchhoff tensor $\Pi = \Pi(a, \mathbf{F})$ satisfies

$$\Pi_{ij}(a, \mathbf{F}) = \frac{\partial W}{\partial \mathbf{F}_{ij}}(a, \mathbf{F}), \quad \forall a \in \Omega, \tag{17.18}$$

for every 3×3 positive definite matrix $\mathbf{F}$, where $W = W(a, \mathbf{F})$ is a given function, called the stored energy function (which was already introduced in the context of linear elasticity in Chapter 16 and denoted by w).

The virtual power theorem, applied in the reference configuration (see [C88], Section 2.6, for a direct derivation, starting from the equations of equilibrium), yields the following variational formulations:

$$\int_\Omega \Pi \cdot \nabla v \ da = \int_\Omega f_a \cdot v \ da + \int_{\Gamma_T} g_a \cdot v \ d\Gamma, \tag{17.19}$$

for every sufficiently smooth vector field $v = v(a)$ on Ω, which vanishes on Γ_u. We defined f_a in (17.2); g_a is defined in a similar way: g is the surface density, in the deformed configuration, of the traction forces applied on Γ_T and g_a is their surface density in the reference configuration, that is, according to Chapter 5, Section 4, $g(x)d\Gamma_x = g_a(a)d\Gamma_a$ or equivalently

$$g_a(a) = g(x)\frac{d\Gamma_x}{d\Gamma_a},$$

with the expression of $d\Gamma_x/d\Gamma_a$ given by formula (5.18) (where it is called $d\Gamma_t/d\Gamma_0$).

We will actually restrict ourselves to a special class of volume and surface forces according to the following definition.

Definition 17.2. (i) *The applied volume forces with density* $f_a = f_a(a) = f(\Phi(a))\det F(a)$ *in the reference configuration are called* conservative *if the integral*

$$\int_\Omega f_a \cdot v \, da$$

is the Gâteaux derivative of a functional $F : \{\Psi : \overline{\Omega} \to \mathbb{R}^3\} \mapsto F(\Psi) = \int_{\Omega} \overline{F}(a, \Psi(a))da,$ *namely*

$$F'(\Phi)v = \int_{\Omega} f_a \cdot v da, \qquad (17.20)$$

where the function $\overline{F}$ *is called the potential of the applied volume forces and* Φ *is the deformation.*

(ii) *The applied surface forces with density* g_a *in the reference configuration are called* conservative *if the integral*

$$\int_{\Gamma_T} g_a \cdot v d\Gamma$$

is the Gâteaux derivative of a functional $G : \{\Psi : \overline{\Omega} \to \mathbb{R}^3\} \mapsto G(\Psi) = \int_{\Gamma_T} \overline{G}(a, \Psi(a), \nabla\Psi(a))da,$ *namely*

$$G'(\Phi)v = \int_{\Gamma_T} g_a \cdot v d\Gamma, \qquad (17.21)$$

where the function G *is called the potential of the applied surface forces and* Φ *is the deformation.*

For instance, the centrifugal force in a body rotating with a constant angular velocity is a conservative volume force when expressed in a frame rotating with it. Similarly, an applied surface force which is a deadload, i.e., the density is independent of the particular deformation Φ considered, is conservative.

One advantage of considering conservative applied volume and surface forces is that we can prove that the solutions of the boundary value problem (17.10) are critical points of the functional $L(\Psi)$ defined by

$$L(\Psi) = \int_{\Omega} W(a, \mathbf{F})da - (F(\Psi) + G(\Psi)), \qquad (17.22)$$

for smooth mappings $\Psi : \overline{\Omega} \to \mathbb{R}^3$. That is

$$L'(\Phi)v = 0, \qquad (17.23)$$

for every smooth field $v : \overline{\Omega} \to \mathbb{R}^3$ that vanishes on Γ_0 (see [C88], p. 142).

The functional

$$\mathcal{W}(\Phi) = \int_{\Omega} W(a, \mathbf{F})da \qquad (17.24)$$

is called the strain energy; the functional L is called the total energy.

Using the principles of material indifference and isotropy, we then deduce (see [C88], Section 4.2) that the stored energy takes the form

$$W = W(a, C), \quad C = \mathbf{F}^{\mathrm{T}}\mathbf{F}. \qquad (17.25)$$

Furthermore, one then finds, in view of (17.4) and (17.18), that

$$P_{ij} = 2\frac{\partial W}{\partial C_{ij}}. \tag{17.26}$$

Isotropy also implies that the stored energy function satisfies the following additional relation:

$$W(a, C) = W(a, QCQ), \quad \forall Q \in O(3), \tag{17.27}$$

where $O(3)$ is the orthogonal group.

Finally, we assume the existence of a stress-free natural state such that

$$W(a, I) = 0, \tag{17.28}$$

$$\Pi(a, I) = 0, \tag{17.29}$$

I being the identity tensor.

For such materials, the equilibrium problem becomes a calculus of variations problem. See the studies in the book by Ciarlet [C88], and in the references therein, in particular a number of references by J. M. Ball.

17.4. Hyperelastic materials in biomechanics

Saint Venant-Kirchhoff materials

We recall that a Saint Venant-Kirchhoff material is a material for which the stress tensor P is of the form (17.9). One then deduces that a Saint Venant-Kirchhoff material is hyperelastic, with a stored energy function W of the form

$$W = \frac{\lambda}{2}(trC)^2 + \mu trC^2. \tag{17.30}$$

In particular, this shows that the definition of a Saint Venant-Kirchhoff material is intrinsic (i.e. frame invariant) and that such a material is isotropic.

Hyperelastic materials in biomechanics

In order to model the behavior of soft tissues including muscles, the following types of stored energy functions have been proposed (see, e.g., the review article by Jemiolo and Telega and the references therein):

$$W = W(C) = \frac{\lambda}{\mu}(e^{\mu\psi(\overline{C}_I, \overline{C}_{II}, \overline{C}_{III})} - 1). \tag{17.31}$$

Here $\overline{C}_I$ and $\overline{C}_{II}$ are the first and second invariants of the tensor $\overline{C}$; $\overline{C}$ and $\mathbf{F}$ are defined by

$$\overline{C} = \overline{\mathbf{F}}^T \overline{\mathbf{F}}, \quad \overline{\mathbf{F}} = C_{III}^{1/3} \mathbf{F}, \tag{17.32}$$

so that $det\ \overline{\mathbf{F}} = det\ \overline{C} = \overline{C}_{III} = 1$, and λ and μ are coefficients which depend on the material.

Assuming the existence of a strain-free natural state, we deduce, at first approximation, from (17.28)–(17.29), for the materials considered in biomechanics, e.g., soft tissues, that

$$\psi(3, 3, 1) = 0, \tag{17.33}$$

$$\Pi(I) = 0. \tag{17.34}$$

Having a stored energy function (17.30), we can show that the corresponding Cauchy stress tensor has the form

$$\sigma = \beta_0 I + \beta_1 \overline{B}_D + \beta_{-1} \overline{B}_D^{-1}, \tag{17.35}$$

where

$$\overline{B} = \overline{\mathbf{F}}\overline{\mathbf{F}}^T, \quad \overline{B}_D = \overline{B} - \frac{1}{3}\overline{C}_I I, \quad \overline{B}_D^{-1} = \overline{B}^{-1} - \frac{1}{3}\overline{C}_{II} I, \tag{17.36}$$

and

$$\beta_0 = \lambda e^{\mu\psi} \frac{\partial \Psi}{\partial C_{III}}, \quad \beta_1 = \frac{2}{C_{III}} \lambda e^{\mu\psi} \frac{\partial \Psi}{\partial \overline{C}_I}, \quad \beta_{-1} = -\frac{2}{C_{III}} \lambda e^{\mu\psi} \frac{\partial \psi}{\partial \overline{C}_{II}}. \tag{17.37}$$

It is reasonable here to consider functions ψ of the form

$$\psi = \mu_1(\overline{C}_I - 3) + \mu_2(\overline{C}_{II} - 3) + \mu_3(C_{III} - 1) + \mu_4(\overline{C}_I\overline{C}_{II} - 1) + \cdots, \tag{17.38}$$

where the constants μ_i depend on the material. A particular case is given by incompressible materials, i.e., materials such that $C_{III} = 1$. In that case, we take

$$W = \frac{\lambda}{2\mu}(e^{\mu(\overline{C}_I - 1)} - 1), \tag{17.39}$$

where λ is called the shear modulus. For such materials, the constitutive law (17.35) takes the form

$$\sigma = -pI + \lambda e^{\mu(\overline{C}_I - 3)} \overline{B}. \tag{17.40}$$

More generally, we can consider functions ψ in (17.30) of one of the following forms:

$$\psi = \psi(\overline{C}_I, \overline{C}_{II}),$$
$$\psi = \psi_1(\overline{C}_I, \overline{C}_{II}) + \psi_2(C_{III}),$$
$$\psi = \psi_1(\overline{C}_I, \overline{C}_{II}, C_{III}).$$

Remark 17.3: For a compressible material, the following stored energy function is considered:

$$W = \frac{\lambda}{e}(e^{\mu_1(\overline{C}_I-3)+\mu_2(\overline{C}_{II}-3)+\mu_3(C_{III}-1)^2} - 1). \tag{17.41}$$

To go further:

Several of the books cited in the first part contain additional developments in solid mechanics; in particular, in French, the books by Duvaut, Germain and Salençon and, in English, the books by Gurtin and Spencer, the book by Landau and Lifschitz (more physically oriented), and the book by Truesdell, in a more formal style.

More advanced books on linear or nonlinear elasticity are those by Antman, Ciarlet, Ericksen, Gurtin, Sokolnikoff and Truesdell (1977). Concerning viscoelasticity, the reader can refer to the book by Renardy, Hrusa and Nohel (mathematical aspects) and, concerning plasticity, that by Hodge (physical aspects) or that by Temam (mathematical aspects).

Homogenization is a subject that has developed greatly in relation with new composite materials. Some mathematically oriented books are those by Bensoussan, Lions and Papanicolaou, and by Jikov, Kozlov and Oleinik, and the introductory (and advanced) book by Cioranescu and Donata.

PART IV

INTRODUCTION TO WAVE PHENOMENA

CHAPTER EIGHTEEN

Linear wave equations in mechanics

Before studying, in Chapters 19 and 20, some nonlinear wave equations oc-
curing in mechanics, we consider in this chapter several linear wave equations
arising in mechanics and study some fundamental aspects of the corresponding vibration phenomena.

In Section 18.1, we start by recalling the fundamental wave equations that
appeared in the previous chapters: the equations of linear acoustics that appeared in Chapter 8 in the context of fluid mechanics and the Navier equation
that appeared in Chapter 13 in the context of linear elasticity. We also special-
ize these equations to specific phenomena such as sound pipes and vibrating
cords and membranes.

In Section 18.2, we show how to solve the one-dimensional wave equation
considered in the whole space $\mathbb{R}$. Then, in Section 18.3, we are interested in
bounded intervals, leading us to introduce the normal (self-vibration) modes,
which also depend on the boundary conditions; some typical examples of
boundary conditions are considered and the corresponding eigenmodes made
explicit. In Section 18.4, we show how to solve the wave equation in a general bounded domain of $\mathbb{R}^3$ by using the corresponding eigenmodes again;
however, in this case, the solution is not complete because we cannot compute the eigenmodes, in general. Finally, in the last section of this chapter,
Section 18.5, we give some indications on other important vibration phenomena such as superposition of waves, beats, and wave packets.

18.1. Returning to the equations of linear acoustics
and of linear elasticity

Returning to the wave equations of linear acoustics

The equations of linear acoustics have been introduced in Chapter 8,
Section 8.4. They are deduced from the equations of compressible

fluids under the small motions assumption by using asymptotic expansions.

A typical equation of linear acoustics is

$$\frac{\partial^2 u}{\partial t^2} - c^2 \Delta u = 0, \qquad (18.1)$$

where c denotes the velocity of the sound in the air near rest and u denotes
the density or the pressure. A similar equation has also been obtained for the
velocity, which is denoted here by $\vec{u}$ (linearized motions around equilibrium):

$$\frac{\partial^2 \vec{u}}{\partial t^2} - c^2 \text{grad div } \vec{u} = 0. \qquad (18.2)$$

Equation (18.1) is called the wave equation; it is a hyperbolic equation
(see the appendix to this book). As we will see, it leads to wave propagation
phenomena in which the waves propagate with velocity c. This equation is
a model equation for linear wave phenomena, both in mathematics and in
physics.

A noteworthy particular case is that of sound pipes.

a) Particular case: sound pipes

We assume in this experiment that air moves in a long cylindrical pipe of axis
Ox, that the velocity $\vec{u}$ is parallel to Ox, $\vec{u} = (u, 0, 0)^T$, and that this velocity
depends only on x and t. Equation (18.2) then reduces to the one-dimensional
wave equation, namely

$$\frac{\partial^2 u}{\partial t^2} - c^2 \frac{\partial^2 u}{\partial x^2} = 0. \qquad (18.3)$$

Remark 18.1: As we will see, Eq. (18.3) also characterizes the longitudinal
vibrations of an elastic string (and many other "simple" vibration phenomena).

Returning to the Navier equation of linear elasticity

In Chapter 13, we have derived, for an elastic homogeneous medium, and
under the small deformations assumption, the equation for displacements,
which is the Navier equation:

$$\rho \frac{\partial^2 \vec{u}}{\partial t^2} - \mu \Delta \vec{u} + (\lambda + \mu) \text{grad div } \vec{u} = \vec{f}; \qquad (18.4)$$

here $\vec{u}$ is the displacement, ρ is the density (assumed constant), $\vec{f}$ denotes
the volume forces, and λ and μ are the Lamé coefficients of the medium
(we recall that a consequence of the second principle of thermodynamics
is that $\mu > 0$ and $3\lambda + 2\mu \geq 0$; hence, $\lambda + \mu \geq 0$). Equation (18.4) was

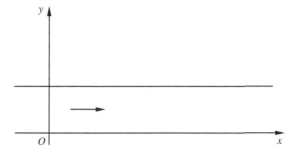

Figure 18.1 The sound pipe.

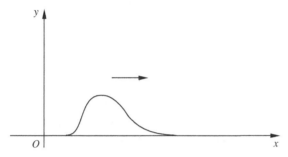

Figure 18.2 The vibrating cord.

obtained by linearizing the equation of conservation of momentum. The Navier equation is also an important linear wave equation; we present some noteworthy particular cases.

Particular cases

a) The vibrating cord

We consider a thin elastic infinite cord supported along the Ox axis. We assume that the cord moves only in the Oy direction so that

$$\vec{u} = \begin{pmatrix} 0 \\ u \\ 0 \end{pmatrix} ;$$

because the cord is thin, we can assume that u depends only on x and t. We then have div $\vec{u} = 0$, and the vector equation (18.4) reduces to a single scalar equation for u:

$$\frac{\partial^2 u}{\partial t^2} - \nu \frac{\partial^2 u}{\partial x^2} = 0, \tag{18.5}$$

where $\nu = \mu/\rho$. We thus recover Eq. (18.3) with c^2 replaced here by ν.

b) The vibrating membrane

Another important particular case of Eq. (18.4) is that of a thin vibrating membrane (a drum, for instance). The membrane fills the domain Ω of the plane Oxy and moves in the orthogonal direction Oz; therefore,

$$\vec{u} = \begin{pmatrix} 0 \\ 0 \\ u \end{pmatrix}.$$

The membrane being thin, we can assume that u depends only, to first approximation, on x and y (and t).

Again, automatically, we have div $\vec{u} = 0$, and the vector equation (18.4) reduces to a scalar equation for u:

$$\frac{\partial^2 u}{\partial t^2} - \nu \Delta_2 u = 0, \quad \text{in } \Omega. \tag{18.6}$$

We have set $\nu = \mu/\rho$, and Δ_2 denotes here the Laplace operator with respect to the variables x and y ($\Delta_2 = \partial^2/\partial x^2 + \partial^2/\partial y^2$).

Assuming, for instance, that the boundary of the membrane is fixed (the case of the drum), we supplement Eq. (18.6) with a Dirichlet-type boundary condition as follows:

$$u = 0 \quad \text{on the boundary } \partial\Omega \text{ of the membrane.} \tag{18.7}$$

Flexion of an elastic string

We studied in Chapter 14, in the context of linear elasticity, the simple bending of a cylindrical beam; this study was made, in the context of statics, for the stress-tensor field.

If the string is thin, and very elongated, we can study its dynamical flexion in a direction orthogonal to the string. With proper simplifying assumptions and a modeling that we will not undertake here, we are led to introduce a function $u = u(x, t)$ that completely characterizes the flexing motion, the string being supported along the Ox axis. The wave equation obtained for u differs from Eqs. (18.3) and (18.5); it contains the fourth-order derivative of u in x, and it reads

$$\rho \frac{\partial^2 u}{\partial t^2} + EI \frac{\partial^4 u}{\partial x^4} = 0, \tag{18.8}$$

where E is the Young's modulus (assumed constant) and I is the inertia moment of the sections in the Ox direction.

18.2. Solution of the one-dimensional wave equation

We consider in this section the vibrations of an infinite cord governed by Eq. (18.5). We can also consider an infinite sound pipe (Eq. (18.3)), but the cord provides a better visualization of the solution.

Because the cord is infinite, no boundary conditions need to be taken into account. We thus study the one-dimensional wave equation

$$\frac{\partial^2 u}{\partial t^2} - c^2 \frac{\partial^2 u}{\partial x^2} = 0, \tag{18.9}$$

where $c = \sqrt{v}$. We supplement this equation with the following initial conditions (at $t = 0$) for position and velocity:

$$u(x, 0) = u_0(x), \quad \frac{\partial u}{\partial t}(x, 0) = u_1(x), \tag{18.10}$$

where the functions u_0 and u_1 are given and are, as usual, sufficiently regular, say in $C^1(\mathbb{R})$.

To solve this equation, we perform the change of independent variables

$$r = x + ct, \quad s = x - ct,$$

and set

$$\bar{u}(r, s) = u(x, t).$$

A straightforward calculation shows that Eq. (18.8) reduces to

$$\frac{\partial^2 \bar{u}}{\partial r \, \partial s} = 0, \tag{18.11}$$

the solution of which is easily found:

$$\bar{u}(r, s) = f(r) + g(s),$$

where the functions f and g are arbitrary. Hence, u may be written in the form

$$u(x, t) = f(x + ct) + g(x - ct). \tag{18.12}$$

Remark 18.2: Equation (18.12) describes the general form of the solutions of Eq. (18.9). If $g = 0$, $u(x, t) = f(x + ct)$ describes a wave that moves to the left at velocity c without changing shape. If $f = 0$, $u(x, t) = g(x - ct)$ is a wave moving to the right with the same velocity. The general solution is thus the sum of two waves moving in opposite directions, at velocity c, without changing shape.

When Eq. (18.9) is supplemented by the initial conditions of Eq. (18.10), we can compute f and g using the initial conditions. We assume that the functions f and g are also of class C^1. We deduce from Eq. (18.10) that

$$f(x) + g(x) = u_0(x), \quad \forall\, x \in \mathbb{R}, \tag{18.13}$$

and

$$f'(x) - g'(x) = \frac{1}{c} u_1(x), \quad \forall\, x \in \mathbb{R}, \tag{18.14}$$

which yields, by differentiating Eq. (18.13) with respect to x,

$$f'(x) = \frac{1}{2}\left[u_0'(x) + \frac{1}{c}u_1(x) \right],$$

$$g'(x) = \frac{1}{2}\left[u_0'(x) - \frac{1}{c}u_1(x) \right].$$

It then follows that

$$f(x) = \frac{1}{2}u_0(x) + \frac{1}{2c}\int_0^x u_1(s)\,ds + k_1,$$

$$g(x) = \frac{1}{2}u_0(x) - \frac{1}{2c}\int_0^x u_1(s)\,ds + k_2,$$

where k_1 and k_2 are two constants. We deduce from Eq. (18.13) that $k_1 + k_2 = 0$, and we can actually take $k_1 = k_2 = 0$. We finally obtain

$$u(x,t) = \frac{1}{2}u_0(x + ct) + \frac{1}{2}u_0(x - ct)$$

$$+ \frac{1}{2c}\int_0^{x+ct} u_1(s)\,ds - \frac{1}{2c}\int_0^{x-ct} u_1(s)\,ds;$$

that is,

$$u(x,t) = \frac{1}{2}u_0(x + ct) + \frac{1}{2}u_0(x - ct) + \frac{1}{2c}\int_{x-ct}^{x+ct} u_1(s)\,ds.$$

18.3. Normal modes

The previous method of resolving the wave equation does not apply to the case of a bounded domain. From the physical point of view, the waves described in Remark 18.2 reach the boundary, reflect, and are superposed with the waves moving towards the boundary; the method of Section 18.2 is totally inappropriate.

A method well suited to our problem consists in looking first for the eigen-modes (also called normal or self-vibration modes) of the form

$$u(x, y, z, t) = V(x, y, z) \exp(i\omega t). \tag{18.15}$$

In this section, we will only describe the eigenmodes for some of the examples considered in Section 18.1. The eigenmodes are actually solutions of an infinite dimensional eigenvalue–eigenfunction problem, but we will not address this aspect here.

Cord fixed at endpoints

We consider here a vibrating cord fixed at its endpoints $x = 0$ and $x = L$ (see Figure 18.3).

Consequently, we associate with the wave equation

$$\frac{\partial^2 u}{\partial t^2} - c^2 \frac{\partial^2 u}{\partial x^2} = 0 \tag{18.16}$$

characterizing the vibrations of the cord (c^2 is equal to v in that case, see Section 18.1) the following Dirichlet-type boundary conditions:

$$u(0, t) = u(L, t) = 0, \quad \forall t \geq 0. \tag{18.17}$$

We look for a solution of the form $u(x, t) = U(x) \exp(i\omega t)$, and we obtain the following equation for the function U:

$$U'' + \frac{\omega^2}{c^2} U = 0.$$

This is a second-order ordinary differential equation with constant coefficients. We deduce that

$$U(x) = a \sin\left(\frac{\omega}{c} x + \varphi\right),$$

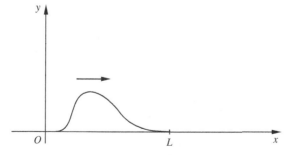

Figure 18.3 Cord fixed at endpoints.

where a and $\varphi \in \mathbb{R}$. Then, Eq. (18.17) yields $U(0) = U(\varphi) = 0$, and it follows that

$$\varphi = 0,$$

$$\frac{\omega L}{c} = k\pi, \quad k \in \mathbb{N}^*.$$

The vibration eigenmodes are thus given by

$$u_k(x, t) = U_k(x) \exp(i\omega_k t),$$

where

$$U_k(x) = a_k \sin \frac{k\pi}{L} x, \quad k \in \mathbb{N}^*,$$

and

$$\omega_k = \frac{k\pi c}{L}, \quad k \in \mathbb{N}^*$$

is the corresponding eigenfrequency.

We can easily verify that the eigenfunctions U_k satisfy the orthogonality property[1]

$$\int_0^L U_i(x)U_j(x)\,dx = 0, \quad \forall\, i \neq j. \tag{18.18}$$

One can actually show that they form an orthogonal basis of the space $L^2(0, L)$ (this can be proven by using the spectral properties of symmetric unbounded operators; see also below).

Sound pipe

The wave equation is the same as in Section 18.3 with c denoting the velocity of sound. We associate, for instance, with this equation the boundary conditions

$$u(0, t) = \frac{\partial u}{\partial x}(L, t) = 0, \quad \forall\, t \geq 0, \tag{18.19}$$

corresponding to wind at rest at the entry of the pipe and to a free (open) exit of the pipe.

Proceeding as in Section 18.3, we find

$$U(x) = a \sin\left(\frac{\omega}{c}x + \varphi\right),$$

---

[1] This is a particular case of a more general result recalled below.

and because, according to Eq. (18.19), $U(0) = U'(L) = 0$, it follows that

$$\varphi = 0,$$

$$\frac{\omega L}{c} = (2k + 1)\frac{\pi}{2}, \quad k \in \mathbb{N}.$$

The vibration eigenmodes are thus the functions

$$u_k(x, t) = U_k(x) \exp(i\omega_k t),$$

where

$$U_k(x) = a_k \sin\left(\frac{(2k + 1)\pi}{2L}x\right), \quad k \in \mathbb{N},$$

and

$$\omega_k = \frac{(2k + 1)\pi c}{2L}, \quad k \in \mathbb{N},$$

is the corresponding eigenfrequency.

The eigenfunctions $U_k(x)$ satisfy the orthogonality property of Eq. (18.18) and they also form an orthogonal basis of $L^2(0, L)$.

Remark 18.3: The boundary conditions of Eq. (18.19) also correspond to a vibrating cord, which is fixed at $x = 0$, and free at $x = L$.

Membrane fixed at its boundary

We recall the equation characterizing the vertical vibrations of a membrane filling the domain $\Omega \subset \mathbb{R}^2$:

$$\frac{\partial^2 u}{\partial t^2} - c^2 \Delta_2 u = 0, \tag{18.20}$$

where $c^2 = \nu = \mu/\rho$. We assume that the membrane is fixed on its boundary, which is expressed by the boundary conditions

$$u = 0, \quad \text{on } \partial\Omega. \tag{18.21}$$

The vibration eigenmodes are the solutions of the form

$$u(x, y, t) = U(x, y) \exp(i\omega t). \tag{18.22}$$

Replacing, in Eq. (18.20), u by its expression given by Eq. (18.22), we obtain the following equation for U:

$$\Delta_2 U + \frac{\omega^2}{c^2} U = 0, \quad \text{in } \Omega, \tag{18.23}$$

and, owing to Eq. (18.21),

$$U = 0, \quad \text{on } \partial\Omega. \tag{18.24}$$

Thus, ω^2/c^2 is an eigenvalue of the operator $-\Delta_2$ associated with the Dirichlet-type boundary conditions, and U is the associated eigenfunction. One can prove (we refer the reader interested in more details to the book of Courant and Hilbert (1953)) that this operator possesses an infinite family of eigenvalues $0 < \lambda_1 \le \lambda_2 \le \ldots \le \lambda_n \le \ldots$ satisfying $\lambda_n \to +\infty$ as $n \to +\infty$. The corresponding eigenfunctions U_n form *an orthogonal basis* of the Hilbert space $L^2(\Omega)$.

Remark 18.4: Actually, the eigenmodes described in Section 18.3 correspond to similar eigenvalue and eigenfunction problems that we could solve explicitly in one dimension. In two dimensions, Eqs. (18.23) and (18.24) cannot be solved explicitly by analytical methods except for simple domains such as rectangles or circles.

Remark 18.5: If it is assumed, for example, that Ω is a drum, the eigenvalues λ_i give the accoustic signature of the drum. The inverse problem, namely that of recovering the shape of Ω given its self-vibration frequencies λ_i, is an outstanding problem in mathematical physics mostly solved in recent years. Mathematicians and physicists refer to it by the question: Can we hear the shape of a drum?

The self-vibration eigenmodes for the membrane are thus given by

$$u_k(x, y, t) = U_k(x, y) \exp(i\omega_k t), \quad k \in \mathbb{N}^*, \tag{18.25}$$

the associated eigenfrequencies being defined by

$$\omega_k = \sqrt{\lambda_k} c, \quad k \in \mathbb{N}^*.$$

Let U_i and U_j be two eigenfunctions associated with two distinct eigenvalues λ_i and λ_j. Then, using the Green formula and considering Eq. (18.24), we see that

$$\left(\lambda_i - \lambda_j\right) \int_\Omega U_i U_j \, dx \, dy = - \int_\Omega (U_j \Delta_2 U_i - U_i \Delta_2 U_j) \, dx \, dy$$
$$= \int_\Omega (\nabla_2 U_j \nabla_2 U_i - \nabla_2 U_i \nabla_2 U_j) \, dx \, dy$$
$$= 0,$$

where ∇_2 is the gradient with respect to the variables x and y. We assume here that the functions U_i and U_j, as well as the domain Ω, are sufficiently regular.

We thus recover the orthogonality property that has already been described:

$$\int_{\Omega} U_i U_j \, dx \, dy = 0, \tag{18.26}$$

for all eigenfunctions U_i and U_j associated with distinct eigenvalues λ_i and λ_j. Furthermore, we can choose the eigenfunctions U_k associated with the same eigenvalue to be orthogonal (in the sense of Eq. (18.26)) to one another (the eigenspace corresponding to a specific eigenvalue has finite dimension). We thus obtain an orthogonal (or orthonormal) family of eigenvectors; one can show that the family is complete; that is, its linear combinations are dense in $L^2(\Omega)$. The eigenfunctions form a basis of $L^2(\Omega)$.

18.4. Solution of the wave equation

Our aim in this section is to solve the wave equation (18.1), supplemented with the boundary and initial conditions, by using spectral expansions, that is to say expansions of the form

$$u(x, y, z, t) = \sum_{k=1}^{\infty} u_k(t) U_k(x, y, z), \tag{18.27}$$

where the $U_k(x, y, z)$ are the eigenfunctions of the stationary problem (as defined in the previous section). We will assume throughout this section that the expansion defined by Eq. (18.27) is convergent and differentiable term by term.

We note that, if a solution of Eq. (18.1) is of the form $u(x, y, z, t) = f(t)U(x, y, z)$, then

$$\frac{f''}{c^2 f} = \frac{\Delta U}{U}, \tag{18.28}$$

and the common value of these two quantities is a constant (denoted $-\lambda$), since the left-hand side depends only on t, while the right-hand side is independent of t.

Under these circumstances, and assuming that U vanishes at the boundary, $\sqrt{-\lambda}$ is an eigenfrequency for $U = U_k$ and, in Eq. (18.27), we can thus take for f_k,

$$f_k(t) = c_k \cos(\omega_k t) + d_k \sin(\omega_k t), \tag{18.29}$$

where $\omega_k = \sqrt{\lambda_k} c$ is the eigenfrequency associated with the eigenmode U_k.

We first consider the particular cases described in the previous sections, and then give some indications on the general three-dimensional problem.

General vibrations of a cord fixed at its endpoints

We consider the boundary value problem

$$\frac{\partial^2 u}{\partial t^2} - c^2 \frac{\partial^2 u}{\partial x^2} = 0, \quad 0 < x < L, \ t > 0, \tag{18.30}$$

$$u(0) = u(L) = 0, \tag{18.31}$$

that we supplement with the initial conditions

$$u(x, 0) = u_0(x), \quad \frac{\partial u}{\partial t}(x, 0) = u_1(x). \tag{18.32}$$

We saw in the previous section that the eigenfunctions are given (we take here $a_k = 1$) by

$$U_k(x) = \sin \frac{k\pi}{L} x, \quad k \in \mathbb{N}^*. \tag{18.33}$$

We thus look for a solution of the form

$$u(x, t) = \sum_{k=1}^{\infty} \sin\left(\frac{k\pi}{L} x\right) (c_k \cos \omega_k t + d_k \sin \omega_k t), \tag{18.34}$$

where $\omega_k = (k\pi c)/L$, $k \in \mathbb{N}^*$. The problem consists thus of finding the coefficients c_k and d_k. We can compute the coefficients c_k and d_k using the initial conditions. Indeed, taking $t = 0$ in Eq. (18.34), we find

$$u_0(x) = \sum_{k=0}^{\infty} c_k \sin\left(\frac{k\pi}{L} x\right). \tag{18.35}$$

Furthermore,

$$\frac{\partial u}{\partial t}(x, t) = \sum_{k=1}^{\infty} \omega_k \sin\left(\frac{k\pi}{L} x\right)(-c_k \sin \omega_k t + d_k \cos \omega_k t);$$

hence, for $t = 0$

$$u_1(x) = \sum_{k=1}^{\infty} \omega_k d_k \sin\left(\frac{k\pi}{L} x\right).$$

It finally follows from the orthogonality property of Eq. (18.18) that

$$c_k = \frac{1}{\int_0^L U_k^2(x)\,dx} \int_0^L u_0(x) U_k(x)\,dx,$$

$$d_k = \frac{1}{\omega_k \int_0^L U_k^2(x)\,dx} \int_0^L u_1(x) U_k(x)\,dx.$$

This finishes the resolution of the wave equation in this case.

Remark 18.6: We can obtain a similar result for sound pipes.

General vibrations of a membrane fixed on its boundary

We consider the two-dimensional wave equation

$$\frac{\partial^2 u}{\partial t^2} - c^2 \Delta_2 u = 0, \quad \text{in } \Omega, \tag{18.36}$$

$$u = 0, \quad \text{on } \partial\Omega, \tag{18.37}$$

where Ω is the domain filled by the membrane. We supplement this equation with the initial conditions

$$u(x, y, 0) = u_0(x, y), \quad \frac{\partial u}{\partial t}(x, y, 0) = u_1(x, y). \tag{18.38}$$

Again, we look for a function u under the form of a spectral expansion

$$u(x, y, t) = \sum_{k=1}^{\infty} U_k(x, y)(c_k \cos \omega_k t + d_k \sin \omega_k t),$$

where the eigenfunctions U_k and the eigenfrequencies ω_k are as has been defined in Section 18.3.

As above, we have

$$u_0(x, y) = \sum_{k=1}^{\infty} c_k U_k(x, y),$$

and

$$u_1(x, y) = \sum_{k=1}^{\infty} d_k \omega_k U_k(x, y).$$

Hence, because the U_k are orthogonal,

$$c_k = \frac{1}{\int_\Omega U_k^2 \, dx \, dy} \int_\Omega u_0 U_k \, dx \, dy,$$

$$d_k = \frac{1}{\omega_k \int_\Omega U_k^2 \, dx \, dy} \int_\Omega u_1 U_k \, dx \, dy.$$

When the functions U_k are known, we obtain, in principle, the general solution of Eqs. (18.36) to (18.38). The main difficulty for the resolution reduces to the determination of the eigenfunctions U_k (which is not an easy task, except for "simple" domains Ω).

The three-dimensional case

In the general case, we essentially follow the same procedure as in the previous examples and look for a function u in the form of a spectral expansion

$$u(x, y, z, t) = \sum_{k=1}^{\infty} U_k(x, y, z)(c_k \cos \omega_k t + d_k \sin \omega_k t). \qquad (18.39)$$

Here, $(U_k)_{k \in \mathbb{N}^*}$ will be an orthogonal family of eigenvectors corresponding to the eigenvalues of the operator $-\Delta$ associated with the boundary conditions of the problem; ω_k is the associated eigenfrequency.

The problem will then be reduced to the computation of the eigenmodes U_k. As mentioned above, these eigenmodes cannot be computed explicitly in general. We observe, however, that in certain cases we can still make a separation of variables for the search of the eigenfunctions. Let us consider for instance the case of a cylindrical domain Ω of section $S \subset \mathbb{R}^2$ and delimited by the planes $z = 0$ and $z = L$ (i.e., $\Omega = S \times (0, L)$). We further assume that the wave equation is supplemented with the Dirichlet-type boundary conditions. In this case, we can write

$$U(x, y, z) = f(z)g(x, y), \qquad (18.40)$$

and the relation $-\Delta U = \lambda U$ gives

$$\frac{-f''}{f} = \frac{\Delta_2 g}{g} + \lambda = k, \qquad (18.41)$$

where k is a constant independent of x, y, and z. It then follows that $k = n^2$, $n \in \mathbb{N}^*$,

$$f(z) = \sin n^2 z, \quad n \in \mathbb{N}^*, \qquad (18.42)$$

and g is solution of the problem

$$\Delta_2 g + (\lambda - n^2)g = 0, \quad \text{in } S, \qquad (18.43)$$

$$g = 0, \quad \text{on } \partial S. \qquad (18.44)$$

The problem is then reduced to a two-dimensional problem. The problem is thus simplified by reducing the space dimension, but it is not solved completely.

18.5. Superposition of waves, beats, and packets of waves

What follows is valid in a more general context, but we can think, for visualization purposes, of waves propagating on a long vibrating cord – preferably away from the ends of the cord to avoid reflection problems.

It follows from Section 18.3 that the vibrations of the cord are superpositions of elementary waves of the form

$$a_k \sin \lambda_k x \cos \omega_k t, \quad \text{and} \quad b_k \sin \lambda_k x \sin \omega_k t.$$

The first ones can be written as

$$\frac{a_k}{2} \left\{ \cos \left(\lambda_k x - \omega_k t + \frac{\pi}{2} \right) - \cos \left(\lambda_k x + \omega_k t + \frac{\pi}{2} \right) \right\},$$

and the second ones can be also written in a similar fashion (see also Section 18.2). Hence, we will consider waves of the form $a \cos(\lambda x - \omega t + \varphi)$, which are customarily written in the complex form $Re\, ae^{i(\lambda x - \omega t + \varphi)}$; ω is the frequency of the wave, $T = 2\pi/\omega$ is the period (in time), λ represents the period (in space) or wavelength, and $k = 2\pi/\lambda$ is the wave number; a measures the intensity of the wave; finally, φ is called the phase of the wave and $v = \omega/\lambda$ is the phase speed.

The superposition of waves with the same frequency and wavelength, but with different phases, produces a wave with the same frequency and wavelength but with a different phase. Indeed

$$a_1 e^{i(\lambda x - \omega t + \varphi_1)} + a_2 e^{i(\lambda x - \omega t + \varphi_2)} = ae^{i(\lambda x - \omega t + \varphi)},$$

with

$$ae^{i\varphi} = a_1 e^{i\varphi_1} + a_2 e^{i\varphi_2};$$

hence,

$$a = \{(a_1 \cos \varphi_1 + a_2 \cos \varphi_2)^2 + (a_1 \sin \varphi_1 + a_2 \sin \varphi_2)^2\}^{1/2},$$
$$\tan \varphi = \frac{a_1 \sin \varphi_1 + a_2 \sin \varphi_2}{a_1 \cos \varphi_1 + a_2 \cos \varphi_2}.$$

Let us now consider the superposition of two waves with different frequencies; we assume, for the sake of simplicity, that they have the same intensity and phase,

$$u_1 = a \cos(\lambda x - \omega t), \quad u_2 = a \cos(\lambda' x - \omega' t).$$

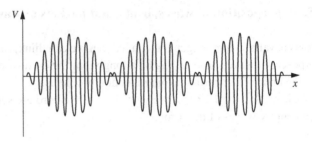

Figure 18.4 Linear superposition of two sinusoidal waves producing a long wave of modulated amplitude.

Then, setting $\delta\lambda = \lambda' - \lambda$, $\delta\omega = \omega' - \omega$,

$$u = u_1 + u_2 = 2a \cos\left[\frac{1}{2}(\delta\lambda x - \delta\omega t)\right]$$

$$\times \cos\left[\left(\lambda + \frac{\delta\lambda}{2}\right)x - \left(\omega + \frac{\delta\omega}{2}\right)t\right]. \qquad (18.45)$$

An interesting case is that in which $\delta\lambda$ and $\delta\omega$ are small, $\delta\lambda \ll \lambda$, and $\delta\omega \ll \omega$; *we then obtain the beats.* The new wave $u = u_1 + u_2$ described by Figure 18.4 is a wave of modulated amplitude. It consists of a wave with high frequency $\omega + \delta\omega/2$ and large wavenumber $\lambda + \delta\lambda/2$, of modulated intensity, and that is contained in an envelope that is itself a wave of low frequency and small wave number; its amplitude is thus

$$\mathcal{U} = 2a \cos\left[\frac{1}{2}(\delta\lambda x - \delta\omega t)\right].$$

The superposed wave is itself periodic with slow variation and is of period $4\pi/\delta\omega$ and wavelength $4\pi/\delta\lambda$. Here appears the beat phenomena used, for instance, in music to tune an instrument by comparing a vibration to a reference frequency.

Each long wave is formed from a packet of short waves. The short waves propagate with a phase speed

$$v = \frac{2\omega + \delta\omega}{2\lambda + \delta\lambda},$$

and the envelope (which contains the wave group or packet), propagates with the phase velocity

$$\frac{\delta\omega}{\delta\lambda}.$$

When $\delta\omega$ and $\delta\lambda$ tend to 0, this velocity approaches the velocity

$$v_g = \frac{d\omega}{d\lambda},$$

called the group velocity.

Exercises

1. Assume that u_0 and u_1 belong to $C^2(\mathbb{R})$. Show that the function u given in Section 18.2 is indeed a solution of the wave equation (18.9)–(18.10), i.e. $u \in C^2(\mathbb{R} \times [0, +\infty))$,

$$\frac{\partial^2 u}{\partial t^2} - c^2 \frac{\partial^2 u}{\partial x^2} = 0 \text{ in } \mathbb{R} \times]0, +\infty[,$$

$$\lim_{(x,t) \to (x_0,0)} u(x, t) = u_0(x_0),$$

$$\lim_{(x,t) \to (x_0,0)} \frac{\partial u}{\partial t}(x, t) = u_1(x_0), \forall x_0 \in \mathbb{R}.$$

2. Give the explicit solution of the wave equation on the half line:

$$\frac{\partial^2 u}{\partial t^2} - c^2 \frac{\partial^2 u}{\partial x^2} = 0, \text{ in } \mathbb{R}_+ \times]0, +\infty[$$

$$u(x, 0) = u_0(x), \frac{\partial u}{\partial t}(x, 0) = u_1(x), x \in \mathbb{R}_+,$$

where u_0 and u_1 belong to $C^2(\mathbb{R}_+)$.

3. Consider the wave equation on a bounded domain $\Omega \subset \mathbb{R}^n$:

$$\frac{\partial^2 u}{\partial t^2} - \Delta u = f, \text{ in } \Omega \times (0, T),$$

$$u = 0, \text{ on } \partial\Omega,$$

$$u(x, 0) = u_0(x), \text{ in } \Omega,$$

$$\frac{\partial u}{\partial t}(x, 0) = u_1(x), \text{ in } \Omega.$$

Show that this problem possesses at most one solution u such that $u \in C^2(\overline{\Omega} \times (0, T))$.

4. (Finite propagation speed) Assume that $u \in C^2(\mathbb{R}^n \times [0, +\infty))$ solves the wave equation

$$\frac{\partial^2 u}{\partial t^2} - \Delta u = 0, \text{ in } \mathbb{R}^n \times]0, +\infty[.$$

Assume that u and $\partial u/\partial t$ vanish in the ball $B = B(x_0, t_0)$ of center x_0 and radius t_0. Then show that u vanishes in the cone

$$\mathcal{C} = \left\{(x, t) \in \mathbb{R}^n \times (0, +\infty), \ |x - x_0| \le t_0 - t, \ 0 \le t \le t_0\right\}.$$

5. Consider equation (18.8) modeling the flexion of an elastic string

$$\frac{\partial^2 u}{\partial t^2} + \frac{EI}{\rho}\frac{\partial^4 u}{\partial x^4} = 0.$$

We assume that the string is fixed at its endpoints $x = 0$ and $x = L$ so that

$$u(0, t) = u(L, t) = 0, \forall t \geq 0,$$

$$\frac{\partial u}{\partial x}(0, t) = \frac{\partial u}{\partial x}(L, t) = 0, \forall t \geq 0,$$

and we look for solutions of the form

$$u(x, t) = U(x)\exp(i\omega t)$$

a)

- (i) Show that u is the solution of the ordinary differential equation

$$U^{(4)} - m^4 U = 0,$$

  where $m^4 = \dfrac{\omega^2 \rho}{EI}$.
- (ii) Deduce that U is of the form

$$U(x) = A\cos mx + B\sin mx + Cchmx + Dshmx.$$

b) We set $\sigma = mL$. Show that

$$U = A[(\sin\sigma - sh\sigma)(\cos mx - chmx) - (\cos\sigma - ch\sigma)(\sin mx - shmx)],$$

with

$$\cos\sigma\,ch\sigma = 1,$$

and compute the eigenfrequencies σ.

6. Same exercise as above but assuming that the string is simply supported by its endpoints.

7. Same exercise as above but assuming that the endpoint $x = 0$ is fixed and that the endpoint $x = L$ is free.

CHAPTER NINETEEN

The soliton equation:
the Korteweg–de Vries equation

Solitons are a type of nonlinear wave whose discovery and study are relatively recent. Solitons appear in numerous wave propagation phenomena, and they should eventually play a major role in telecommunications by optical fibers. Two well-known equations possess soliton solutions: the Korteweg–de Vries equation and the nonlinear Schrödinger equation; we will present these equations in Chapters 19 and 20, respectively. These equations may be obtained by passing to the limit in various equations (e.g., fluid mechanics, electromagnetism). Let us mention here two remarkable properties of solitons. On the one hand, these are waves that propagate by keeping a constant shape; on the other hand, when two solitons that propagate at different velocities meet, they interact for some time and then recover their initial shapes and continue their propagation with their own initial velocity.

Solitons appear in particular in the small-amplitude motions of the surface of a shallow fluid; they are then governed by the Korteweg–de Vries equation. In this chapter, we write different water-wave equations and show how the Korteweg–de Vries equation may be deduced from the Euler equation of incompressible fluids; we finally present the soliton solutions of the Korteweg–de Vries equation (KdV).

Solitons were discovered by the Englishman John Scott Russell. He noticed the appearance of waves with constant shape propagating over large distances in shallow canals, and he used to follow them on horseback along the banks of the canal (J.S. Russell, 1884). The theoretical (analytical) study of these waves was then undertaken by Korteweg and de Vries, who introduced the equation bearing their names (Korteweg and de Vries (1895)).

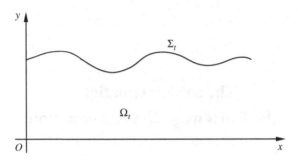

Figure 19.1 The domain Ω_t.

19.1. Water-wave equations

In this section, we study two-dimensional flows of an incompressible irrotational fluid filling the region $\Omega = \Omega_t$ defined by $0 < x < \ell, 0 < y < h(x, t)$. The surface $\Sigma = \Sigma_t$ of the liquid with equation $y = h(x, t)$ is a free surface (see Figure 19.1).

Because the flow is incompressible and irrotational, there exists (see Chapter 8) a function $\varphi = \varphi(x, t)$ such that

$$u = \operatorname{grad} \varphi \quad \text{and} \quad \operatorname{div} u = \Delta\varphi = 0;$$

that is to say,

$$\varphi_{,xx} + \varphi_{,yy} = 0. \tag{19.1}$$

To determine the boundary conditions satisfied by φ on the free surface Σ, we start from the Euler equation

$$\frac{\partial u}{\partial t} + \operatorname{curl} u \wedge u + \operatorname{grad}\left(\frac{p}{\rho_0} + \frac{u^2}{2}\right) = f,$$

and, because $\operatorname{curl} u = 0, u = \operatorname{grad} \varphi$, and $f = -\operatorname{grad}(gy)$, we deduce the Bernoulli equation

$$\varphi_{,t} + \frac{p}{\rho_0} + \frac{1}{2}\left(\varphi_{,x}^2 + \varphi_{,y}^2\right) + gy = \text{const.}$$

On the surface Σ_t, the pressure is equal to the outer atmospheric pressure p_0, and we then have

$$\varphi_{,t} + \frac{1}{2}\left(\varphi_{,x}^2 + \varphi_{,y}^2\right) + gh(x, t) = c(t), \quad \text{on } \Sigma_t.$$

Because φ is defined up to the addition of a function of t (which is spatially constant), we can replace $\varphi(x, y, t)$ in the preceding equality by $\varphi(x, y, t) - \int_0^t c(s)\,ds$, and we then obtain

$$\varphi_{,t} + \frac{1}{2}\left(\varphi_{,x}^2 + \varphi_{,y}^2\right) + gh = 0, \quad \text{on } \Sigma_t. \tag{19.2}$$

On the other hand, by the nonpenetration condition, the normal velocity of Σ_t is the same as the normal velocity of the fluid. A point of Σ_t that is at $[x, h(x, t)]$ at time t will be at $[x, h(x, t + \Delta t)]$ at time $t + \Delta t$; we deduce from this that the velocity of Σ_t is $(0, h_{,t})$. Because $(-h_{,x}, 1)$ is normal to Σ_t and $u = \operatorname{grad}\varphi$, it follows that

$$-\varphi_{,x}h_{,x} + \varphi_{,y} = h_{,t}, \quad \text{on } \Sigma_t,$$

which we rewrite in the form

$$h_{,t} + \varphi_{,x}h_{,x} - \varphi_{,y} = 0, \quad \text{on } \Sigma_t. \tag{19.3}$$

Finally, on the lower horizontal wall, the nonpenetration condition $u \cdot n = 0$ gives

$$\frac{\partial \varphi}{\partial y} = \varphi_{,y} = 0, \quad \text{at } y = 0. \tag{19.4}$$

Small amplitude waves in shallow water: nondimensional form of the equations

We now assume that the equation of Σ_t is of the form $h = h_0 + \eta$, where h_0 is a constant and $\eta = O(a)$ with a denoting the amplitude of the wave (see Figure 19.2).

In what follows, we will assume that $a \ll h_0$ and $h_0 \ll \ell$, which means that we are interested in waves of small amplitude in shallow water.

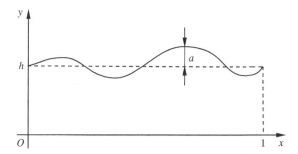

Figure 19.2 Small-amplitude waves in shallow water.

To obtain the nondimensional form of the wave equations, we set

$$x = \ell x', \quad y = h_0 y', \quad t = \frac{\ell t'}{c_0}, \quad c_0 = \sqrt{g h_0},$$

$$\eta = a \eta', \quad \varphi = g \frac{\ell a}{c_0} \varphi', \quad \alpha = \frac{a}{h_0}, \quad \beta = \frac{h_0^2}{\ell^2}.$$

We then delete the primes in the equations obtained by this change of variables. The equation of Σ_t thus becomes $y = 1 + \alpha \eta$, and Eqs. (19.1)–(19.4) can be rewritten as

$$\beta \varphi_{,xx} + \varphi_{,yy} = 0, \quad 0 < x < 1, \quad 0 < y < 1 + \alpha \eta,$$

$$\eta_{,t} + \alpha \varphi_{,x} \eta_{,x} - \frac{1}{\beta} \varphi_{,y} = 0, \quad \text{at } y = 1 + \alpha \eta,$$

$$\varphi_{,t} + \frac{1}{2} \alpha \varphi_{,x}^2 + \frac{1}{2} \frac{\alpha}{\beta} \varphi_{,y}^2 + \eta = 0, \quad \text{at } y = 1 + \alpha \eta,$$

$$\varphi_{,y} = 0 \quad \text{at } y = 0. \tag{19.5}$$

19.2. Simplified form of the water-wave equations

Our aim in this section is to solve, in an approximate way (for small β), the equation

$$\beta \varphi_{,xx} + \varphi_{,yy} = 0,$$

$$\varphi_{,y} = 0, \quad \text{at } y = 0.$$

To do so, we look for a solution having a series expansion with respect to y of the form:

$$\varphi = \varphi(x, y, t) = \sum_{n=0}^{\infty} f_n(x, t) y^n.$$

Thus, $\varphi_{,y} = \sum_{n=0}^{\infty} n y^{n-1} f_n(x, t)$, and we obtain, for $y = 0$,

$$f_1(x, t) = 0.$$

Then the Laplace equation $\beta \varphi_{,xx} + \varphi_{,yy} = 0$ yields

$$\beta \sum_{n=0}^{\infty} y^n \frac{\partial^2 f_n}{\partial x^2} + \sum_{n=2}^{\infty} n(n-1) y^{n-2} f_n = 0;$$

that is to say

$$\sum_{n=0}^{\infty} \left[\beta \frac{\partial^2 f_n}{\partial x^2} + (n+1)(n+2) f_{n+2} \right] y^n = 0.$$

It then follows that

$$\beta \frac{\partial^2 f_n}{\partial x^2} + (n+1)(n+2) f_{n+2} = 0, \quad n \geq 0.$$

We can solve these equations recursively for even n and for odd n separately.

Because $f_1 = 0$, we deduce that $f_3 = 0$ and, by induction, $f_{2j+1} = 0$, $\forall \, j \in \mathbb{N}$.

Then, for even n, we obtain

$$\beta \frac{\partial^2 f_0}{\partial x^2} + 2 f_2 = 0,$$

$$\beta \frac{\partial^2 f_2}{\partial x^2} + 12 f_4 = 0;$$

hence,

$$f_2 = -\frac{\beta}{2} \frac{\partial^2 f_0}{\partial x^2}$$

$$f_4 = \frac{\beta^2}{24} \frac{\partial^4 f_0}{\partial x^4};$$

we can prove by induction that, similarly,

$$f_{2j} = \frac{(-1)^j}{(2j)!} \beta^j \frac{\partial^{2j} f_0}{\partial x^{2j}}.$$

By now setting $f_0(x, t) = f(x, t)$, we obtain the following expression for φ:

$$\varphi = \varphi(x, y, t) = \sum_{j=0}^{\infty} (-1)^j \frac{y^{2j}}{(2j)!} \beta^j \frac{\partial^{2j} f}{\partial x^{2j}}.$$

Asymptotic expansions

Next, we substitute this expression for φ in the equation for η, and we make asymptotic expansions for small β (and possibly also for small α). We have, with $y = 1 + \alpha \eta$:

$$\varphi = f - \beta \frac{y^2}{2} f_{,xx} + \beta^2 \frac{y^4}{24} f_{,xxxx} + O(\beta^3),$$

$$-\frac{1}{\beta} \varphi_{,y} = y f_{,xx} - \beta \frac{y^3}{6} f_{,xxxx} + O(\beta^2),$$

$$\varphi_{,x} = f_{,x} - \beta \frac{y^2}{2} f_{,xxx} + O(\beta^2),$$

and, because

$$\alpha \eta_{,x} f_{,x} + (1 + \alpha\eta) f_{,xx} = [(1 + \alpha\eta) f_{,x}]_{,x},$$

the first equation in η gives

$$\eta_{,t} + [(1 + \alpha\eta) f_{,x}]_{,x}$$

$$- \left[\frac{1}{6}(1 + \alpha\eta)^3 f_{,xxxx} + \frac{1}{2}\alpha(1 + \alpha\eta)^2 \eta_{,x} f_{,xxx} \right] \beta + O(\beta^2) = 0.$$

(19.6)

Furthermore,

$$\varphi_{,t} = f_{,t} - \beta \frac{y^2}{2} f_{,xxt} + O(\beta^2),$$

and we deduce from the second equation for η the following equation:

$$\eta + f_{,t} - \frac{1}{2}\beta y^2 f_{,xxt} + \frac{1}{2}\alpha f_{,x}^2$$

$$- \frac{1}{2}\alpha\beta y^2 f_{,x} f_{,xxx} + \frac{1}{2}\alpha\beta y^2 f_{,xx}^2 + O(\beta^2) = 0. \qquad (19.7)$$

Simplified equations (based on asymptotic expansions)

A first simplification of Eqs. (19.6) and (19.7) consists of neglecting the terms of order β (i.e., $O(\beta)$). Setting $w = f_{,x}$ and differentiating Eq. (19.7) with respect to x, we then obtain

$$\eta_{,t} + [(1 + \alpha\eta)w]_{,x} = 0, \qquad (19.8)$$

$$w_{,t} + \eta_{,x} + \alpha w w_{,x} = 0. \qquad (19.9)$$

These equations are called nonlinear shallow water equations.

Another simplification consists of keeping the terms in β but of neglecting the terms of order $\alpha\beta$ (i.e., $O(\alpha\beta)$). We thus obtain

$$\eta_{,t} + [(1 + \alpha\eta)w]_{,x} - \frac{\beta}{6} w_{,xxx} + O(\alpha\beta, \beta^2) = 0, \qquad (19.10)$$

$$\eta_{,x} + w_{,t} + \alpha w w_{,x} - \frac{1}{2}\beta w_{,xxt} + O(\alpha\beta, \beta^2) = 0. \qquad (19.11)$$

These last equations are a variant of the so-called Boussinesq equations.

Remark 19.1: We saw that

$$\varphi_{,x} = w - \beta \frac{y^2}{2} w_{,xx} + O(\beta^2).$$

Consequently, by averaging over y (over the depth), we note that the averaged velocity $\tilde{u}$ satisfies

$$\tilde{u} = w - \frac{1}{6}\beta w_{,xx} + O(\alpha\beta, \beta^2),$$

which gives, upon inverting:

$$w = \tilde{u} + \frac{1}{6}\beta \tilde{u}_{,xx} + O(\alpha\beta, \beta^2).$$

By then replacing w with this expression, we find

$$\eta_{,t} + [(1 + \alpha\eta)\tilde{u}]_{,x} + \frac{\beta}{6}\tilde{u}_{,xxx} + O(\alpha\beta, \beta^2) = 0,$$

$$\tilde{u}_{,t} + \alpha\tilde{u}\tilde{u}_{,x} + \eta_{,x} - \frac{1}{3}\beta\tilde{u}_{,xxt} + O(\alpha\beta, \beta^2) = 0.$$

19.3. The Korteweg–de Vries equation

Our aim in this section is to deduce the Korteweg–de Vries equation from the preceding equations. This equation, very important for historical reasons because it has led to the discovery of the concept of the soliton, can actually be deduced from any one of the equations derived in the previous sections by considering the waves propagating to the right.

Hence, Eqs. (19.9) and (19.10) give, if we take $\alpha = \beta = 0$,

$$\eta_{,t} + w_{,x} = 0,$$

$$w_{,t} + \eta_{,x} = 0,$$

which yields

$$\eta_{,tt} + w_{,xt} = 0,$$

and

$$\eta_{,tt} - \eta_{,xx} = 0.$$

As we saw in Chapter 18, all the solutions of this equation propagating to the right are of the form

$$\eta(x, t) = a(x - t) = \eta(x - t, 0),$$

which yields $\eta_t + \eta_x = 0$ and, for $\psi = \eta - w$,

$$\psi_{,t} = \eta_{,t} - w_{,t} = \eta_{,t} + \eta_{,x} = 0,$$

$$\psi_{,x} = \eta_{,x} - w_{,x} = \eta_{,x} + \eta_{,t} = 0;$$

hence,

$$\psi = \eta - w = 0,$$
$$\eta_{,t} + \eta_{,x} = 0.$$

Next, we look for a correction, to first order in α and β, of the form

$$w = \eta + \alpha A + \beta B + O(\alpha^2 + \beta^2)$$

and

$$\eta_{,t} + \eta_{,x} = O(\alpha, \beta).$$

We then deduce from Eqs. (19.10) and (19.11) that

$$\eta_{,t} + \eta_{,x} + \alpha(A_{,x} + 2\eta\eta_{,x}) + \beta\left(B_{,x} - \frac{1}{6}\eta_{,xxx}\right) + O(\alpha^2 + \alpha\beta + \beta^2) = 0,$$

$$\eta_{,t} + \eta_{,x} + \alpha(A_{,t} + \eta\eta_{,x}) + \beta\left(B_{,t} - \frac{1}{2}\eta_{,xxx}\right) + O(\alpha^2 + \alpha\beta + \beta^2) = 0.$$

But $\eta_{,t} = -\eta_{,x} + O(\alpha, \beta)$, which implies that, to first order, for the terms involving η,

$$\frac{\partial}{\partial t} \equiv -\frac{\partial}{\partial x},$$

and thus we obtain the following equations:

$$\eta_{,t} + \eta_{,x} + \alpha(A_{,x} + 2\eta\eta_{,x}) + \beta\left(B_{,x} - \frac{1}{6}\eta_{,xxx}\right) + O_2 = 0, \quad (19.12)$$

$$\eta_{,t} + \eta_{,x} + \alpha(A_{,t} + \eta\eta_{,x}) + \beta\left(B_{,t} - \frac{1}{2}\eta_{,xxx}\right) + O_2 = 0, \quad (19.13)$$

where $O_2 = O(\alpha^2 + \alpha\beta + \beta^2)$. The compatibility of Eqs. (19.12) and (19.13) and the asymptotic expansions require then that

$$A_{,x} + 2\eta\eta_{,x} = A_{,t} + \eta\eta_{,x},$$
$$B_{,x} - \frac{1}{6}\eta_{,xxx} = B_{,t} - \frac{1}{2}\eta_{,xxx}.$$

The first equation can be rewritten as

$$A_{,t} - A_{,x} = \eta\eta_{,x}.$$

We then perform the change of variables

$$r = x - t, \quad s = x + t,$$
$$x = \frac{1}{2}(r + s), \quad t = \frac{1}{2}(s - r),$$

and, for every function $\varphi(x, t)$, we write $\varphi(x, t) = \bar{\varphi}(r, s)$. We obtain

$$\bar{\eta}_{,s} = \eta_{,x} x_{,s} + \eta_{,t} t_{,s};$$

that is,

$$\bar{\eta}_{,s} = \frac{1}{2}\eta_{,x} + \frac{1}{2}\eta_{,t} = O(\alpha, \beta),$$

because $\eta_{,t} + \eta_{,s} = O(\alpha, \beta)$. Similarly,

$$\bar{\eta}_{,r} = \frac{1}{2}\eta_{,x} - \frac{1}{2}\eta_{,t}$$
$$= \eta_{,x} + O(\alpha, \beta).$$

Moreover,

$$A_{,t} = -\bar{A}_{,r} + \bar{A}_{,s},$$
$$A_{,x} = \bar{A}_{,r} + \bar{A}_{,s};$$

hence,

$$A_{,t} - A_{,x} = -2\bar{A}_{,r}.$$

Therefore,

$$-2\bar{A}_{,r} = \bar{\eta}\bar{\eta}_{,r} + O(\alpha, \beta),$$

that is to say

$$\bar{A}_{,r} = -\frac{1}{4}(\bar{\eta}^2)_{,r} + O(\alpha, \beta).$$

We can then take, to order $O(\alpha, \beta)$,

$$A = -\frac{\eta^2}{4}.$$

Proceeding similarly for the equation for B,

$$B_{,t} - B_{,x} = -\frac{2}{3}\eta_{,xxx},$$

we obtain

$$-2\bar{B}_{,r} = -\frac{2}{3}\bar{\eta}_{,rrr} + O(\alpha, \beta);$$

that is,

$$\bar{B}_{,r} = \frac{1}{3}\bar{\eta}_{,rrr} + O(\alpha, \beta),$$

which yields

$$B = \frac{1}{3}\tilde{\eta}_{,rr} + O(\alpha, \beta),$$

$$B = \frac{1}{3}\eta_{,xx} + O(\alpha, \beta).$$

We will then take

$$B = \frac{1}{3}\eta_{,xx}.$$

Replacing A and B by these values in Eqs. (19.12) and (19.13), we obtain

$$\eta_{,t} + \eta_{,x} + \frac{3}{2}\alpha\eta\eta_{,x} + \frac{1}{6}\beta\eta_{,xxx} + O(\alpha^2, \beta^2, \alpha\beta) = 0,$$

that is, to first order,

$$\eta_{,t} + \eta_{,x} + \frac{3}{2}\alpha\eta\eta_{,x} + \frac{1}{6}\beta\eta_{,xxx} = 0, \qquad (19.14)$$

which is almost the Korteweg–de Vries equation.

To obtain the slightly simpler Korteweg–de Vries equation, we now perform the change of variables

$$x = t' + x',$$

$$t = t';$$

setting

$$\eta(x, t) = \tilde{\eta}(x', t'),$$

we deduce from Eq. (19.14) the equation

$$\tilde{\eta}_{t'} + \frac{3}{2}\alpha\tilde{\eta}\tilde{\eta}_{,x'} + \frac{1}{6}\beta\tilde{\eta}_{,x'x'x'} = 0. \qquad (19.15)$$

Finally, we rescale x', t', and $\tilde{\eta}$ and set

$$x' = kx,$$

$$t' = \ell t,$$

$$\tilde{\eta} = u,$$

where k and ℓ will be chosen later. We find

$$\frac{1}{\ell}u_{,t} + \frac{3}{2}\frac{\alpha}{k}uu_{,x} + \frac{1}{6}\frac{\beta}{k^3}u_{,xxx} = 0.$$

We then choose k and ℓ such that

$$\frac{3}{2}\frac{\alpha\ell}{k} = 1, \quad \frac{1}{6}\frac{\beta\ell}{k^3} = 1;$$

hence,

$$\frac{\ell}{k} = \frac{2}{3\alpha}, \quad \frac{\ell}{k^3} = \frac{6}{\beta},$$

and

$$k = \frac{1}{3}\sqrt{\frac{\beta}{\alpha}},$$

$$\ell = \frac{2}{9}\sqrt{\frac{\beta}{\alpha^3}},$$

and we finally obtain the Korteweg–de Vries (KdV) equation

$$u_{,t} + uu_{,x} + u_{,xxx} = 0. \tag{19.16}$$

19.4. The soliton solutions of the KdV equation

We conclude this chapter by the derivation of the soliton solutions of the KdV equation. These solutions correspond to waves that propagate with a constant velocity and keep a given shape; we will then have

$$u(x,t) = \Psi(\xi) = \Psi(x - Vt),$$

where $\xi = x - Vt$. Substituting this expression of Ψ in Eq. (19.16), we obtain, for Ψ, the following ordinary differential equation:

$$(\Psi - V)\frac{d\Psi}{d\xi} + \frac{d^3\Psi}{d\xi^3} = 0.$$

It follows by integration that

$$\frac{1}{2}\Psi^2 - V\Psi + \frac{d^2\Psi}{d\xi^2} = k_1, \tag{19.17}$$

where k_1 is a constant. We multiply both sides of this equation by $d\Psi/d\xi$ and integrate again to obtain

$$\frac{1}{6}\Psi^3 - \frac{1}{2}V\Psi^2 + \frac{1}{2}\left(\frac{d\Psi}{d\xi}\right)^2 = k_1\Psi + k_2, \tag{19.18}$$

k_2 being another constant. The constants k_1 and k_2 are determined by the boundary conditions: the wave is flat at infinity, and thus Ψ and its derivatives vanish at infinity. This gives $k_2 = 0$ in Eq. (19.18); then, by letting $\xi \to \infty$ in Eq. (19.17), we find $k_1 = 0$.

Equation (19.18), in which $k_1 = k_2 = 0$, can be solved by using the elementary transcendental functions. We find the solution

$$\Psi = \Psi(\xi) = 3V \left[1 + \sinh^2 \left(\frac{1}{2}\sqrt{V}\xi \right) \right]^{-1} = \frac{3V}{\cosh^2 \left(\frac{1}{2}\sqrt{V}\xi \right)};$$

therefore,

$$u(x, t) = 3V \left[1 + \sinh^2 \left(\frac{1}{2}\sqrt{V}(x - Vt) \right) \right]^{-1}$$

$$= \frac{3V}{\cosh^2 \left[\frac{1}{2}\sqrt{V}(x - Vt) \right]} = 3V \operatorname{sech}^2 \left[\frac{1}{2}\sqrt{V}(x - Vt) \right].$$

It is easy to check that this function u solves Eq. (19.16). We notice that u almost vanishes as soon as ξ is larger than a few units (e.g., $\sqrt{V}|\xi| = \sqrt{V}|x - Vt| \geq 20$).

Exercises

1. Consider the Sine-Gordon equation

$$\frac{\partial^2 \theta}{\partial t^2} - c_0^2 \frac{\partial^2 \theta}{\partial x^2} + \omega_0^2 \sin \theta = 0.$$

This equation is obtained, e.g., when one considers a mechanical transmission consisting of elastically coupled pendulums.

a) We first assume that the angle θ is small. Then, one approximates the Sine-Gordon equation by the Klein-Gordon equation

$$\frac{\partial^2 \theta}{\partial t^2} - c_0^2 \frac{\partial^2 \theta}{\partial x^2} + \omega_0^2 \theta = 0.$$

Under what condition (dispersive relation) is

$$\theta = \theta_0 \cos(kx - wt)$$

solution of the Klein-Gordon equation?

b)
- (i) Show that, under the change of variables $t \mapsto \omega_0 T, x \mapsto \frac{\omega_0}{c_0}x$, the Sine-Gordon equation can be rewritten in the form

$$\frac{\partial^2 \theta}{\partial t^2} - \frac{\partial^2 \theta}{\partial x^2} + \sin \theta = 0.$$

- (ii) We look for solutions of the above equation in the form

$$\theta = \theta(s) = \theta(x - ut), \quad -1 < u < 1.$$

Show that

$$(1 - u^2)\theta_{,ss} = \sin\theta,$$

and that,

$$\theta_{,s} = \pm\sqrt{\frac{2(c - c_\infty\theta)}{1 - u^2}},$$

where c is an integration constant.

- (iii) Compute the value of c if we assume that we look for localized solutions (i.e., such that θ and its derivatives vanish as $s \longrightarrow \pm\infty$).
- (iv) Give the expression of the soliton solutions.

2. Consider the following system of equations which describes small amplitude and long waves in a water channel

$$\eta_{,t} + u_{,x} + (u\eta)_{,x} + u_{,xxx} - \eta_{,xxt} = 0,$$

$$u_{,t} + \eta_{,x} + uu_{,x} + \eta_{,xxx} = 0.$$

We set $\xi = x + x_0 - ct$ and we look for solutions of the above system of the form

$$\eta(x, t) = \eta(\xi), \quad u(x, t) = \alpha\eta(\xi), \quad \alpha \neq 0,$$

where η is localized.

a) Show that $c = \dfrac{2 - \alpha^2}{\alpha}$ and that

$$(\alpha^2 - 1)\eta_{,\xi} + \alpha^2\eta\eta_{,\xi} + \eta_{,\xi\xi\xi} = 0.$$

b) Consider the equation

$$a\eta_{,\xi} - b\eta_{,\xi\xi\xi} = \eta\eta_{,\xi}, \quad ab > 0.$$

Show that

$$\eta = 3a \; sech^2(\frac{1}{2}\sqrt{\frac{a}{b}}(\xi + \xi_0)),$$

where ξ_0 is an arbitrary constant, is the solution of the above equation.

c) Deduce that

$$\eta(x, t) = \eta_0 sech^2(\lambda(x + x_0 - ct)),$$

$$u(x, t) = \pm\sqrt{\frac{3}{\eta_0 + 3}} \; \eta_0 sech^2(\lambda(x + x_0 - ct)),$$

where

$$\eta_0 > 0, \quad c = \frac{3 + 2\eta_0}{\pm\sqrt{3(3 + \eta_0)}}, \quad \lambda = \frac{1}{2}\sqrt{\frac{\eta_0}{\eta_0 + 3}},$$

is a soliton solution of the initial system.

3. Give a soliton solution of the following Whitham's system:

$$\eta_{,t} + u_{,x} + (u\eta)_{,x} - \frac{1}{6}u_{,xxx} = 0,$$

$$u_{,t} + \eta_{,x} + uu_{,x} - \frac{1}{2}u_{,xxt} = 0,$$

such that $\eta(x, t) = \eta(\xi) = \eta(x + x_0 - ct), \quad u(x, t) = \alpha\eta(\xi).$

CHAPTER TWENTY

The nonlinear Schrödinger equation

The purpose of this chapter is to introduce another equation describing nonlinear wave phenomena: the nonlinear Schrödinger equation (NLS), which should not to be confused with the linear Schrödinger equation from quantum mechanics (see below).

As indicated in Chapter 18, this equation, like the KdV equation, has been discovered rather recently. The two equations appear in, and are used for, wave phenomena of various types. In particular, the NLS equation, like the KdV equation, can describe water-wave phenomena, and it can also be deduced from the Euler equation of perfect fluids under appropriate hypotheses.

However, the NLS equation also describes phenomena that are very important nowadays: the propagation of waves in wave guides in relation to the design of optical long-distance communications lines and all-optical signal-processing devices for reliable and high-bit-rate transmission of information.

Owing to the importance of the subject, and to diversify the mathematical techniques developed in this book, we will in this chapter derive the NLS equation from the Maxwell equations in the context of wave guides rather than deduce them from the Euler equations in the context of fluid mechanics.

We start in Section 20.1 by recalling the Maxwell equations, and we introduce a new phenomenon that is essential for optic fibers, namely polarization, which corresponds to, and describes, the electromagnetically anisotropic behavior of the medium. In Sections 20.2 and 20.3, we complete the modeling of electromagnetic wave propagation in polarized media by introducing the constitutive laws of the medium that link the polarization and the electric field together (these constitutive laws are similar in many respects to those that appeared in Chapters 5 and 10). The nonlinear Schrödinger equation is then introduced in Section 20.4. Finally, in Section 20.5, we give the soliton solutions of the undamped NLS equation. The developments of Sections 20.1 and 20.2 are general and contain basic elements used in a large variety of photoelectric phenomena in wave guides. The next sections are specific to optic fibers, and take into account the isotropy of the medium and its geometry (section very small by comparison with the length).

20.1. Maxwell equations for polarized media

We are interested in this section in the propagation of signals in polarized media such as optical fibers. These phenomena, of electromagnetic nature, are governed by the Maxwell equations introduced in Chapter 10 to which we add the polarization.

We thus start by recalling the Maxwell equations as they appeared in Chapter 10. The electric and magnetic fields E, H, and the electric and magnetic inductions D, B, satisfy, in the medium, the equations

$$\frac{\partial B}{\partial t} + \text{curl } E = 0, \tag{20.1}$$

$$\text{curl } H = J + \frac{\partial D}{\partial t}, \tag{20.2}$$

$$\text{div } D = q, \tag{20.3}$$

$$\text{div } B = 0, \tag{20.4}$$

where q is the charge density. In the case of optical fibers, and more generally in the absence of free charge in the medium, we have $J = 0$ and $q = 0$; hence,

$$\text{curl } H = \frac{\partial D}{\partial t}, \tag{20.5}$$

$$\text{div } D = 0. \tag{20.6}$$

As in Chapter 10, the inductions D and B are related to the fields E and H by the constitutive laws of the electromagnetic medium. It is at this level that polarization appears and that the first differences with Chapter 10 occur. We introduce the induced electric and magnetic polarization vectors P and M. The constitutive laws, similar to Eqs. (10.14) of Chapter 10, then read

$$D = \varepsilon_0 E + P, \tag{20.7}$$

$$B = \mu_0 H + M, \tag{20.8}$$

where ε_0 and μ_0 are, respectively, the permittivity and the permeability of vacuum. Henceforth, we will take $M = 0$ because the magnetic polarization vanishes for nonmagnetic media such as optical fibers; hence,

$$B = \mu_0 H. \tag{20.9}$$

Taking the curl of Eq. (20.1), it follows, in view of Eq. (20.9) that

$$\text{curl curl } E = -\frac{\partial}{\partial t}\text{curl } B$$

$$= -\frac{\partial}{\partial t}\mu_0 \text{ curl } H.$$

Therefore, because $J = 0$, we find

$$\text{curl curl } E = -\mu_0 \frac{\partial^2 D}{\partial t^2},$$

and, finally, we deduce from Eq. (20.7) the following wave-type equation:

$$\text{curl curl } E = -\frac{1}{c^2}\frac{\partial^2 E}{\partial t^2} - \frac{1}{\varepsilon_0 c^2}\frac{\partial^2 P}{\partial t^2}, \tag{20.10}$$

where we have set $\mu_0 \varepsilon_0 = 1/c^2$, c denoting the speed of light in the vacuum.

Equation (20.10) is the fundamental equation for the propagation of the electric field E in optical fibers. It must be supplemented with suitable constitutive relations between E and the induced electric polarization vector P: we develop this question in the following sections. For the time being, we give different forms of Eq. (20.10).

In particular, we will use the Fourier transform with respect to time defined for E by

$$\mathcal{F}E(x, \omega) = \hat{E}(x, \omega) = \int_{-\infty}^{+\infty} E(x, t) \exp(i\omega t)\, dt.$$

By Fourier transform, Eq. (20.10) becomes

$$\text{curl curl } \hat{E} = \frac{\omega^2}{c^2}\hat{E} + \frac{\omega^2}{\varepsilon_0 c^2}\hat{P}.$$

In isotropic media, E and D are proportional (with a scalar proportionality coefficient independent of t). It follows that $\hat{E}$ and $\hat{D}$ are also proportional; then, div $D = 0$ yields div $E = 0$, and div $\hat{E} = 0$, so that

$$\text{curl curl } E = \text{grad div } E - \Delta E = -\Delta E,$$
$$\text{curl curl } \hat{E} = -\Delta \hat{E},$$

and the last equation becomes

$$-\Delta \hat{E} = \frac{\omega^2}{c^2}\left(\hat{E} + \frac{1}{\varepsilon_0}\hat{P}\right). \tag{20.11}$$

In some circumstances, it is important to consider nonisotropic media for which $D = [\varepsilon]E$, where $[\varepsilon]$ is a tensor of order two; in this case div $E \neq 0$, but this issue will not be addressed here.

20.2. Equations of the electric field: the linear case

In the most general case, the derivation of the relations between the electric field E and the polarization P relies on quantum mechanics; quantum mechanics is needed in particular when the optical frequency is close to resonance with the medium.[1] Far from the resonances, which is in particular the case for optical fibers with wavelengths contained between 0.5 and 2 μm, we can use simpler phenomenological relations between P and E – for instance a relation corresponding to the beginning of a Taylor expansion in E around 0. This expansion is justified by the fact that E is small (in a proper nondimensional form), which expresses, from the physical point of view, the fact that the electromagnetic forces are weak by comparison with the cohesion forces.

We will then have an expansion of the form

$$P = \varepsilon_0 \big\{ \mathcal{L}^{(1)}(E) + \mathcal{L}^{(2)}(E \otimes E) + \mathcal{L}^{(3)}(E \otimes E \otimes E) + \cdots$$

$$+ \mathcal{L}^{(j)}(E \underbrace{\otimes \cdots \otimes}_{j \text{ times}} E) + \cdots \big\}, \qquad (20.12)$$

where the $\mathcal{L}^{(j)}$ are linear functional operators. A phenomenological study similar to that of rheology, described in Chapter 5, must then be performed here, and one must take into account, in particular, the isotropy (invariance by a change of orthonormal system), causality (the past does not depend on the future), and temporal invariance (invariance by translation in time). Through an analysis not developed here, these principles lead to the following form for the terms $\mathcal{L}^{(j)}$ (we limit ourselves to $j = 1, 2, 3$):

$$\mathcal{L}^{(1)}(E)(t) = \int_{-\infty}^{+\infty} \chi^{(1)}(t - \tau) \cdot E(\tau) \, d\tau,$$

$$\mathcal{L}^{(2)}(E)(t) = \int_{-\infty}^{+\infty} \int_{-\infty}^{+\infty} \chi^{(2)}(t - \tau_1, t - \tau_2) \cdot E(\tau_1) \otimes E(\tau_2) \, d\tau_1 \, d\tau_2,$$

$$\mathcal{L}^{(3)}(E)(t) = \int_{-\infty}^{+\infty} \int_{-\infty}^{+\infty} \int_{-\infty}^{+\infty} \chi^{(3)}(t - \tau_1, t - \tau_2, t - \tau_3)$$

$$\times E(\tau_1) \otimes E(\tau_2) \otimes E(\tau_3) \, d\tau_1 \, d\tau_2 \, d\tau_3,$$

where $\cdot$ denotes a contracted tensorial product and $\chi^{(j)}$ is a tensor of order

---

[1] One then needs the *linear* Schrödinger equation of quantum mechanics. As indicated in the introduction, this equation is totally distinct from the NLS equation below as far as its role and the use that is made of it are concerned.

$j + 1$ called the multiimpulsional response tensor; its Fourier transform in time is the susceptibility tensor. By causality, the $\chi^{(j)}$ vanish for $t < 0$.

For symmetry and isotropy reasons, the term $\mathcal{L}^{(2)}$ vanishes.[2] If we thus limit ourselves to the term of order 3, there remains

$$P = P_{\mathrm{L}} + P_{\mathrm{NL}}, \qquad (20.13)$$

where P_{L} is the linear part; for reasons of isotropy, we can take $\chi^{(1)}$ diagonal, $\chi^{(1)} = \chi^{(1)}I$, $\chi^{(1)}$ scalar, and thus

$$P_{\mathrm{L}}(x, t) = \varepsilon_0 \int_{-\infty}^{+\infty} \chi^{(1)}(x, t - \tau)E(x, \tau)\, d\tau. \qquad (20.14)$$

It follows, by taking a Fourier transform in time, that

$$\hat{P}_{\mathrm{L}}(x, \omega) = \varepsilon_0 \hat{\chi}^{\,1}(x, \omega)\hat{E}(x, \omega). \qquad (20.14')$$

In Eqs. (20.14) and (20.15), x denotes, as usual, the point of $\mathbb{R}^3$ with coordinates x_1, x_2, x_3 ($x_3 = z$ hereafter).

We assume that, in the nonlinear term P_{NL} (which represents lower-order effects, E being small), the response is instantaneous, which is expressed by

$$\chi^{(3)}(x, t, t', t'') = \chi^3(x)\delta_t \delta_{t'} \delta_{t''},$$

where δ_t denotes here the Dirac mass at 0 for the variable t, and $\chi^{(3)}$ is a tensor of order four independent of t. It follows that

$$P_{\mathrm{NL}}(x, t) = \varepsilon_0 \chi^{(3)}(x) \cdot E(x, t) \otimes E(x, t) \otimes E(x, t). \qquad (20.15)$$

The expression of $\chi^{(3)}$ and its dependence on x will be further discussed below.

Linearized equation

In a first step, we will neglect the nonlinear term P_{NL} and study the linearized form of Eq. (20.10). Because

$$D = \varepsilon_0 E + P \simeq \varepsilon_0 E + P_{\mathrm{L}},$$

it follows, owing to Eq. (20.14), that

$$\hat{D} = \varepsilon_0 \hat{E} + \hat{P}_{\mathrm{L}},$$
$$\hat{D} = \varepsilon_0 \big(1 + \hat{\chi}^{(1)}\big)\hat{E}. \qquad (20.16)$$

---

[2] Every function, tensor of odd order, invariant by central symmetry, vanishes; from this, we deduce that $\mathcal{L}^{(2)} = 0$.

Equation (20.11) becomes

$$\begin{cases} \Delta \hat{E} + \varepsilon(\omega)\dfrac{\omega^2}{c^2}\hat{E} = 0, \\ \varepsilon(\omega) = 1 + \chi^{(1)}(\omega). \end{cases} \qquad (20.17)$$

Let us note here that, because $\hat{\chi}^{(1)}(\omega)$ is complex, the same is true for $\varepsilon(\omega)$. By definition

$$\varepsilon = \left(n + \frac{i\alpha c}{2\omega}\right)^2 = n^2 - \frac{\alpha^2 c^2}{4\omega^2} + i\frac{n\alpha c}{\omega}, \qquad (20.18)$$

where $n + i\alpha c/2\omega$ is the complex refraction index of the medium, $n = n(\omega)$ is the real part of this index; $\alpha = \alpha(\omega)$ denotes the absorption coefficient of frequency ω, and c denotes as usual the speed of light in vacuum. Thus,

$$\alpha = \frac{\omega}{nc}\,\mathrm{Im}\big[\hat{\chi}^{(1)}(\omega)\big],$$

where Im denotes the imaginary part (the real part will be denoted by Re); this relation is exact. Furthermore, we have $\hat{\chi}^{(1)} \ll 1$, so that, to first approximation,

$$n + \frac{i\alpha c}{2\omega} = \varepsilon^{1/2} \simeq 1 + \frac{1}{2}\hat{\chi}^{(1)},$$

that is

$$n \simeq 1 + \frac{1}{2}\,\mathrm{Re}\big[\hat{\chi}^{(1)}(\omega)\big].$$

Thus, and always to first approximation, we will have

$$\varepsilon \simeq n^2 \simeq 1.$$

We now turn to the resolution of Eq. (20.17) for a long fiber of section $\Omega \subset \mathbb{R}^2_{x_1 x_2}$ (let us observe here that we will arrive at a problem similar to that of Eq. (20.17) in the nonlinear case). To do so, we look for a solution of the form

$$\hat{E}(x, \omega) = A(\omega)\, F(x_1, x_2)\, \exp(i\beta z),$$

where we have set $z = x_3$. Dividing by $\exp(i\beta z)$, we deduce from Eq. (20.17) that

$$A\Delta_2 F - A\beta^2 F + \varepsilon(\omega)\frac{\omega^2}{c^2} AF = 0,$$

where Δ_2 denotes the two-dimensional Laplacian with respect to the variables x_1 and x_2 and where A is a normalization factor. We arrive at the following eigenvalue problem for F:

$$\begin{cases} \Delta_2 F + \kappa^2 F = 0, & x_1, x_2 \in \Omega, \\ F = 0, & \text{on the boundary } \partial\Omega \text{ of } \Omega. \end{cases} \qquad (20.19)$$

We make here the assumption that E (and thus F) vanishes on $\partial\Omega$ (we refer the reader interested in more details to Agrawal (1989)).

Therefore κ^2 is an eigenvalue of the Laplacian in Ω associated with the Dirichlet boundary condition on $\partial\Omega$. This corresponds to an infinite dimensional spectral problem (several similar problems have appeared in Chapter 17). The possible values of κ^2 form a sequence of nonnegative real numbers λ_n

$$0 < \lambda_1 \le \lambda_2 \le \dots, \quad \lambda_n \to \infty \quad \text{for } n \to \infty.$$

Choosing for κ^2 one of these eigenvalues, $\beta = \beta(\omega)$ will be defined for each frequency ω by the relation

$$\varepsilon(\omega)\frac{\omega^2}{c^2} - \beta(\omega)^2 = \kappa^2. \tag{20.20}$$

Finally, it remains to normalize the eigenvector F (defined up to a multiplicative constant). If we choose, for instance, F such that

$$\int_\Omega F^2(x_1, x_2)\, dx_1\, dx_2 = 1,$$

$A(\omega)$ will then be called the intensity factor.

20.3. General case

We return here to the study of Eq. (20.10) in the nonlinear case; that is to say, we no longer neglect the nonlinear term P_{NL}. We assume, for the sake of simplicity, that the waves propagate in the direction $x_3 = z$ and that the polarization is in the direction Ox_1 (with unit vector e_1), and thus

$$\begin{cases} E(x, t) = E_1(x, t)e_1, \\ P_{\mathrm{L}}(x, t) = P_{1\mathrm{L}}(x, t)e_1, \\ P_{\mathrm{NL}}(x, t) = P_{1\mathrm{NL}}(x, t)e_1. \end{cases} \tag{20.21}$$

On the other hand, for glass fibers, the frequencies ω are close to a central frequency ω_0 ($|\omega - \omega_0| \ll \omega_0$). We then set

$$E_1(x, t) = \mathrm{Re}[\mathcal{E}(x, t)\, \exp(-i\omega_0 t)], \tag{20.22}$$

$$P_{1\mathrm{L}}(x, t) = \mathrm{Re}[\mathcal{P}_{\mathrm{L}}(x, t)\, \exp(-i\omega_0 t)], \tag{20.23}$$

where $\mathcal{E}$ and $\mathcal{P}_{\mathrm{L}}$ are complex; in view of Eq. (20.14), it is natural to require that $\mathcal{P}_{\mathrm{L}}$ and $\mathcal{E}$ satisfy

$$\mathcal{P}_{\mathrm{L}}(x, t) = \varepsilon_0 \int_{-\infty}^{+\infty} \hat{\chi}^{(1)}(t - t')\, \mathcal{E}(x, t')\, \exp(i\omega_0(t - t'))\, dt'. \tag{20.24}$$

Moreover, because Eq. (20.24) is equivalent to

$$\mathcal{P}_{\mathrm{L}}(x, \cdot) = \varepsilon_0 \left\{ \hat{\chi}^{(1)}(\cdot) \, \exp(i\omega_0 \cdot) \right\} \star \mathcal{E}(x, \cdot),$$

where $\star$ is the convolution product, we have

$$\widehat{\mathcal{P}_{\mathrm{L}}}(x, \omega) = \varepsilon_0 \mathcal{F} \left\{ \hat{\chi}^{(1)}(\cdot) \, \exp(i\omega_0 \cdot) \right\} (\omega) \, \hat{\mathcal{E}}(x, \omega)$$

(where $\mathcal{F}G = \widehat{G}$). Furthermore,

$$\mathcal{F} \left\{ \hat{\chi}^{(1)}(\cdot) \, \exp(i\omega_0 \cdot) \right\} (\omega) = \int_{-\infty}^{+\infty} \hat{\chi}^{(1)}(t) \, \exp(i\omega_0 t) \, \exp(i\omega t) \, dt$$

$$= \hat{\chi}^{(1)}(\omega + \omega_0),$$

and thus

$$\widehat{\mathcal{P}_{\mathrm{L}}}(x, \omega) = \varepsilon_0 \hat{\chi}^{(1)}(\omega + \omega_0) \, \hat{\mathcal{E}}(x, \omega),$$

which finally gives, by inverse Fourier transform:

$$\mathcal{P}_{\mathrm{L}}(x, t) = \frac{1}{2\pi} \int_{-\infty}^{+\infty} \hat{\chi}^{(1)}(\omega' + \omega_0) \, \hat{\mathcal{E}}(x, \omega') \, \exp(-i\omega' t) \, d\omega'$$

$$= (\text{Setting } \omega' + \omega_0 = \omega) \qquad\qquad (20.25)$$

$$= \frac{1}{2\pi} \int_{-\infty}^{+\infty} \hat{\chi}^{(1)}(\omega) \, \hat{\mathcal{E}}(x, \omega - \omega_0) \, \exp(-i(\omega - \omega_0)t) \, d\omega.$$

We then deduce from Eqs. (20.15) and (20.21) that

$$P_{\mathrm{1NL}} = \varepsilon_0 \chi_{1111}^{(3)} (E_1)^3,$$

where $\chi_{1111}^{(3)} = \chi_{x_1, x_1, x_1, x_1}^{(3)}$ is the component on $e_1 \otimes e_1 \otimes e_1 \otimes e_1$ of the fourth-order tensor $\chi^{(3)}$. Relation (20.22) then furnishes

$$E_1 = \mathrm{Re} \, a = \frac{1}{2}(a + \bar{a}),$$

where

$$a = \mathcal{E}(x, t) \, \exp(-i\omega_0 t),$$

which gives

$$E_1^3 = \frac{1}{8} \, \mathrm{Re}(a^3 + 3a^2\bar{a} + 3a\bar{a}^2 + \bar{a}^3)$$

$$= \frac{1}{8} \, \mathrm{Re}[\mathcal{E}^3 \exp(-3i\omega_0 t) + 3\mathcal{E}^2\bar{\mathcal{E}} \, \exp(-i\omega_0 t)$$

$$+ 3\mathcal{E}\bar{\mathcal{E}}^2 \exp(i\omega_0 t) + \bar{\mathcal{E}}^3 \exp(3i\omega_0 t)].$$

The terms of the form $\exp(\pm 3i\omega_0 t)$ correspond to very fast oscillations and are negligible by comparison with the terms of the form $\exp(\pm i\omega_0 t)$. Then there remains, to first approximation

$$E_1^3 \simeq \frac{3}{4}\mathcal{E}\bar{\mathcal{E}}\,\mathrm{Re}[\mathcal{E}\exp(-i\omega_0 t)] = \frac{3}{4}|\mathcal{E}|^2 E_1.$$

Consequently,

$$P_{1\mathrm{NL}} \simeq \mathrm{Re}(\mathcal{P}_{\mathrm{NL}}\exp(-i\omega_0 t)), \tag{20.26}$$

where

$$\begin{cases} \mathcal{P}_{\mathrm{NL}} \simeq \varepsilon_0 \varepsilon_{\mathrm{NL}}\mathcal{E}, \\ \varepsilon_{\mathrm{NL}} \simeq \dfrac{3}{4}\chi_{1111}^{(3)}|\varepsilon(x,t)|^2. \end{cases} \tag{20.27}$$

To write Eq. (20.11), we need to take the Fourier transform in time of Eqs. (20.26) and (20.27). We thus specify (assume) – which is reasonable – that, for an optic fiber, $\chi_{1111}^{(3)}$ and $|\mathcal{E}(x,t)|^2$ are very close to their time mean values. Everything thus behaves, to first approximation, as if $\varepsilon_{\mathrm{NL}}$ were independent of time, and thus

$$\hat{\mathcal{P}}_{\mathrm{NL}} \simeq \varepsilon_0 \varepsilon_{\mathrm{NL}}\hat{\mathcal{E}}. \tag{20.28}$$

By applying the Fourier transform to Eq. (20.7), we then see that div $\hat{D}$ is again proportional to div $\hat{E}$; we thus have again div $\hat{E} = 0$, curl curl $\hat{E} = -\Delta\hat{E}$. Equation (20.10) then gives

$$\Delta\hat{\mathcal{E}} + \varepsilon(\omega)k^2\hat{\mathcal{E}} = 0, \tag{20.29}$$

where

$$\varepsilon(\omega) = 1 + \hat{\chi}^1(\omega) + \varepsilon_{\mathrm{NL}},$$

and $k = \omega/c$; $\varepsilon_{\mathrm{NL}}$ is independent of ω.[3]

Now our aim is to solve Eq. (20.29) in an approximate way and, at the end of the calculations, the nonlinear Schrödinger equation will appear. To do so, we need some preliminary computations and some remarks concerning $\varepsilon(\omega)$.

As in the linear case, we have

$$\varepsilon = \left(\tilde{n} + \frac{i\tilde{\alpha}}{2k_0}\right)^2,$$

---

[3] We have actually implicitly set it to its value at ω_0: The dispersive effects (corresponding to $\omega \neq \omega_0$) are taken into account in the linear part P_{L}, whereas the nonlinear part does not take into account the dispersive effects.

where $\tilde{n}$ and $\tilde{\alpha}$ are defined as in the linear case, and $k_0 = \omega_0/c$; however, the different quantities depend now on the intensity $|\mathcal{E}|^2$. It is usual to write $\tilde{n}$ and $\tilde{\alpha}$ in the form

$$\tilde{n} = n + n_2|\mathcal{E}|^2, \quad \tilde{\alpha} = \alpha + \alpha_2|\mathcal{E}|^2, \tag{20.30}$$

where n and α have been defined in the previous section (linear case). It then follows that

$$\frac{\tilde{n}\tilde{\alpha}}{k_0} = \text{Im}\left(1 + \hat{\chi}^{(1)} + \frac{3}{4}\chi_{1111}^{(3)}|\mathcal{E}|^2\right)$$

$$= \frac{n\alpha}{k_0} + \frac{3}{4}\text{Im}\chi_{1111}^{(3)}|\mathcal{E}|^2;$$

hence,

$$\tilde{\alpha} = \alpha\frac{n}{\tilde{n}} + \frac{k_0}{\tilde{n}}\frac{3}{4}\text{Im}\chi_{1111}^{(3)}|\mathcal{E}|^2.$$

We will have, as a first approximation, $\tilde{n} \approx n$ (and also $\omega \approx \omega_0$), so that

$$\alpha_2 = \frac{3}{4}\frac{\omega_0}{nc}\ \text{Im}\chi_{1111}^{(3)}.$$

Moreover,

$$\varepsilon = \tilde{n}^2 - \frac{\tilde{\alpha}^2}{4k_0^2} + i\frac{\tilde{\alpha}\tilde{n}}{k_0}$$

$$= 1 + \hat{\chi}^{(1)} + \frac{3}{4}\chi_{1111}^{(3)}|\mathcal{E}|^2,$$

and hence,

$$\varepsilon = 1 + \text{Re}\ \hat{\chi}^{(1)} + \frac{3}{4}\text{Re}\left(\chi_{1111}^{(3)}|\mathcal{E}|^2\right) = \tilde{n}^2 - \frac{\tilde{\alpha}^2}{4k_0^2}.$$

On the other hand,

$$(n + n_2|\mathcal{E}|^2)^2 = n^2 + 2nn_2|\mathcal{E}|^2 + n_2^2|\mathcal{E}|^4$$

$$\simeq n^2 + 2nn_2|\mathcal{E}|^2,$$

where we have neglected, as a first approximation, the term $n_2^2|\mathcal{E}|^4$. Because $1 + \text{Re}\chi^{(1)} \simeq n^2$ and $\tilde{\alpha} \ll \tilde{n}$, it follows that

$$\frac{3}{4}\text{Re}\chi_{1111}^{(3)} = 2nn_2,$$

and thus, to first approximation,

$$n_2 \simeq \frac{3}{8n}\text{Re}\chi_{1111}^{(3)}.$$

This last quantity measures the nonlinearity of the fiber.

20.4. The nonlinear Schrödinger equation

We are now in a position to solve (in an approximate way) Eq. (20.29). To do so, we look for a solution of the form

$$\hat{\mathcal{E}}(x, \omega - \omega_0) = F(x_1, x_2)\, \hat{A}(z, \omega - \omega_0)\, \exp(i\beta_0 z),$$

where we have again set $z = x_3$. We insert this expression of $\hat{\mathcal{E}}$ in Eq. (20.29) and obtain, upon dividing by $\exp(i\beta_0 z)$:

$$\hat{A}\Delta_2 F - \hat{A} F \beta_0^2 + \frac{\partial^2 \hat{A}}{\partial z^2} F + 2i\beta_0 \frac{\partial \hat{A}}{\partial z} F + \varepsilon(\omega)\frac{\omega^2}{c^2}\hat{A} F = 0.$$

Under the assumption of slow variation along the fiber, the term $\partial^2 \hat{A}/\partial z^2$ is neglected by comparison with $\beta_0\, \partial \hat{A}/\partial z$. If we divide by $\hat{A}F$, it then follows that

$$\frac{\Delta_2 F}{F} + \varepsilon(\omega)\frac{\omega^2}{c^2} = \beta_0^2 - \frac{2i\beta_0}{\hat{A}}\frac{\partial \hat{A}}{\partial z}.$$

Let $\tilde{\beta}^2 = \tilde{\beta}(z, \omega)^2$ be the common value of these two quantities (the right-hand side depends only on z and ω). Thus,

$$\Delta_2 F + \left[\varepsilon(\omega)\frac{\omega^2}{c^2} - \tilde{\beta}^2\right] F = 0, \tag{20.31}$$

and

$$2i\beta_0 \frac{\partial \hat{A}}{\partial z} + \left(\tilde{\beta}^2 - \beta_0^2\right)\hat{A} = 0. \tag{20.32}$$

We first solve Eq. (20.31), which we supplement, as before, with the Dirichlet-type boundary condition. The corresponding eigenvalue problem is a bit different from that encountered in the linear case because the coefficient of F also depends on x_1 and x_2 (and z). One can show, using perturbation theory for eigenvalue problems,[4] that F is unchanged at first order, that is

$$\Delta_2 F + \kappa^2 F = 0, \quad \text{in } \Omega,$$
$$F = 0, \quad \text{on } \partial\Omega,$$

to first approximation, as in the linear case. Similarly, we will have

$$\tilde{\beta}(\omega) = \beta(\omega) + \delta\beta,$$

where $\beta(\omega)$ has been computed in Section 20.2 (i.e., when $P_{\mathrm{NL}} = 0$). Thus,

$$\kappa^2 = \varepsilon(\omega)k^2 - \tilde{\beta}^2,$$

---

[4] See details and references in Agrawal (1989).

and

$$\varepsilon = (n + \delta n)^2 \simeq n^2 + 2n \cdot \delta n$$

$$= \left(n + n_2 |\mathcal{E}|^2 + \frac{i\alpha}{2k_0} + i\frac{\alpha_2 |\mathcal{E}|^2}{2k_0} \right)^2$$

$$\simeq n^2 + 2nn_2 |\mathcal{E}|^2 + n\frac{i\alpha}{k_0}$$

$$+ \text{ lower-order terms,}$$

which yields, to first approximation,

$$\delta n \simeq n_2 |\mathcal{E}|^2 + \frac{i\alpha}{2k_0}.$$

Moreover,

$$\tilde{\beta}^2 \simeq \beta^2 + 2\beta \cdot \delta\beta.$$

The eigenfunction F is thus, to first approximation, a solution of

$$\Delta_2 F + \left[k_0^2 (n^2 + 2n \cdot \delta n) - \beta^2 - 2\beta \cdot \delta\beta \right] F = 0,$$

where δn depends on x_1 and x_2, but n, β and $\delta\beta$ do not depend on x_1 and x_2. Multiplying the previous equation by F and integrating over Ω, we find

$$\int_\Omega |\nabla F|^2 \, dx_1 \, dx_2 = \int_\Omega \left[k_0^2 (n^2 + 2n \cdot \delta n) - \beta^2 - 2\beta \cdot \delta\beta \right] F^2 \, dx_1 \, dx_2.$$

Because, owing to Eqs. (20.19) and (20.20),

$$\int_\Omega |\nabla F|^2 \, dx_1 \, dx_2 \simeq \int_\Omega \left(k_0^2 n^2 - \beta^2 \right) F^2 \, dx_1 \, dx_2$$

to first order, we deduce that

$$\beta \cdot \delta\beta \int_\Omega |F|^2 \, dx_1 \, dx_2 \simeq k_0^2 n \int_\Omega \delta n |F|^2 \, dx_1 \, dx_2,$$

and thus

$$\delta\beta \simeq \frac{k_0^2 n}{\beta} \frac{\int_\Omega \delta n F^2 \, dx_1 \, dx_2}{\int_\Omega F^2 \, dx_1 \, dx_2}.$$

Now, κ^2 is an eigenvalue of the linear problem and $\varepsilon \simeq n^2$; hence, $n^2 k_0^2 - \beta^2 = \kappa^2$, which gives

$$\delta\beta \simeq \frac{k_0^2 n}{\sqrt{n^2 k_0^2 - \kappa^2}} \frac{\int_\Omega \delta n F^2 \, dx_1 \, dx_2}{\int_\Omega F^2 \, dx_1 \, dx_2}. \qquad (20.33)$$

Once the functions F and $\tilde{\beta}$ are determined as indicated before, we turn to the resolution of Eq. (20.32). Because $\tilde{\beta} \simeq \beta_0$, we have $\tilde{\beta}^2 - \beta_0^2 \simeq 2\beta_0(\tilde{\beta} - \beta_0)$, and thus we obtain, to first approximation, the equation

$$\frac{\partial \hat{A}}{\partial z} = i(\beta + \delta\beta - \beta_0)\hat{A}.$$

Because $\omega \simeq \omega_0$, we write the beginning of the Taylor expansion:

$$\beta(\omega) \simeq \beta_0 + (\omega - \omega_0)\beta_1 + \frac{1}{2}(\omega - \omega_0)^2\beta_2,$$

where

$$\beta_n = \left.\frac{d^n\beta}{d\omega^n}\right|_{\omega=\omega_0}.$$

We infer from these equations that

$$\frac{\partial \hat{A}}{\partial z}(z, \omega - \omega_0) = i\left[(\omega - \omega_0)\beta_1 + \frac{1}{2}(\omega - \omega_0)^2\beta_2 + \delta\beta\right]\hat{A}(z, \omega - \omega_0),$$

and, by applying the inverse Fourier transform, we finally obtain the following equation for A:

$$\frac{\partial A}{\partial z} = -\beta_1\frac{\partial A}{\partial t} - \frac{i}{2}\beta_2\frac{\partial^2 A}{\partial t^2} + i\delta\beta A.$$

The term $\delta\beta$ includes the loss effects of the fiber and of the nonlinearity. We deduce from the preceding relations, and in particular from Eq. (20.33), that

$$\delta\beta \simeq \gamma|A|^2 + \frac{i\alpha}{2},$$

where

$$\gamma = \frac{n_2\omega_0}{c\,A_{\text{eff}}},$$

$$A_{\text{eff}} = \frac{(\int_\Omega |F|^2 dx_1\,dx_2)^2}{\int_\Omega |F|^4\,dx_1\,dx_2},$$

and where F is the eigenfunction computed; we thus find A_{eff} once the eigenmode F has been determined.

Finally, A is a solution of the equation

$$\frac{\partial A}{\partial z} + \beta_1\frac{\partial A}{\partial t} + \frac{i}{2}\beta_2\frac{\partial^2 A}{\partial t^2} + \frac{\alpha}{2}A = i\gamma|A|^2 A.$$

To conclude, we perform the change of variables $z = z'$, $t = t' + \beta_1 z'$. The previous equation then becomes, with the primes omitted:

$$\frac{\partial A}{\partial z} + \frac{i}{2}\beta_2 \frac{\partial^2 A}{\partial t^2} - i\gamma |A|^2 A + \frac{\alpha}{2}A = 0, \qquad (20.34)$$

which is the **nonlinear Schrödinger equation**.

Remark 20.1: The linear Schrödinger equation (known and studied long before the NLS equation) consists of Eq. (20.34) with $\gamma = \alpha = 0$ and its analogue in space dimension two or three. The term involving γ is the nonlinear term specific to the present study. This is the term that gives rise to solitons, $\gamma > 0$. The term involving α, $\alpha > 0$, is a damping term.

20.5. Soliton solutions of the NLS equation

In the absence of damping ($\alpha = 0$), the nonlinear Schrödinger equation (20.34) possesses soliton solutions like the KdV equation, namely, waves propagating without changing their shape along the z axis. We conclude this study by giving, as in the previous chapter, the expression of the soliton solutions. We assume, for the sake of simplicity, that $\beta_2 = -1$ (β_2 is negative for physical reasons) and $\gamma = 1$. We observe that we can arrive at these values by a mere change of scale for A, z, and t.

For these values of β_2, γ, and α, it is easy to verify that the following function is an **exact solution** of Eq. (20.34) for every value of the real constant b and of the complex constant a of modulus 1:

$$A(t, z) = \frac{a \, \exp\left(\frac{1}{2}iz\right)}{\cosh(t - b)}. \qquad (20.35)$$

To go further:

Continuations of Part 4 are numerous: acoustics, general wave phenomena, soliton theory.

It is worth mentioning, as basic references, the books by Courant and Hilbert and by Roseau; concerning acoustics and aeroacoustics, see the books by Goldstein, Lightill, or Rayleigh. The physical or mathematical aspects of musical sounds are developed in the books by Benade, Fletcher and Rossing, Roederer, and Sundberg.

As explained in the introduction to the fourth part, solitons, which originally appeared as a curiosity which could occur in certain channels, turned out to be a fundamental phenomenon for the propagation of digital signals in optic fibers and wave guides. The books by Agrawal, Boyd, and Remoissenet are examples of basic or advanced books on this very important subject.

Exercises

1. Consider the following weakly dispersive nonlinear wave equation which appears in the modeling of elastic networks:

$$u_{,tt} - \frac{\delta^2}{LC_0}u_{,xx} = \frac{\delta^4}{12LC_0}u_{,xxxx} + bu_{,tt}^2.$$

We look for a soliton solution $u(x, t) = U(x - vt)$ of this equation.

a) Write the ordinary differential equation satisfied by U.

b) We look for localized solutions of this equation (i.e., such that U and its derivatives tend to zero at $+$ and $-\infty$). Show that U is a solution of

$$(v^2 - v_0^2)U = bv^2U^2 + \delta^2\frac{v_0^2}{12}U_{,ss}.$$

c) Show that

$$\frac{2bv^2}{3}U^3 - (v^2 - v_0^2)U^2 + \delta^2\frac{v_0^2}{12}(U_{,s})^2 = 0.$$

d) Show that

$$U^2 = \frac{3(v^2 - v_0^2)}{2bv^2}sech^2[\frac{\sqrt{3(v^2 - v_0^2)}}{v_0}\frac{(x - vt)}{\delta}]$$

is a soliton solution of the equation.

2. Consider the following NLS equation which appears in the modeling of electric networks:

$$i\frac{\partial\Psi}{\partial t} + \frac{\partial^2\Psi}{\partial\xi^2} + q|\Psi|^2\Psi = 0, q > 0.$$

We write Ψ in the form $\Psi = a(\xi, t)e^{i[\theta(\xi,t)+nt]}$, where a and θ are real functions and n is a real constant.

a) Write the equations satisfied by a and θ.

b) We look for a solution of the form $a = a(\xi - ct)$ and $\theta = \theta(\xi - ct)$.

• (i) Show that

$$a_{,ss} + a[c\theta_s - n - (\theta_{,s})^2] + qa^3 = 0,$$

$$a\theta_{,ss} + 2a_{,s}\theta_{,s} - ca_{,s} = 0.$$

- (ii) Show that

$$a^2(2\theta_{,s} - c) = A,$$

where A is an integration constant, and that

$$a_{,ss} = (n - \frac{c^2}{4})a + \frac{A^2}{4a^3} - qa^3.$$

- (iii) Show that

$$(a_{,s})^2 = (n - \frac{c^2}{4})a^2 - \frac{A^2}{4a^2} - q\frac{a^4}{2} + \kappa,$$

where κ is an integration constant, so that

$$4(a\,a_{,s})^2 = 4(n - \frac{c^2}{4})a^4 - A^2 - 2qa^6 + 4\kappa a^2.$$

- (iv) Deduce that $S = a^2$ is a solution of

$$\frac{1}{2q}S^2_{,s} = E + BS + V_m^2 S^2 - S^3,$$

where $E = -\frac{A^2}{2q}$, $B = \frac{4\kappa}{2q}$ and $V_m = \frac{2(n - c^2/4)}{q}$.

- (v) We look for localized solutions (i.e., such that S and its derivatives vanish at $\pm\infty$), so that $E = B = 0$. Show that $S = V_m sech\left(\frac{\sqrt{2q}}{2}V_m s\right)$ is a solution.

- (vi) Show finally that $\Psi = V_m sech\left[\sqrt{\frac{2q}{2}}V_m(\xi - ct)\right]\exp\left[i\frac{c}{2}(\xi - ct) + int\right]$ is a solution of the initial equation.

The partial differential equations
of mechanics

Although we chose to refrain from introducing functional analysis and the theory of partial differential equations (PDEs) in this book, it is desirable to have an overview and make a few comments on the tremendously rich and diverse set of PDEs that we have introduced. In fact, some important PDEs that we introduced are well understood, whereas others are still at the frontier of science as far as their mathematical theory is concerned, which deals with "well-posedness in the sense of Hadamard." This means existence and uniqueness of solutions in suitable function spaces and continuous dependence on the data.

For simplicity, let us restrict ourselves to space dimension two. Several interesting and important PDEs are of the form

$$a\frac{\partial^2 u}{\partial x^2} + b\frac{\partial^2 u}{\partial x \partial y} + c\frac{\partial^2 u}{\partial y^2} = 0. \tag{A.1}$$

Here a, b, c may depend on x and y or be constants, and then Eq. (A.1) is linear; they may depend also on u, $\partial u/\partial x$, and $\partial u/\partial y$, in which case the equation is nonlinear.

Such an equation is

- elliptic when (where) $b^2 - 4ac < 0$,
- hyperbolic when (where) $b^2 - 4ac > 0$,
- parabolic when (where) $b^2 - 4ac = 0$.

Among the simplest linear equations we have seen already are the elliptic equation

$$\Delta u = 0, \tag{A.2}$$

which appeared in Chapter 8, Section 8.2, and in Chapter 14, Section 14.6, and the hyperbolic equation

$$\frac{\partial^2 u}{\partial t^2} - \frac{\partial^2 u}{\partial x^2} = 0, \tag{A.3}$$

(t instead of y), which appeared in Chapter 8, Section 8.4, and throughout Chapter 17. Finally, we saw also the linear parabolic equation

$$\frac{\partial u}{\partial t} - \frac{\partial^2 u}{\partial x^2} = 0, \tag{A.4}$$

which corresponds to the heat equation [Eq. (6.6) of Chapter 6] when the fluid is at rest [in space dimension one, with suitable values of the quantities appearing in Eq. (6.6) of Chapter 6].

Much more difficult, and still raising many unsolved mathematical problems are the Navier–Stokes and Euler equations described in Chapters 5 and 7 and corresponding (somehow) to parabolic nonlinear equations.

Even more difficult and barely touched from the mathematical point of view is the equation of transsonic flows [Eq. (8.13) of Chapter 8], which is a mixed second-order equation: it is elliptic in the subsonic region, hyperbolic in the supersonic region, and parabolic on the sonic line $M = 1$. This equation is furthermore nonlinear.

Finally, let us recall that still different nonlinear PDEs were introduced in Chapters 18 and 19, the Korteweg–de Vries and the nonlinear Schrödinger equations. These are nonlinear wave equations very different from Eqs. (A.1) to (A.4) and reasonably well understood from the mathematical point of view; they produce and describe an amazing physical wave phenomenon, the soliton.

Hints for the exercises

CHAPTER 1

1. $U = \partial\Phi(a, t, t_0)/\partial t$. Compute U_1, U_2, and U_3 in terms of $x_1, x_2, x_3, x = \Phi(a, t, t_0)$.
2.

$$\frac{1}{2}\nabla|U|^2 + (\text{curl } U) \wedge U = (U \cdot \nabla)U.$$

3. $U = \partial\Phi(a, t, t_0)/\partial t$ and $x = (x_1, x_2, x_3) = \Phi(a, t, t_0) = (\Phi_1(a, t, t_0),$ $\Phi_2(a, t, t_0), \Phi_3(a, t, t_0))$; note that $\Phi_1\partial\Phi_1/\partial t + \Phi_2\partial\Phi_2/\partial t = 0$.
4. Note that $U_1 = -\omega x_2$ and $U_2 = \omega x_1$ and that $x_1^2 + x_2^2 = a_1^2 + a_2^2$.
5. Use proper coordinates with a proper choice of the reference frame: e.g., if $\vec{A}, \vec{B} \neq \vec{0}$, then choose $\vec{A}$ parallel to the first vector of the frame, and $\vec{B}$ in the plane $0x_1x_2$.
7. (i) Note that, if $v_i = u$ and $v_j = 0$ when $j \neq i$, then div $v = \partial u/\partial x_i$.
   (ii) Use (i) for $\partial u/\partial x_i$ and then sum over i.
   (iii) Use (ii).
   (iv) Note that div $(u \wedge v) = v$ curl $u - u$ curl v.

CHAPTER 2

1. Use the fundamental law $m\gamma = F$.
2. Consider the equations $D = \{(x, y) \in \mathbb{R}^2, y \geq 0 \text{ and } x^2 + y^2 \leq R^2\}$ and use polar coordinates.
3. Write the fundamental law in polar coordinates.
4. Use Green's formula $\int_{\partial\Omega} f\vec{n}d\Sigma = \int_\Omega \nabla f\, dx$.

CHAPTER 3

1. Compute the eigenvalues and eigenvectors of the matrix (σ_{ij}).
3. Note that $\int_A \sigma_{ij,j}\theta_i dx = -\int_A \sigma_{ij}\partial\theta_i/\partial x_j dx + \int_{\partial A}(\sigma \cdot n)\cdot\theta d\Sigma$ and show that $\int_A \sigma_{ij}\,\partial\theta_i/\partial x_j dx = 0$.
4. In the basis with axes the principal directions, one of the $\sigma_{ij} \neq 0$ and all the others vanish.
5. Note that the stress tensor has the dimension of a force times a length. Use the equilibrium relation $\sigma_{ij,j} = 0$ to compute c_1 and c_2 in a)(i). Note finally that the surface density of forces is given by $F = \sigma \cdot n$ on S_0 and S_1.

CHAPTER 4

1. a) Consider for each material point the forces exerted by each spring and pay attention to the direction of these forces (choose an oriented axis). For the system consisting of the two material points, the virtual power theorem is given by Theorem 4.2.
   b) Consider, in addition to the forces in a), the action of gravity.
2. We have $\mathcal{P}_a = \mathcal{P}_{ext} + \mathcal{P}_{int}$, $\mathcal{P}_{ext} = \int_{\partial\Omega}(\sigma \cdot n)\cdot V d\Gamma + \int_\Omega f \cdot V dx$ and $\mathcal{P}_{int} = -\int_\Omega \sigma\cdot\varepsilon(V)dx$, $\varepsilon(V)_{ij} = \frac{1}{2}(V_{i,j}+V_{j,i})$. Then, note that $\sigma \cdot n = F$ on $\Gamma_1 (\sigma \cdot n$ is unknown on Γ_0).

CHAPTER 5

1. $\gamma = \partial^2\Phi/\partial t^2$, with $x = \Phi(a, t, t_0)$ and $U = \partial\Phi/\partial t$.
2. Note that $\omega x = \Omega \wedge x$, where ω is the rotation tensor.
3. Note that U_1 (resp. U_2, U_3) only depends on (x_2, x_3) (resp. (x_1, x_3), (x_1, x_2)). Set $D_{13} = f$ and $D_{23} = g$, where f and g only depend on x_1 and x_2. Show that $\partial f/\partial x_2 - \partial g/\partial x_1 = -\partial^2 U_2/\partial x_1\partial x_3 = \partial^2 U_1/\partial x_2\partial x_3$, so that $\partial f/\partial x_2 - \partial g/\partial x_1$ does not depend on x_1 and x_2.
4. a) $\omega x = \Omega \wedge x$, with $\omega = 0$ (rotation rate tensor).
   b) $\nabla U = D + \omega$, $D_I = -Tr D$, $D_{II} = \frac{1}{2}[Tr(D)^2 - Tr(D^2)]$, $D_{III} = \det D$.

CHAPTER 6

1. a) Note that $d\,\dfrac{p}{\rho} = \dfrac{1}{\rho}d\,p - \dfrac{p}{\rho^2}d\rho$.

b) Note that it follows from (6.12) that

$$T = \frac{\partial e}{\partial s} \text{ and } \frac{p}{\rho^2} = \frac{\partial e}{\partial \rho}.$$

3. a) Note that $\sigma = -pN$ and $q = 0$.
   c) Prove that $c^2 = \gamma \dfrac{p}{\rho}$.
4. a) Combine the conservation of enthalpy and mass equations and of momentum and mass.
   b) Eliminate the velocity in the relation derived in a).
5. If we call v_{it} and v_{in} the tangential and normal components of the velocity v_i with respect to the shock, then

$$v_{1t} = V_1 \cos \varepsilon, \; V_{1n} = V_1 \sin \varepsilon, \; v_{2t} = V_2 \cos \beta,$$
$$v_{2n} = V_2, \; V_i^2 = v_i^2, i = 1, 2.$$

CHAPTER 7

1. a) Note that $\rho = $ const., div $u = 0$, and $\gamma_i = u_j u_{i,j}$.
   b) We have $\sigma^D = \sigma - \frac{1}{3}(Tr\sigma)I$ and $\varepsilon^D = \varepsilon - \frac{1}{3}(Tr\varepsilon)I$ and note that $u_{j,ji} = (\text{div} u)_{,i} = 0$.
   c) $\mathcal{P} = \int_\Omega \rho\gamma \cdot vdx, \mathcal{P}_{ext} = \int_\Omega f \cdot vdx + \int_{\partial\Omega} F \cdot vd\Gamma$ and $\mathcal{P}_{int} = \int_\Omega \sigma_{ij,j}v_i dx - \int_{\partial\Omega} \sigma_{ij}n_j v_i d\Gamma$.
2. $\dfrac{dp}{dx_3} = -\rho g$, with $\rho = \dfrac{p}{RT}$. Thus, $\dfrac{1}{p}\dfrac{dp}{dx_3} = -\dfrac{g}{RT} = f(x_3)$.
3. Differentiate the Euler equation $\dfrac{\partial u}{\partial t} + u\dfrac{\partial u}{\partial x} = -\dfrac{1}{\rho}\dfrac{\partial p}{\partial x}$ with respect to t to obtain

$$\frac{\partial^2 u}{\partial t^2} + \frac{\partial}{\partial x}(u\frac{\partial u}{\partial t}) = -k\frac{\partial}{\partial x}(\frac{1}{\rho}\frac{\partial\rho}{\partial t}).$$

Then, use the continuity equation $\dfrac{\partial\rho}{\partial t} + \dfrac{\partial}{\partial x}(\rho u) = 0$ and use again the Euler equation (with $p = k\rho$).

CHAPTER 8

1. a) Check that div $U = 0$ and that γ is a gradient (since it follows from the Euler equation that $\gamma = -\nabla(\dfrac{P}{\rho} + gx_3)$, $0x_3$ pointing downwards).
   Write $\gamma = \frac{1}{2}\nabla U^2 + (\text{curl } U) \wedge U$ and show that, in cylindrical coordinates, curl $U \wedge U = -[\psi\psi' + \dfrac{\phi}{r}(r\phi)']\vec{r}$.

b) Prove that $x_1 dx_1 + x_2 dx_2 (= \frac{1}{2} dr^2) = 0$ and that $x_1 dx_2 - x_2 dx_1 = \frac{\phi r}{\psi} dx_3$ (where (r, θ, x_3) denote the cylindrical coordinates), so that, on the streamlines, $dx_3 = (\frac{r\psi}{\phi}) d\theta$, with $\frac{r\psi}{\phi} = $ Const.

c) Note that, in cylindrical coordinates, $U_r = 0$, $U_\theta = \phi(r)$, $U_3 = \psi(r)$, so that curl $U = (0, -\psi'(r), (\phi/r) + \phi')$. Then, $\gamma + \nabla(P/\rho + gx_3) = 0$, with $\gamma = \frac{1}{2} \nabla U^2$.

2. a) Use polar coordinates (set $z = re^{i\theta}$).

  b) The equation for the streamlines read

$$\frac{r^2 + 1}{r^2 - 1} \tan\theta = \lambda,$$

which gives, in cartesian coordinates,

$$(x^2 + y^2 + 1)y = \lambda x(x^2 + y^2 - 1).$$

Take then $\lambda = 0$ and $\lambda = +\infty$.

3. At $t = 0$, the streamlines are given by $dy = -dx/\tan(kx)$. Study the solutions on $]0, \pi/k[$ and draw the picture.

4. a) Note that $\vec{U} = \nabla\Phi = \Phi'(r)\vec{e}_r$. Apply Bernoulli's theorem, with $p = f(\rho)$ (the gaz being perfect) to obtain $f(\rho)/\rho = c(r)$.

  b) Use the continuity equation and show that div $(\rho\vec{U}) = \frac{1}{2} d(r^2 \rho U)/dr$ (use spherical coordinates).

  c) Note that $r^2\rho U = $ const. Study the variation of the function $M \mapsto r^2(M)$ and show that it has a minimum at $M = 1$.

5. Apply Bernoulli's theorem between a point A at the free surface in the container and a point B at the opening. We have, if h denotes the height of fluid between the opening and the free surface in the container

$$\frac{v_B^2}{2g} = h + \frac{v_A^2}{2g},$$

and, neglecting v_A (since the section of the container is large with respect to that of the opening), we find v_B.

CHAPTER 9

1. a) Show that p_0 does not depend on x_1 and that $dp_0/dx_2 = -\rho_0 g(\psi - T_*)$. Then, show that $\kappa \psi'' = 0$ and deduce that $\psi = \frac{T_1 - T_0}{L} x_2 + T_0$.

b) (i) Note that $[(v \cdot \nabla)v] \cdot v = v_i v_{j,i} v_j = \frac{1}{2} v_i (v_j)^2_{,i}$. Then integrate by parts and use the incompressibility condition. Proceed similarly for (ii), (iii). Use Green's formula to integrate by parts and take into account the incompressibility condition.

c) Use equations (1′) and (3′) and b) to show that

$$\frac{1}{2}\frac{d}{dt}\left[\int_\Omega (\rho_0 |v|^2 + c_v \theta^2)dx\right] = \mu \int_\Omega \Delta v \cdot v dx + \kappa \int_\Omega \Delta \theta \cdot \theta dx$$

$$-c_v \int_\Omega \left[(\rho_0 g + \tfrac{T_1-T_0}{L})v_2 \theta\right] dx.$$

Then, integrate by parts.

2. b) We obtain, for $k \neq 0$

$$-\frac{v 4\pi^2 |k|^2}{L^2} u_k + \frac{2i\pi k}{L} p_k = f_k, \quad k \cdot u_k = 0.$$

Taking the scalar product of the first equation by k, we find p_k thanks to the incompressibility condition, and the u_k then follow.

CHAPTER 10

1. Note that it follows from (10.26) and (10.28) that $E = (1/\sigma_m)J - \mu_m u \wedge H$. Insert then this expression into (10.23).
2. a) Multiply by H the equation for H and integrate over Ω.
   b) Use (iv) Exercise 7, Chapter 1.
   c) Use (10.24) (with $D = 0$) and (10.26) (with $u = 0$).
3. a) See the proof of Kelvin's theorem (Theorem 8.4).
   b) Proceed as in Exercise 2.
   c) Proceed as in Exercise 2, b) to show that

$$\frac{d\mathcal{M}}{dt} = \mu_m \int_\Omega (u \wedge H) \cdot \operatorname{curl} H dx - \mu_m \int_{\partial\Omega} [H \wedge (u \wedge H)] \cdot n d\sigma.$$

Then use (10.24) (with $D = 0$) and Exercise 5, (i) and (ii), Chapter 1, noting that $u \cdot n = 0$ on $\partial\Omega$.

CHAPTER 11

1. a) Study the variations of the function $f(\theta) = e^{z\theta} - \alpha\theta$. Note that it reaches its minimum for $\theta = \frac{1}{z}\log(\frac{\alpha}{z})$ and that this minimum is strictly negative for $\frac{\alpha}{z} > e$.

b) Note that the equation linearized around θ_i, $i = 1, 2$, reads

$$\frac{d\theta}{dt} = \beta z \theta e^{z\theta i} - \sigma\theta,$$

i.e., noting that $e^{z\theta i} = \alpha\theta_i$,

$$\frac{d\theta}{dt} = \beta\alpha(z\theta_i - 1)\theta.$$

The solution of this equation reads

$$\theta = \theta(0)e^{\beta\alpha(z\theta_i - 1)t}.$$

Then, note that $z\theta_2 - 1 > 0$. Keeping the equation in the form

$$\frac{d\theta}{dt} = \beta(ze^{z\theta_i} - \alpha)\theta,$$

note that $ze^{z\theta_i} - \alpha < 0$

2. a) Make the change of variable $x' = x/L$. Then note that the problem reduces to

$$\theta'' + L^2\beta e^{z\theta} = 0, \ 0 < x < 1,$$
$$\theta(0) = \theta(1) = 0.$$

Multiply the reduced equation by θ' and integrate to obtain, noting that $\theta'(1) = 0$,

$$(\theta')^2 = \frac{2L^2\beta}{z}(e^{z\theta(1)} - e^{z\theta}).$$

Thus

$$\theta' = \pm\sqrt{\frac{2L^2\beta}{z}e^{z\theta(1)}} \ \sqrt{1 - e^{z(\theta - \theta(1))}}.$$

Set $u = z(\theta - \theta(1))$. Then

$$\frac{du}{\sqrt{1 - e^u}} = \pm \sqrt{2L^2\beta z e^{z\theta(1)}}\, dx.$$

Integrating, we find

$$\int \frac{du}{\sqrt{1 - e^u}} = \pm\sqrt{2L^2\beta z e^{z\theta(1)}}\, x + c.$$

Setting $v = \sqrt{1 - e^u}$, we have $dv = -\dfrac{e^u du}{2\sqrt{1 - u}}$, so that $\dfrac{du}{\sqrt{1 - e^u}} =$

$-\dfrac{2dv}{1 - v^L}$. Therefore

$$\int \frac{du}{\sqrt{1 - e^u}} = -2\int \frac{dv}{1 - v^2} = -\int \left(\frac{1}{1 - v} + \frac{1}{1 + v}\right) dv,$$

and

$$\log \left| \frac{1-v}{1+v} \right| = \pm\sqrt{2zL^2\beta e^{z\theta(1)}}\, x + c,$$

which yields

$$\frac{1-v}{1+v} = ce \pm \sqrt{2zL^2\beta e^{z\theta(1)}}\, x.$$

Finally, compute c by taking $x = 1$ and show that

$$\theta(x) = \theta(1) - \frac{2}{z} \log \left[ch \left(\sqrt{\frac{zL^2\beta e^{z\theta(1)}}{2}}(x-1) \right) \right].$$

b) Note that we must have $\theta(0) = 0$ and study the variations of the function $f(x) = -\frac{2}{z} \log \left[ch \left(\sqrt{\frac{zd}{2}}e^{zx/2} \right) \right]$. We have $f'(x) = 1 - \sqrt{\frac{z\alpha}{2}}e^{zx/2}th \left[e^{zx/2}\sqrt{\frac{\alpha z}{2}} \right]$. Then note that f' is strictly decreasing.

3. a) If $Le = 1$, then $v_0(T - q\alpha)'' + c(T - q\alpha)' = 0$, and one can thus express α in terms of T.

b) Consider the system linearized around $(T_\alpha, 0)$:

$$T' = p$$
$$p' = \frac{1}{v_0}(-cp - F'(T_\alpha)T).$$

Then show that the eigenvalues of the matrix

$$\begin{pmatrix} 0 & 1 \\ -F'(T_\alpha) & -\dfrac{c}{\kappa} \end{pmatrix}$$

have opposite signs (note that $F'(T_0) < 0$)

CHAPTER 13

1. a) Note that, necessarily, χ is of the form

$$\chi = \int_0^{x_2} ds \int_0^s \sigma_{11}(x_1, \xi)d\xi + x_2 F(x_1) + G(x).$$

Furthermore, it follows from the equilibrium equations $\sigma_{ij,j} = 0$ that $\sigma_{11,11} = \sigma_{22,22}$. This yields

$$\frac{\partial^2\chi}{\partial x_1^2} = \sigma_{22}(x_1, x_2) - \sigma_{22}(x_1, 0) - x_2\sigma_{22,2}(x_1, 0) + x_2 F''(x_1) + G''(x_1).$$

Now, since $\sigma_{11,1} = -\sigma_{12,2}$,

$$\frac{\partial^2 \chi}{\partial x_1 \partial x_2} = -\sigma_{12}(x_1, x_2) + \sigma_{12}(x_1, 0) + F'(x_1).$$

When specifying F and G, note that $\sigma_{12,1} = \sigma_{22,2}$. Finally, note that

$$\Delta(\Delta\chi) = \sigma_{11,22} - 2\sigma_{12,12} + \sigma_{22,11}.$$

  b) Show that f is solution of the ordinary differential equation

$$f'''' - 2\omega^2 f'' + \omega^4 f = 0,$$

where $' = d/dx_2$.
2. a) The Navier equation reads in that case

$$(\lambda+\mu)\nabla(\text{div}\,\vec{\chi_0} + \text{div}\,\vec{B} +\nabla\chi)+\mu(\Delta\vec{\chi_0}+\Delta\vec{B}+\Delta\nabla\chi)+ \vec{f} = 0,$$

hence the result, $\vec{\chi_0}$ being a solution of the Navier equation. Then,

$$\Delta\chi + \frac{\lambda+\mu}{\lambda+2\mu}\,\text{div}\,\vec{B} = \text{const} = a,$$

and show that $\Delta(\vec{r}\cdot\vec{B}) = 2\,\text{div}\,\vec{B}$, and, noticing that $\Delta(a\vec{r}^2) = 6a$, deduce that

$$\Delta(\chi + \frac{\lambda+\mu}{2(\lambda+2\mu)}\vec{r}\cdot\vec{B} - \frac{a}{6}\vec{r}^2) = 0,$$

so that $\chi + \frac{\lambda+\mu}{2(\lambda+2\mu)}\vec{r}\cdot\vec{B} - \frac{a}{6}\vec{r}^2$ is a harmonic function that we can rewrite as $-\frac{\lambda+\mu}{2(\lambda+2\mu)}\vec{B_0}$, with $\vec{B_0}$ harmonic.
  b) Note that $\vec{\chi_1} = \vec{\chi_0} + b\vec{r}$ is solution of the Navier equation.
3. The relations $\sigma_{ij,j} + f_i = 0$ yield $f = 0$.
The Beltrami equations

$$(1 + v)\Delta\sigma_{ij} + (Tr\sigma),_{ij} = 0$$

then yield (for $i = j = 1$ and $i = j = 3$) that $k = 0$ and $v = -2$, which is impossible since, necessarily, $v \geq 0$.
4. The equilibrium equations $\sigma_{ij,j} = 0$ yield that (for $i = 1$ and 3) $\partial F/\partial x_2$ and $\partial F/\partial x_1$ are independent of x_3. Then, note that only the derivatives of F appear in σ. The Beltrami equations (for $i = 1, j = 3$, and $i = 2, j = 3$) finally yield that ΔF is affine.

CHAPTER 14

1. a) The equilibrium equations yield $2A + D + F = 0$ and the Beltrami equations yield $2(1 + v)(A + B) + F = 0$.

   b) Consider a point M with cylindrical coordinates $(a \cos \theta, a \sin \theta, x_3)$ on the lateral surface. Thus, $\vec{n} = (\cos \theta, \sin \theta, 0)$ and use the relation $T = \sigma n = 0, \forall \ \theta$, to obtain

$$A = B + D$$
$$C + Aa^2 = 0.$$

On the base $x_3 = \ell, n = (0, 0, 1)$, and we find

$$\mathcal{R} = ((A + B)a^4 \frac{\pi}{4} + \pi a^2, 0, 0)$$

and

$$\mathcal{M} = (0, -(E + F\ell)\frac{\pi a^4}{4}, 0).$$

We finally obtain 6 equations for the 6 unknowns A,B,C,D,E, and F.

2. b) Note that $\varepsilon_{ij} = \frac{1 + v}{E}\sigma_{ij} - \frac{v}{E}(Tr\sigma)\delta_{ij}$.

   c) Note that $\varepsilon = \frac{1}{2}(u_{i,j} + u_{j,i})$.

3. a) In cylindrical coordinates, $\operatorname{div} u = \frac{1}{r}\frac{\partial}{\partial r}(rU(r))$ and $\operatorname{curl} u = 0$. The Navier equation then yields, noting that $\Delta u = -\operatorname{curl} \operatorname{curl} u + \nabla \operatorname{div} u$,

$$\nabla[\frac{1}{r}\frac{\partial}{\partial r}(rU)] = 0.$$

   b) We have $U(r) = \frac{ar}{2} + \frac{b}{r}$. The constitutive law

$$\sigma = 2\mu\varepsilon + \lambda(Tr\varepsilon)I,$$

yields, noting that $\varepsilon = \frac{1}{2}(u_{i,j} + u_{j,i})$ and $u = U(r)e_r, e_r = \left(\frac{x_1}{r}, \frac{x_2}{r}, 0\right)$,

$$\sigma_{11} = \lambda a + 2\mu\left[\frac{U}{r} + \frac{\partial}{\partial r}(\frac{U}{r})\frac{x_1^2}{r}\right],$$

$$\sigma_{12} = 2\mu\frac{x_1 x_2}{r}\frac{\partial}{\partial r}\left(\frac{U}{r}\right).$$

Consider finally the boundary conditions (see, e.g., Section 14.3) to find

$$a = \frac{1}{\lambda + \mu} \frac{P_2 R_2^2 - P_1 R_1^2}{R_2^2 - R_1^2},$$

$$d = \frac{(P_2 - P_1)R_1^2 R_2^2}{2\mu(R_1^2 - R_2^2)}.$$

CHAPTER 16

1. a) A picture can help. Take, e.g., $a = 0, b = 2$ and consider the values $\varepsilon = 2^{-n}, n \in \mathbb{N}$.

   b) Note that $\int_\alpha^\beta \sin(2\pi \frac{x}{\varepsilon})dx = -\frac{\varepsilon}{2\pi} \left[\cos(\pi \frac{x}{\varepsilon})\right]_\alpha^\beta$.

   c) Note that $\int_a^b f_2^2 dx = \frac{b-a}{2} + \frac{\varepsilon}{8\pi} \left[\sin\left(\frac{4\pi a}{\varepsilon}\right) - \sin\left(\frac{4\pi b}{\varepsilon}\right)\right]$ (use the relation $\sin^2 y = \frac{1 - \cos 2y}{2}$).

CHAPTER 18

2. Set

$$\tilde{u}(x, t) = \begin{cases} u(x, t) & \text{if } x \geq 0, t \geq 0, \\ -u(-x, t) & \text{if } x \leq 0, t \geq 0, \end{cases}$$

and define $\tilde{u}_0$ and $\tilde{u}_1$ similarly. Then, consider the equation satisfied by $\tilde{u}$ (for simplicity, assume that $\tilde{u}_0$ and $\tilde{u}_1$ belong to $\mathcal{C}^2(\mathbb{R})$).

3. Consider the "energy"

$$\mathcal{E}(t) = \frac{1}{2}\int_\Omega \left[\left(\frac{\partial u}{\partial t}\right)^2 + |\nabla u|^2\right] dx, \quad 0 \leq t \leq T,$$

and show that $\dfrac{d\mathcal{E}}{dt} = 0$.

4. Set

$$\mathcal{E}(t) = \frac{1}{2}\int_B \left[\left(\frac{\partial u}{\partial t}\right)^2 + |\nabla u|^2\right] dx.$$

Use (1.8) in Proposition 1.3 to show that

$$\frac{d\mathcal{E}}{dt} = \int_B \frac{\partial u}{\partial t} \left(\frac{\partial^2 u}{\partial t^2} - \Delta u \right) dx + \int_{\partial B} \frac{\partial u}{\partial v} \frac{\partial u}{\partial t} d\Sigma$$

$$-\frac{1}{2} \int_{\partial B} \left[\left(\frac{\partial u}{\partial t} \right)^2 + |\nabla u|^2 \right] dx.$$

Then, use the Cauchy-Schwarz inequality to obtain

$$|\frac{\partial u}{\partial v} \frac{\partial u}{\partial t}| \le \frac{1}{2} \left(\frac{\partial u}{\partial t} \right)^2 + \frac{1}{2} |\nabla u|^2,$$

and conclude that $\dfrac{d\mathcal{E}}{dt} \le 0$. Here, v denotes the unit outer normal vector to ∂B.

5. a) ii) Note that the solutions of the characteristic equation

$$r^4 - m^4 = 0,$$

are $\pm m$ and $\pm im$.

b) Use the boundary conditions. The solutions of the equation $\cos\sigma \; ch\sigma = 1$ can be written in the form σ_p, $p \in \mathbb{N}^*$ (a picture can help: draw, on the same picture, the curves $\eta = \cos\xi$ and $\eta = \dfrac{1}{ch\xi}$), hence the eigenfrequencies

$$\omega_p = \frac{\sigma_p^2}{L^2} \sqrt{\frac{EI}{\rho}}, \quad p \in \mathbb{N}^*.$$

(Note that $\omega^2 = m^4 \dfrac{EI}{\rho} = \dfrac{\sigma^4}{L^4} \dfrac{EI}{\rho}$.)

6. In that case, we have $u(0, t) = \dfrac{\partial^2 u}{\partial x^2}(0, t) = u(L, t) = \dfrac{\partial^2 u}{\partial x^2}(L, t) = 0$. This yields $U = B \sin mx$ and the eigenfrequencies are given by $\sigma_p = p\pi$, $p \in \mathbb{N}^*$.

7. In that case, $u(0, t) = \dfrac{\partial u}{\partial x}(0, t) = \dfrac{\partial^2 u}{\partial x^2}(L, t) = \dfrac{\partial^3 u}{\partial x^3}(L, t) = 0$, $\forall t \ge 0$, and we find

$$U = A \left[(\cos\sigma + ch\sigma)(\cos mx - chmx) \right.$$
$$\left. + (\sin\sigma - sh\sigma)(\sin mx - shmx) \right], ch\sigma \cos\sigma + 1 = 0.$$

CHAPTER 19

1. b) Multiplying the first equation in (ii) by $\theta,_s$ and integrating, we obtain the second equation. For localized solutions, we have $c = 1$, and we obtain, by integration

$$\pm\frac{s - s_0}{\sqrt{1 - u^2}} = ln(tan\frac{\theta}{4});$$

hence

$$\theta = 4 \ arctan\left[\exp\left(\pm\frac{s - s_0}{\sqrt{1 - u^2}}\right)\right].$$

2. a) Insert the expressions for η and u in the system (with $\eta,_x = \eta,_\xi$ and $\eta,_t = -c\eta,_\xi$) and note that the two equations obtained must be compatible.

   c) Set $\eta_0 = \dfrac{3(1 - \alpha^2)}{\alpha^2}$ and use b).

3. Proceed as in Exercise 2, a) to show that $\alpha = \pm\frac{2\sqrt{15}}{5}$, $c = \mp\frac{\sqrt{15}}{15}$ and that η satisfies the equation

$$-\frac{7}{5}\eta,_\xi + \frac{1}{5}\eta,_{\xi\xi\xi} = \frac{12}{5}\eta\eta,_\xi,$$

and use Exercise 2, b) to show that

$$\eta(x, t) = -\frac{7}{4}sech^2(\lambda(x + x_0 - ct)),$$

$$u(x_1, t) = \mp\frac{7\sqrt{15}}{10} \ sech^2(\lambda(x + x_0 - ct)),$$

where $\lambda = \frac{\sqrt{7}}{2}$.

CHAPTER 20

1. a) Set $s = x - vt$ and note that $\dfrac{\partial}{\partial x} = \dfrac{d}{ds}$ and $\dfrac{\partial}{\partial t} = -v\dfrac{\partial}{\partial x}$.

   b) Integrate the equation twice and note that the integration constants vanish.

   c) Multiply the equation in b) by $2dU,_s$ and integrate. Note also that the integration constant vanishes.

2. a) Separate the real and imaginary parts to obtain

$$\theta_{,\xi\xi} - a(\theta_{,t} + n) - a(\theta_{,\xi})^2 + qa^3 = 0,$$
$$a\theta_{,\xi\xi} + 2a_{,\xi}\theta_{,\xi} + a_{,t} = 0.$$

   b) (i) Set $s = \xi - ct$.
      (ii) Integrate the second equation in (i), then inject the value obtained for $\theta_{,s}$ into the first equation.
      (iii) Multiply the equation in (ii) by $2a_{,s}$ and integrate.

References

PART ONE

General references

Duvaut, G., *Mécanique des Milieux Continus*, Masson, Paris, 1990 (in French).

Germain, P., *Cours de Mécanique des Milieux Continus*, Vol. 1, Masson, Paris, 1973 (in French).

Germain, P., *Mécanique*, Vols. 1 & 2, Cours de l'Ecole Polytechnique, Ellipses, Paris, 1986 (in French).

Goldstein, H., *Classical Mechanics*, 2d ed., Addison–Wesley, Reading, MA, 1980.

Gurtin, E., *An Introduction to Continuum Mechanics*, Academic Press, Boston, 1981.

Lai, W.M., D. Rubin and E. Krempl, Introduction to Continuum Mechanics, Pergammon Press, New York, 1974.

Landau, L., and E. Lifschitz, *Mechanics of Continuum Media*, Butterworth–Hernemann, New York, 1976.

Salençon, J., *Mécanique des Milieux Continus*, Vols. 1 & 2, Ellipses, Paris, 1988 (in French).

Segel, L.A., *Mathematics Applied to Continuum Mechanics*, Dover Publications, New York, 1987.

Spencer, A.J.M., *Continuum Mechanics*, Longman Scientific and Technical, New York, 1975.

Truesdell, C., *A First Course in Rational Continuum Mechanics*, Vol. 1, Academic Press, New York, 1977.

Ziegler, H., *mechanics*, Vol. 2, *Dynamics of Rigid Bodies and Systems* (English translation), Addison–Wesley, Reading, MA, 1965.

More advanced or more specialized books

Zemansky, M.W., *Heat and Thermodynamics*, 5th ed., McGraw–Hill, New York, 1968.

PART TWO

General references

Batchelor, G.K., *An Introduction to Fluid Dynamics*, Cambridge University Press, Cambridge, 1988.

Chorin, A.J., and J.E. Marsden, *A Mathematical Introduction to Fluid Mechanics*, Springer–Verlag, Heidelberg, 1979.

Lamb, H., *Hydrodynamics*, 6th ed., Cambridge University Press, London and New York, 1932. Reprinted New York, Dover, 1945.

More advanced or more specialized books

Cabannes, H., *Theoretical Magnetofluiddynamics*, Academic Press, New York, 1970.

Candel, S., *Mécanique des Fluides*, Dunod, Paris, 1995 (in French).

Chandrasekhar, S., *Hydrodynamic and Hydromagnetic Stability*, Dover Publications, New York, 1981.

Doering, R., and J.D. Gibbon, *Applied Analysis of the Navier–Stokes Equations*, Cambridge University Press, Cambridge, 1995.

Drazin, P.G., and W.H. Reed, *Hydrodynamic Stability*, Cambridge University Press, Cambridge, 1981.

Dunwoody, J., *Elements of Stability of Visco-elastic Fluids*, Pitman Research Notes in Mathematical Series, Longman Scientific & Technical, 1989.

Duvaut, G., and J.L. Lions, *Inequations in Mechanics and Physics*, Springer–Verlag, Heidelberg, 1977.

Hinze, J.O., *Turbulence*, 2d ed., McGraw–Hill, New York, 1987.

Ladyzhenskaya, O.A., *The Mathematical Theory of Viscous Incompressible Flow*, Gordon and Breach, New York, 1963.

Landau, L., and E. Lifschitz, *Fluid Mechanics*, Addison–Wesley, New York, 1953.

Lions, J.L., R. Temam, and S. Wang, "Models of the Coupled Atmosphere and Ocean (CAO I & II)," in *Computational Mechanics Advances*, J.T. Oden, ed., 1, Elsevier, Amsterdam, 1993, 5–54 and 55–119.

Majda, A., and A.L. Bertozzi, *Vorticity and Incompressible Flows*, Cambridge University Press, Cambridge, 2001.

Panofsky, W.K.H., and M. Philipps, *Classical Electricity and Magnetism*, Addison-Wesley, London, 1955.

Pedlosky, J., *Geophysical Fluid Dynamics*, 2d ed., Springer–Verlag, Heidelberg, 1987.

Schlichting, H., *Boundary-Layer Theory*, McGraw–Hill, New York, 1968.

Serrin, J., "Mathematical Principles of Classical Fluid Mechanics," in *Handbuch der Physik*, **8**, Springer–Verlag, Berlin, 1959.

Temam, R., *Navier-Stokes Equations, Theory and Numerical Analysis*, AMS-Chelsea Series, American Mathematical Society, Providence, 2001.

Volpert, A.I., V.A. Volpert, and V.A. Volpert, *Traveling Wave Solutions of Parabolic Systems*, American Mathematical Society, Providence, RI, 1994.

Washington, W.M., and C.L. Parkinson, *An Introduction to Three-Dimensional Climate Modeling*, Oxford University Press, Oxford, 1986.

Williams, A., *Combustion Theory*, 2d ed., The Benjamin/Cummings Publishing Company, Menlo Park, CA, 1985.

PART THREE

More advanced or more specialized books and articles

Antman, S., *Nonlinear Problems of Elasticity*, Springer–Verlag, New York, 1995.

Bensoussan, A., J.L. Lions, and G. Papanicolaou, *Asymptotic Analysis for Periodic Structures*, North-Holland, Amsterdam, 1978.

Ciarlet, P.G., *Mathematical Elasticity*, North-Holland, Amsterdam, 1988.

Cioranescu, D., and P. Donata, *An Introduction to Homogenization*, Oxford University Press, New York, 2000.

Ericksen, J.L., *Introduction to the Thermodynamics of Solids*, Rev. Ed., Springer–Verlag, Heidelberg, 1998.

Gurtin, M.E., *Topics in Finite Elasticity*, Society for Industrial and Applied Mathematics, Philadelphia, 1981.

Hodge, P., *Plastic Analysis of Structures*, Reprint, R.E. Krieger Publishing Company, Malabar, Florida, 1981.

Jemiolo, S. and J.J. Telega, Modelling elastic behaviour of soft tissue, I&II, *Engineering Transactions*, 49, 2–3, 2001, pp. 213–240 and 241–281.

Jikov, V.V., S.M. Kozlov, and O.A. Oleinik, *Homogenization of Differential Operators and Integral Functionals*, Springer–Verlag, Heidelberg, 1991.

Ogden, R.W., *Non-linear Elastic Deformations*, Ellis Horwood, 1984 (second edition to appear soon).

Renardy, M., W.J. Hrusa, and J.A. Nohel, *Mathematicals Problems in Viscoelasticity*, Longman Scientific and Technical, New York, 1987.

Sokolnikoff, I.S., *Mathematical Theory of Elasticity*, 2d ed., McGraw–Hill, New York, 1956.

Temam, R., *Mathematical Problems in Plasticity*, Gauthier–Villars, New York, 1985.

Truesdell, C., *Mechanics of Solids* I–III, Springer–Verlag, Heidelberg, 1973.

PART FOUR

More advanced or more specialized books

Agrawal, G.P., *Nonlinear Fiber Optics*, Academic Press, Boston, 1989.

Benade, A.H., *Fundamentals of Musical Acoustics*, Oxford University Press, Oxford, 1976.

Benjamin, T.B., "Lectures on Nonlinear Wave Motion," in *Lectures in Applied Mathematics*, Vol. 15, American Mathematical Society, Providence, RI, 1974, 3–47.

Boyd, R.W., *Nonlinear Optics*, Academic Press, Boston, 1992.

Courant, R., and D. Hilbert, *Methods of Mathematical Physics*, Interscience Publishers, New York, 1953.

Fletcher, N.H., and T.D. Rossing, *The Physics of Musical Instruments*, 2d ed., Springer–Verlag, Heidelberg, 1998.

Goldstein, M.E., *Aeroacoustics*, McGraw–Hill, New York, 1976.

Lightill, J., *Waves in Fluids*, Re-edition, Cambridge University Press, Cambridge, 1978.

Rayleigh, J.W.S., *The Theory of Sound* (1894), first Dover edition published in 1945, Dover, New York.

Remoissenet, M., *Waves Called Solitons*, 2d ed., Springer–Verlag, Heidelberg, 1996.

Roederer, J.G., *The Physics and Psychophysics of Music*, 3d ed., Springer–Verlag, Heidelberg, 1995.

Roseau, M., *Vibrations in Mechanical Systems*, Springer–Verlag, Heidelberg, 1987.

Sundberg, J., *The Science of Musical Sounds*, Academic Press, New York, 1991.

Index

Printed in the United States
By Bookmasters